Emerson's Life in Science

::

Emerson's
Life in Science

::

The Culture of Truth

LAURA DASSOW WALLS

CORNELL UNIVERSITY PRESS

Ithaca and London

First published 2003 by Cornell University Press

Printed in the United States of America

Library of Congress Cataloging-in-Publication Data

Walls, Laura Dassow.
  Emerson's life in science : the culture of truth / Laura Dassow Walls.

    p. cm.
Includes bibliographical references and index.
  ISBN 0-8014-4044-0
  1. Emerson, Ralph Waldo, 1803–1882—Knowledge—Science. 2. Literature and science—United States—History—19th century. 3. Science in literature. I. Title.
  PS1642.S3 W35 2003
  814'.3—dc21

                                        2002151358

Cornell University Press strives to use environmentally responsible suppliers and materials to the fullest extent possible in the publishing of its books. Such materials include vegetable-based, low-VOC inks and acid-free papers that are recycled, totally chlorine-free, or partly composed of nonwood fibers. For further information, visit our website at www.cornellpress.cornell.edu.

Cloth printing 10 9 8 7 6 5 4 3 2 1

Contents

CONTENTS

Acknowledgments

This book has been a part of my life for so many years that everyone I know, including family, friends, colleagues, and students, has contributed some ray of light to whatever illumination I can offer here. Yet all would still be darkness were it not for the gift of time, and for that I owe profound thanks to the American Council of Learned Societies, whose generous support of this project at its crucial, formative stage enabled me to convert vision into reality. A fellowship from the National Endowment for the Humanities allowed me the additional time to complete the final round of revisions. Thanks must also go to Lafayette College and to Provost June Schlueter for the sabbatical leave that allowed the work to continue; to my EXCEL student Kathie Butler, for her help in surveying the dense intellectual landscape; to Neil McElroy and the staff of Skillman Library, who located so many rare science books and placed them so trustingly in my hands; to Bernie Kendler at Cornell University Press, for encouraging me to write the book I really wanted to write, and to my editors Louise E. Robbins and Ann Hawthorne, for their painstaking help in making sure this books says exactly what I wanted it to say.

Among my Emersonian colleagues, my deepest thanks go to Bob Richardson, who supported this book from its inception and whose advice and encouragement were the wind under my wings. His suggestions on reading the manuscript helped me find the courage to write a shorter, stronger, and, I hope, more lyrical book. David Robinson, too, has shown unfailing generosity to a junior colleague, and he, too, read the manuscript in full. The breadth of his knowledge and the depth of his insights gave me additional ways to clarify and deepen my argument, some of which will have to wait for future works. Finally, without the generosity of Ron Bosco, I'm not sure this book would be at all. By sharing in advance the relevant pages of Emerson's *Later Lectures*, not to mention his own deep knowledge of Emerson, Ron helped me see my way, at last, through this project's concluding chapter.

Many others have played significant roles as the germ of this book unfurled into its final form. I thank Lee Sterrenburg and Anka Ryall for their constant friendship, and for serving as my muses when I first started wrestling with "gnomic" science. Jonathan Smith, who also inhabits the

: vii

thinly populated border between literature and the history of science, has many times served as a sounding board. Fred Gregory and Bernie Lightman have welcomed my literary incursions into the history of science and given much good advice. Finally, I thank the greater community of Emerson scholars for making this Thoreauvian feel utterly at home. In particular, my thanks to Larry Buell, Phyllis Cole, Len Gougeon, Barbara Packer, Joel Porte, and Sarah Wider, all of whom have shared their time, their expertise, and, best of all, their passion. Eric Wilson deserves a special mention, for his bottomless energy and inspiration. And above all, my thanks to Joel Myerson, whose wit and graciousness shed a tropic radiance across the austere study of New England Transcendentalism. I count myself lucky for finding my way to the oasis of good fellowship he has done so much to create. Finally, once again I offer thanks to the late Wallace Williams, who long ago, in an office cluttered with Emerson books and manuscripts, pointed the way.

Closer to home, I would like to thank my Lafayette colleagues, whose daily moral support makes the impossible seem a little less so. In particular, my thanks to Susan Blake, Pat Donahue, David Johnson, Deborah Rosen, Lynn Van Dyke, Suzanne Westfall, James Woolley, and Tom Yuster, who have helped me move forward and provided good company along the way. I would also like to thank my students, who cannot know how often they have helped me see old worlds anew, and reminded me why the literature of the past matters so much today.

This brings me to the innermost circle of all. To my husband, Bob, who believes in me most when I'm most skeptical and urges me onward when I hesitate; to my parents, John and Ethel Dassow, who first gave me the tools of thought and convinced me they were mine to use; to my godparents, John and Polly Dyer, who showed me what it meant to live in the love of nature—to all of these, mere thanks are not enough. I dedicate this book to the memory of my father, who, like Emerson, understood the power of knowledge and acquired the knowledge of pain. Dad, long ago you gave me Emerson. Here is my belated gift to you.

Emerson's Life in Science

::

Darwin had keen skills of observation

Prologue

When I finished writing *Seeing New Worlds,* my book on Thoreau and science, I knew I wasn't really finished. I had wondered how Thoreau, who called himself "a mystic—a transcendentalist—& a natural philosopher to boot," could have become so deeply drawn to natural science. It seemed to me that a Romantic writer interested in grasping nature as one great whole had two basic approaches from which to choose. One I named "empirical holism," for it worked toward a vision of that whole through the interconnectedness of individual details, all generating their own pattern or design. By contrast, the other and more dominant approach, "rational holism," embraced the whole in a visionary moment and saw details in that moment's light.[1]

Thoreau, so deeply involved in seeing patterns in the individual details of nature, became my model for the empirical approach. Always aware of a transcendent unity and ever susceptible to visionary moments, he yet adopted as his daily working method a highly disciplined attention to nature's myriad particulars, working literally from the ground up. But this empirical approach is only half the picture. Emerson was hardly insensitive to nature's singular beauties, but temperamentally he responded more to the "rational" side. Hence this book flips the coin—pitches the penny, as Emerson might have said—to look to the other side. Whereas Thoreau's vision opened out to chance and contingency, Emerson's began with the premise that the universe was one because it had been designed and ordered by a single governing intelligence; even pitching pennies was not a random act. Whereas Thoreau saw nature's creatures each stepping to their own music, weaving a collective dance, Emerson saw nature as a grand orchestrated whole, a symphony that gave each part its meaning. Both visions have their satisfactions, and their limitations. This book asks what those were for Emerson, who becomes a lens through which to survey the wider world of the meaning of science to readers and writers of the nineteenth century.

Many commentators have addressed the importance of science to Emerson's thought, though the subject has only begun to be fully explored. One of Emerson's first biographers nicely captured the tone of Emerson's interest in science when he declared, in 1888, "This is an age of science, and science has found no such literary interpreter as Emerson." Yet as Sarah Wider notes, "His claim would soon be disputed, and Emerson's connection with science would all but disappear for the next ninety years." In 1931 Harry Hayden Clark opened his groundbreaking article, "Emerson and Science," with the observation that Emerson's "debt to science" awaited "extensive investigation." As if to stimulate such investigation, he closed with two pages of questions that he admitted would take a "solid book" to address— a book that he predicted would, were it ever written, considerably modify the conventional view of Emerson "as a mystical and irresponsible traveler in realms of Whim." Despite such enticements, a full forty-five years later Gay Wilson Allen opened his article "A New Look at Emerson and Science" by wondering why Clark's important study, though widely cited and praised, had not been assimilated into Emerson scholarship. Allen, too, closed by proposing a book on Emerson and science, which he, too, begged off writing.[2]

More recently, much attention has focused on the moment that David M. Robinson identified as a turning point for Emerson, his visit in 1833 to the Paris Muséum d'Histoire Naturelle. Emerson reacted in his journal with a starburst of prose, culminating in the resolution, made as he sailed home to launch his new career, "I will be a naturalist." As soon as he stepped ashore in Boston, Emerson accepted a speaking engagement at the Boston Society of Natural History, where less than a month later he delivered his first public lecture, "The Uses of Natural History." After three more lectures on science, Emerson diversified his topics, although ever afterward science was never far from his thoughts. It was typical, for instance, that in a lecture on Shakespeare, Emerson turned to Francis Bacon's foundational work on the true method of science, *The New Organon,* to explain how the creative artist unites many disparate parts into an organically complete whole. The creative method of science, by showing how to incorporate mind and nature into a new whole, gave Emerson the key to the universe; it was a fundamental insight that he never tired of repeating.[3]

The claim that Emerson was grounded in science sounds counterintuitive today, on two grounds. First, just as Clark predicted, to recover Emerson's involvement with science overturns what remains of the conventional stereotypes that have dogged critical and popular opinion for 150 years: Emerson as the whimsical and fuzzy-minded idealist, too self-absorbed to attend to the outer world, too blithe to face real evil. My Emerson is a tough-minded survivor, dedicated to facing down tragedy and evil with the keen-edged weapon of scientific truth. This Emerson challenges Stephen Whicher's classic portrait of a man who turns in on himself, relinquishing "revolution" for "acquiescence" and acceptance of "things as they are." On the contrary, for Emerson obedience was but the first step toward victory over limitation and calamity; he deployed with ever greater precision the Baconian strategy of commanding nature by obeying her laws. This Emerson read deeply in the new sciences of his day and involved himself both theoretically and personally, befriending many of the age's leading scientists.[4]

The very intimacy of Emerson's involvement with science leads to the second reason my claims may seem counterintuitive. Emerson, like most intellectuals of his day, was perfectly at ease folding scientific truth into moral truth, reading literature and science together as part of a common intellectual culture. He took scientific literacy so much for granted that his scientific metaphors sink out of sight; worse, from his time to ours, the divorce between "the two cultures" of literature and science has made his deep debt to science virtually invisible. The result is that later generations have allowed this characteristic and crucial aspect of Emerson's thought to go largely unnoticed and unremarked. Moreover, the very goals and assumptions that once yoked poetry and science as allies in a common cultural enterprise are no longer accepted as part of science. Hence the whole Emersonian complex has been swept into "literature," with little recognition of the loss this entails to an understanding of the common roots of literary and scientific thought in America.[5]

My approach has been to use the particular light of Emerson's reading in science to illuminate Emerson's intellectual life. Few can be more aware than I at this point of all that remains unsaid here, for Emerson's reading in science, as in so many other subjects, was wide-ranging and intensive, and whatever he read, he absorbed and refracted in his writing. For a start, one could target any one of the many science writers I discuss (let alone the many I do not) and develop from that one point whole new arguments and deeper critical understandings. That, I hope, is precisely what will happen over the next few years. What I offer in the meantime is an argument about the overall place of science in Emerson's life, thought, and writing, and the place of Emerson in the development of modern science. Hence I have sought to be less definitive than provocative. I hope the story told here will engage nonspecialists interested in the place of science in literature and society, as well as scholars of literature, history, and the history of science. If enough scholars are provoked to look more closely at the questions raised here, then in time we may indeed be able better to define the relationship between Emerson and science—and perhaps even between science, culture, and literature. The Emerson who will emerge from such work is, as I hope to show, radical, daring, and challenging. Part of what he challenges most is the blinkered assumption that culture and science have been, are, or ever could be alien halves of our one global whole.

CHAPTER ONE

The Sphinx at the Crossroads

THE SUN OF SCIENCE

Emerson lived during the golden age of science, and he hoped to translate its dawning radiance into a beacon for the masses. Modern America owes largely to Emerson its faith in science as the bulwark of truth against the tides of history and the storms of war. Yet Emerson has never been known as a "scientist." During his heyday the word barely existed, and to suggest today that his work was scientific strains belief. The degree of that strain marks the distance the "two cultures" of literature and science have drifted apart since 1836, when Emerson could so confidently assert: "The true philosopher and the true poet are one, and a beauty, which is truth, and a truth, which is beauty, is the aim of both." Science for Emerson was not a rarified method or a specialized field of study but the highest form of mental action. For him, the true man of science and the American scholar were one and the same, and science was the dynamic heart of America.[1]

"All science has one aim," Emerson proclaimed: "namely, to find a theory of nature." He knew he could point the way: nature was the embodiment of divine mind. Through knowledge of nature the human mind embodied itself, until at last body and spirit were not the sundered halves of Creation but one great whole, God walking forth anew into the world. Through science the intellect seized the fragmented and unmeaning phenomena of the world and forged them into meaningful, productive wholes. The "method" of science would be the model for thought and action in the world. Yet beneath this sublime confidence lurked a deep uncertainty: were not the phenomena of the world fragmented and without meaning? Was not truth coming apart, into many truths or none? Against this fear, Emerson proposed his marriage of mind and nature by creative reason as the last best hope to make the world whole again, and his proposal carried all the urgency of his conviction that only a collective dedication to the process of making truth would carry America safely into the future.[2]

For unlike scattered and quarreling humanity, nature was self-evidently single, bound into one by the Law that created it and guided it still. Therefore the study of nature seemed to Emerson the one sure path to truth. As a result, science permeated his thought and writing at every level, from its deepest structure to his most casual analogies. The very method of thought,

: 4

of science, and of nature was one and the same: the search for the clairvoy-
ant insight that galvanized the chaos of particulars into their rightful order.
Poetry, too, participated: while science was the vehicle of truth already existing, poetry was the expression of truth in new forms. Moreover, the law
of nature that governed both science and poetry was the same with the
moral law that ruled every individual human being. As Emerson said, "the
whole of nature is a metaphor of the human mind. The laws of moral nature answer to those of matter as face to face in a glass. . . . The axioms of
physics translate the laws of ethics." By this metaphoric transference, the
key working concepts of nineteenth-century science—polarity, magnetism,
Newtonian optics, chemical affinity, electrical circulation, the balance of
nature, progressive development, geologically deep time, anatomical corre-
lation of parts, physical geography—became so much a part of Emerson's
familiar, accepted and unquestioned working vocabulary that they dropped
virtually out of view.[3]

Early nineteenth-century intellectuals called repeatedly for organic connection through both "natural" and "moral" law. This call drove Emerson's own progress from the minister's pulpit to the public lectern, as he
sought to expand the connective energy of divine purpose from the sacred
precincts of the church to the far corners of the secular world. Nature's creative power would dissolve ancient and failing social institutions like fog in
sunlight. For long generations, nature had been merely the supplement to
revelation, but Emerson saw nature as revelation's source. All the old
sources—scripture, the church, the ministry—had shown themselves to be
merely human constructions after all, tainted with vanity, rusted by tradition, and corrupted by history. The only real guide to truth was that sole
realm cleansed of history's impurities and defined not by the treacherous
community of human beings but by the exclusion of humanity altogether.

Students of the Enlightenment recognize the turn to the authority of "the
laws of nature and of nature's God," in Thomas Jefferson's foundational
phrase, as a long-standing tradition, and it is possible to see Romanticism as
an attempt to fulfill the promise made by Enlightenment reason. As Joel
Porte has shown, Emerson was nowhere closer in his affinity to the eighteenth century than in his belief in "moral law," whereby truth and morality were one. The difference between the two eras lay not in Romanticism's
alleged rejection of reason or science but in a shift in connective principle
from outer to inner, or, as Romantics often styled it, from "mechanism" to
"organicism." This shift confirmed and intensified the moral dimension of
natural law by relocating its power from external force to internal suasion.
That is, *natural* law acted not by exercising legislative command from without but by organizing irresistibly from within, and obedience was brought
about not by crude mechanisms of force but by innermost conviction of the
right and true, the "moral," path. This shift had the crucial consequence of
placing the creative mind at the center of moral life. As the organizing

power of the universe was creative and dynamic, so was the organizing power of the human mind. Instead of a passive recipient of messages relayed by the senses, the mind was an active and potent creative agent in its own right, a chip of divinity on everyone's shoulders.[4]

The teenage Emerson already understood this. In 1822 he defined the "Moral Sense" as "a rule coextensive and coeval with Mind." Where does our intuition of right and wrong come from? "Manifestly not from *matter,* which is altogether unmoved by it . . . but from a *Mind,* of which it is the essence. That Mind is God." As this passage suggests, locating moral law in the mind allies it with the imagination, defined by David Robinson as "a power by which man was able to perceive a perfection which transcended the world ordinarily available to his senses." Of course a gap yawns open between the ideal of perfect unity and the fragmented world of our senses, but that gap, far from defeating moral law, *empowers* it; and that dynamic power keeps the soul "in a continual process of flowing toward that unity and away from individual isolation." This dynamic flow, which makes the Many tributary to the One, characterizes Romanticism. As the margins of a leaf connect by veins to the stem, as leaves connect by branches to the tree's living core, so does this principle of dynamic internal creative law draw together writers from Wordsworth and Coleridge to Sir Humphry Davy and John Herschel, from Goethe and Emerson to Edgar Allan Poe and Harriet Beecher Stowe.[5]

This connective, dynamic principle placed science at the heart of antebellum literature and culture. Science became the highest, the most Godlike, culmination of human achievement, sheer intelligence ordering and vivifying the "desultory phenomena" of the material world in the richness and clarity of idea. Intellectuals—both religious and secular—used science to consolidate and advance an argument about the need for total cultural coherence against the specter of atomized incoherence and the baneful influence of materialism. As an 1835 reviewer declared of Coleridge in the conservative *North American Review,* the "false philosophy" of materialism, the "mere mechanical understanding," was "a *malaria* to the general intellect, a brooding fog over the whole mind of the age." Materialism "denies life, and strikes death through the philosophy of the human soul," reducing the "living principles" of the sacred word "to a dead letter." Given this grim diagnosis, the cure was to turn not away from but toward science, the pursuit of general and ultimate principles by which a germinal idea organized scattered and inconsequent facts into progressively more meaningful and coherent wholes. By making this "Method" the quintessence of science, Coleridge had shown science to be the essence of intellectual culture. The creative mind acted by organizing the chaos of experience into meaning, and the scientific mind became the model of creative thought at its most sublimely potent. In Coleridge's words, "every *principle* contains in itself the germs of a prophecy; and as the prophetic power is the essential privilege of science, so the fulfilment of its oracles supplies the outward and (to men in general) the *only* test of its claim to the title." Or, as Emerson himself summarized, also in 1835, "Every principle is an eye to see with."[6]

The horror was life *without* principle. Emerson's 1835 journal entry continued with the melancholy reflection, reminiscent of his poem "Days," that "facts in thousands of the most interesting character are slipping by me every day unobserved, for I see not their bearing, I see not their connexion, I see not what they prove. By & by I shall mourn in ashes their irreparable loss." The creative mind placed the burden of chaos and loss squarely on his own shoulders. By extension, the chaos and creeping skepticism that such powerful theories of total coherence were erected to combat were the fault of society's delegated intellects. It was their job to stem the advancing tide of skepticism through public interventions—sermons, lectures, articles, essays, books—in a campaign of popular education and recruitment. In this system there could be no meaningful knowledge in a vacuum. Moral law directed the flow of energy and imagination away from individual isolation and atomism and toward wholeness and unity, but this flow would disperse into droplets unless everyone was similarly directed, converging through individual instances of active choice toward a society unified in law, "science revealed, in fulness and unity; fulness without perplexity, unity without sameness"—or democracy without mobs.[7]

The alternative bore hard on the imaginations of Emerson's contemporaries. "Cover Nature with darkness, the world with ruin. . . . Then you may safely claim the administration of Chance. DARK DARK DARK DARK." Thus worried Emerson at eighteen. He recurred to this theme several times over the next few years, most memorably shortly after his twentieth birthday, when he related a nightmare vision of a universe in which moral law did not govern: "the Chaos of thought would make life an insupportable Curse." But "rend away the darkness, & restore to man the knowledge of this principle, and you have lit the sun over the world & solved the riddle of life." This young Oedipus, master of the riddle of the Sphinx, imagines his inheritance: "Mind, which is the end and aim of all the Divine Operation, feeds with unsated appetite upon moral & material Nature, that is, upon the order of things which He has appointed. It is perpetually growing wiser and mightier by digesting this immortal food." Without mind governing the world, the world will eat you alive; but *with* mind, you will thrive by eating the world. The imperial mind can slake its hunger only by feeding on the globe itself, assimilating all "moral & material Nature" to its own needs and desires, to its own organizing idea.[8]

Such moments of panic show how urgently Emerson needed to erect and defend a center that would hold all together at a historical moment when everything seemed on the verge of flying apart. This ominous premonition was not limited to the United States. In his omnivorous reading, young Emerson heard such concerns echoed and reechoed. Dugald Stewart concluded anxiously that even the ministrations of reason could not conquer the terrorized people of revolutionary Paris, and he reassured his British readers that Baconian science had erected a bulwark against atheism. Coleridge prescribed an integrated church-state to stave off French-inspired dissolution into tyranny. Sir Humphry Davy asserted (in a textbook on agricultural chemistry) that true English patriotism grew from the cultivation of the soil. Madame de Staël warned that the philosophy of sensualism had nearly plunged "the universe and man back again into darkness." Carlyle lamented that "society, in short, is fast falling in pieces; and a time of unmixed evil is come on us." As each of these writers rushed to affirm, only the rock of principle would hold against the corrosive river of mechanism, materialism, sensualism, skepticism.[9]

Despite the overheated rhetoric of, say, Coleridge or Carlyle, the place of Romantic science in this mix was not to reject or deny the accomplishments of "mechanistic" science, for they were too significant and useful simply to discredit. Instead, mechanistic science had to be renovated and incorporated into a new, reconceived, "higher" science in which the lost unity of self and nature would be restored. The widely admired Copernican reversal attributed to Kant was just then, in the words of de Staël, about to place "the soul of man in the centre, and to make it, in every respect, like the sun, round which external objects trace their circle, and from which they borrow their light." In this way Kant's philosophy would resolve the troubling dualism of mind and body, the source of mechanistic thought, and, happily, do so in a manner not new but implicit since the beginning of science. Sir Francis Bacon himself had understood the great truth that all the branches of science were "reunited" by one "general philosophy." The anatomizing that severed the intellectual faculties from each other brought disease and weakness. Health and restoration lay in the unity which understood that "sensibility, imagination, reason, serve each other. . . . The exact sciences, at a certain height, stand in need of the imagination. She, in her turn, must support herself upon the accurate knowledge of nature."[10]

This "Copernican turn," by which the light of the mind illumined the phenomena of nature and its gravitational force drew them into an ordered system, characterized the early 1800s and generated what historians of science now describe as a second scientific revolution. According to historians of science, "the federation of disciplines we call science" was formed in this period; "Science, in our sense, once held to be more than two thousand years old, is now credited with less than two hundred years of history." And all the great questions asked by science—"What is nature? . . . What are the sciences? On what principles do they rest? What are their limits? How are they to be taught, learned, practiced?"—at this time "came to be perceived as questions of self-understanding."[11]

What, then, were the characteristics of science as Emerson knew them? First, science was moral. Emerson took his line about the axioms of physics translating the laws of ethics directly from de Staël, who affirmed that both the physical and moral universes were thoughts of God, "transfused into two different languages, and capable of reciprocal interpretation. . . . Almost all the axioms of physics correspond with the maxims of morals." Good science was not in competition with religion but, as Bacon had promised, its "handmaid," the means by which man would recover his lost harmony with nature. Every science could claim some kinship with religion, for in the broad and accommodating lap of natural theology, every science could claim to highlight some facet of God's cosmic order. John Herschel, in a book that made a strong impression on Emerson, offered a panoramic survey of the sciences, triumphantly throwing open window after window to reveal a universe ever more dazzling in its connectivity.[12]

Second, science was synthesis. Herschel's rewriting of Bacon updated Bacon's idea that underlying all the phenomena of nature would be found a very few primal generative principles, perhaps even only one. This fundamental conviction secured the era's characteristic commitment to synthesis, which gathered ever more phenomena under ever fewer principles, as well as the equally characteristic detailed empirical researches that by rigorously narrowing the field promised to close in on the ultimate principle at last. Examples could be cited from every field of science: Hans Oersted's researches into electricity and magnetism proved them to be but two forms of a single force; Theodor Schwann's discovery of the cell proved it to be, as he said, the "common developmental principle for the most diverse elementary parts of the organism"; the search by Goethe, Lorenz Oken, Étienne Geoffroy Saint-Hilaire, and Richard Owen for the "single Ideal Plan or Type" underlying the variety of plant and animal forms would show the many to be, finally, one; botanists hoped to replace the clumsy and outworn "artificial" system of Linnaeus with a "natural" system of botanical nomenclature; Justus von Liebig worked throughout his life to unite physics, chemistry, and physiology into one science governed by "the laws of vitality." And so on. A complete review would cover the entire nineteenth century's history of science.[13]

Third, the synthesis offered by science was not static but dynamic and progressive. When the physicist Joseph Henry, in his 1850 address to the American Association for the Advancement of Science as its outgoing president, asked, "What is science? Is it a collection of facts?" he answered with a resounding no. "Science does not consist in a knowledge of facts but of laws. It essentially relates to change, is dynamical rather than statical." Ultimately the principle of connectivity would unite the physical and organic sciences, at least through metaphor. As Walter D. Wetzels notes, an essential part of the Romantic credo was "a 'higher' form of physics in which the metaphor for the universe was not the cosmic clock but the cosmic organism." Drawing the key metaphor from biology rather than from Newtonian physics altered the conception of the universe in a fundamental way.

Whereas Newtonian physical processes are fully reversible, biological processes are not. The trajectory of planets orbiting the sun looks essentially the same in either direction, and indeed, according to the laws of physics, "all natural processes should work equally well forward and backward." Yet organic processes are dramatically irreversible. Unlike planets, organisms trace a one-way trajectory through time, from birth and growth to decay and death. The universe as a cosmic organism is directional, progressive, even purposive. Instead of running on its own like a clock, it requires continual stoking by a creative life force, restoring to God a continuing generative role in his Creation.[14]

Finally, science was still widely diffused both geographically and culturally, not yet strongly centralized around universities or industrial laboratories, nor yet specialized into rigorously defined professional fields. Thus Romantic science, as David Knight has shown, was not a coherent school of thought with a firm roster of converts somehow opposed to "real" science. More simply—and more provocatively—"in this period of political and intellectual ferment . . . some men of science and some writers and painters found that they had things to learn from one another. . . . It is only because we are still dominated by the idea of the Two Cultures that we find this surprising; in the early nineteenth century natural science was a component of the one culture of Western Europe." Emerson, in finding things both to learn from and to teach to men of science, should not be conceived as aligning himself with a special cadre of specialists or as a participant in some Romantic fringe movement sealed off from the genuine article. He was placing himself in the center of intellectual ferment in his time; indeed, he became one of its leading voices.[15]

The operative division in Emerson's day was not the one that emerged later between science and imagination, but the one, in de Staël's words, between "the region of universal ideas," where every science touched upon all the rest, and the "mechanic arts" of labor and material production, ruled over not by the synthesis of intellectual culture but by Adam Smith's principle of the division of labor. It was against this world of trades and manufactures and commercial enterprises that science and literature formed one single intellectual culture, unified by the belief that the various human capacities could be integrated into a harmonious system centered on the creative power of the human soul. As de Staël added, "Literature and science reflect alternate light upon each other; and the connection which exists between all the objects in nature, must also be maintained among the ideas of man." In short, "The universe resembles a poem more than a machine."[16]

However, no one urges a marriage between partners who are not separate, and de Staël, like Emerson, reminded her readers that her vision of unity was urged against a background of growing division. "The scholar has nothing to say to the poet; the poet to the physicist; and even among savans, those who are differently occupied avoid each other, taking no interest in what is out of their own circle. This cannot be when a central philosophy establishes connections of a sublime nature between all our thoughts. The scientific penetrate nature by the aid of imagination. Poets find in the sciences the genuine beauties of the universe." Emerson wanted to be a "central philosopher" in de Staël's sense, establishing sublime connections between poetry and science and reminding each of their need for the other. His warnings to science were quite explicit: "Empirical science is apt to cloud the sight, and, by the very knowledge of functions and processes, to bereave the student of the manly contemplation of the whole. The savant becomes unpoetic." His warnings to poets were rather more implicit, contained in his knotty, fact-dappled prose, which continually roped the poet back to de Staël's "genuine beauties": "Nor has science sufficient humanity, so long as the naturalist overlooks that wonderful congruity which subsists between man and the world; of which he is lord, not because he is the most subtle inhabitant, but because he is its head and heart, and finds something of himself in every great and small thing, in every mountain stratum, in every new law of color, fact of astronomy, or atmospheric influence which observation or analysis lay open."[17]

The finding of self in nature draws men to science; the poet, too, "invests dust and stones with humanity, and makes them the words of the Reason. The imagination may be defined to be, the use which the Reason makes of the material world." Imagination is reason's projection into the material world *beyond* mind, the necessary mediator between "ME" and the "NOT ME" without which reason would be, literally, unrealized. Is reason, the truth-seeing faculty of science, then prior to imagination and the truth-making faculty of poetry? No, for the division is false: science grasps the world through imagination, as much as poetry needs reason to see "all things in their right series and procession." Both poetry and science are formations of reason working through imagination to create new integrated wholes, new forms for eternal truths. In Emerson's view, nature is not "built up" around us so much as "put forth" through us, "as the life of the tree puts forth new branches and leaves through the pores of the old." Both the fine and useful arts, and science, too, are all "a nature passed through the alembic of man," original creative force manifested in different forms, different faces of the same truth. Poetry may, in Plato's words, come "nearer to vital truth," but only if it lifts its own lamp next to that of the scientist and shows that "when the fact is seen under the light of an idea, the gaudy fable fades and shrivels. We behold the real higher law. To the wise, therefore, a fact is true poetry, and the most beautiful of fables."[18]

Facts became poetry during Emerson's revelation at the Paris Muséum d'Histoire Naturelle, but this moment in 1833 marked the culmination of years of preparation. Emerson had already read deeply and meditated keenly on the nature of scientific knowledge, and his inspiration carried through still more years during which he incubated and eventually completed *Nature* (1836), a cosmic progress report that opened the way for renewed development. The dramatic setting of his Paris revelation and the lyrical intensity of his early science lectures have made Emerson's early indebtedness to science quite apparent, but what happened after 1836 is much less clear. Despite his avowal in 1833, Emerson certainly did not "become a naturalist" in the way of, say, his friends Henry Thoreau and Sarah Ripley. Indeed, three years later he warned, "He cannot be a naturalist, until he satisfies all the demands of the spirit. Love is as much its demand, as perception." Emerson's stand would be at the midpoint where love and perception perfect each other, attuning mind and nature to each other, detaching objects from personal relation to see them in the light of thought, and kindling science "with the fire of the holiest affections" to send God forth anew into his Creation. With such a goal, Emerson became interested in science less as a practice than as a discourse, not as a way of doing so much as a way of thinking, seeing, and talking.[19]

Emerson's way of talking about science had breathtaking implications for America. As Leonard Neufeldt points out, "Among literary figures Emerson was virtually alone in his endorsement of the possibilities of technology and science for the individual and the culture." In his treatment of science, Emerson was "more comprehensive than Poe, more ambitious and philosophically rigorous than Thoreau or Hawthorne, and more practical than Whitman." Emerson's inclusive definition of science as, in Neufeldt's words, "a quality of mental action" amounted to something more than the Romantic "marriage" of science and poetry, for it abolished the distinction between them by making each a form of the other. Similarly, as Neufeldt continues, "Emerson rejects the popular distinction among artists of his time between the mechanical and the natural, for machines have as their ultimate origin 'the same Spirit that made the elements at first.' . . . A machine is the wit and will of man combined with the will of nature." Thus technology, for Emerson, was not the conqueror or violator of nature, but nature itself in another form. Conversely, nature was no less than the art or technology of God, the means by which God brought change—progress—into the world. Man—that is, American industrial men—by taking hold of nature's material means and scientific laws for his own purposes, was melding man, God, *and* nature into an exciting new union unlike anything the world had ever seen.[20]

This ideal, however, was not and could not be stable, and Emerson responded through a series of personal and cultural crises to its ever-impending collapse, reiterating, strengthening, adjusting, qualifying, all the while moving ever further into the popular culture of his time. Nor is it fair to assume that Emerson's high-wire act was performed over a safety net. As Neufeldt observes, "It is simply a question of how well the essay, the mind, and the culture can keep the equilibrium even as they continue to move." The very failure of Emerson's persistent but uncompleted work on a "Natural History of the Intellect," which was to have applied the method of science to the study of the mind, suggests that Emerson's engagement with science was, from first to last, not an escape into complacent certainty or unthinking optimism but a fiercely sustained campaign to read order into a universe that persistently threatened to fly apart. Robert Richardson concludes that "the more strongly Emerson felt the case for tragedy, for waste, for chance, for loss, the more he looked to science as well as to his own convictions for proof of underlying order." As Emerson stated in a lecture given during his second trip to England (1847–48), "There is no day so dark but I know that the worst facts will presently appear to me in that high order which makes skepticism impossible." In a later version of this lecture, Emerson explained the nature of that order: "In all sciences the student is discovering that Nature, as he calls it, is always working, in wholes and in every detail, after the laws of the human mind. . . . the Intellect builds the universe and is the key to all it contains." The intellect, then, had both the privilege and the burden of creating the universe in its own image. A dark universe bespoke a dark soul, a soul fallen away from the truth. In the best Realist tradition of science, Emerson refused to believe that the order the mind looked into the universe was not really there.[21]

THE CULTURE OF TRUTH

When the institution of the church failed him by refusing to negotiate the terms by which truth would be constituted, Emerson turned to science instead as a ground and model for free thought constrained by law. Today the word *science* has lost the general overtone of "rational knowledge" that it still carried in Emerson's day, although even then the more specialized usage, "the theoretical and methodical study of *nature*," had begun to harden. Thus, to restrict Emerson's use of the word *science* to the successful experimental method of the objective natural sciences—physics, chemistry, biology—is to impose on Emerson a definition that he would have rejected as an example of the "half-sight" of empirical science. To read Emerson through his own understanding of science is to explore his role in a moral economy of knowledge, its construction, and its circulation.[22]

For Emerson, science included moral and natural philosophy, the natural religion of William Paley's *Natural Theology* and the Bridgewater Treatises, the new natural sciences, and the older natural history presented in such popular treatments as James L. Drummond's *Letters to a Young Naturalist*. These works shared a general concern with the planting, irrigation, and nourishment of an integrated social/natural universe through the circulation of a common core of beliefs. This "cultural agreement," as Donald Pease characterizes it, would convert the dangerous energies of the American and French Revolutions into constructive forces guaranteeing the cohesion and integration of a postrevolutionary civil society. Romantic scientists

Establish the mind Steven Shapin

played a crucial role in settling this cultural agreement. For instance, in 1830 John Herschel opened the book that would establish the modern scientific method by affirming: "There is something in the contemplation of general laws which powerfully persuades us to merge individual feeling, and to commit ourselves unreservedly to their disposal; while the observation of the calm, energetic regularity of nature . . . tends, irresistibly, to tranquillize and re-assure the mind, and render it less accessible to repining, selfish, and turbulent emotions." Or, in Emerson's paraphrase, "The first effect of science is to stablish the mind, to disclose beneficent arrangements, to remove groundless terrors. . . . the survey of nature irresistibly suggests that the world is not a tinderbox left at the mercy of incendiaries." Science turned the isolated self toward both nature and society, calming the terrors of atomism and materialism with the deep stability of natural law.[23]

Emerson's famous turn to nature was, therefore, not to the seemingly unmediated nature of an appreciative stroll in the woods, but to nature as rationalized by such natural philosophers and scientists as Herschel. For Herschel, science was "the knowledge of many, orderly and methodically digested and arranged, so as to become attainable by one." The handle by which the many were to be grasped as one was "law," a metaphor that attributes the regularities of nature to an act of legislation so absolute as to ensure "implicit obedience." This metaphor allows the natural, social, and spiritual worlds to merge, an elision basic to Emerson's enterprise: as David Robinson notes, "The scientist's capacity to move from fact to law was the evidence that the spiritual world showed itself in the material." Since laws encode action in language, the fundamental metaphor of law unites the realms of natural science, cultural politics, *and* social discourse into one broad front, the "one culture" of Anglo-Euro-America before America's Civil War and the post-Darwinian debates.[24]

The source of the power of knowledge was the God of all knowledge, and this notion still underwrites, however distantly, the authority accorded to scientific knowledge today. Cultural critics still echo Emerson's fear that without a unifying and transcendent moral/natural law, all is "DARK DARK DARK DARK." But there is another way in which knowledge can be moral, a way reflected in Emerson's own ideals and practice. Emerson was attracted not just to the texts produced by science, with their technical explications of natural law, but to the moral community articulated by science. As Steven Shapin argues, science searches for truth by establishing who can be trusted, for little of what anyone knows can be personally experienced. Skepticism and distrust cannot constitute an entire system, but can exist only "on the margins of trusting systems." Shapin argues that science is, therefore, a moral endeavor: "Knowledge is a collective good. In securing our knowledge we rely upon others, and we cannot dispense with that reliance. That means that the relations in which we have and hold our knowledge have a

Natural & Social & Spiritual

Knowledge & Moral

: 14

moral character, and the word I use to indicate that moral relation is *trust.*"[25]

This argument may seem to fly in the face of conventional wisdom, which holds, after Darwin, that science is amoral—or, after Hiroshima, downright immoral. Both attitudes exile scientific truth from the lived human world, making of it either a transcendent ideal or an ideological tool. But I would argue, with Shapin, that truth is not located in transcendence, "out there," but "down here," in the painstaking process by which claims to truth are tested, argued, modified, accepted, or rejected, a process that may be in service to ideology but is equally capable of refusing that service. Thus truth has a history after all:

> The history of truth can be a social history because what we know about the world is arrived at, sustained, and recognized through collective action. Against dominant romantic and heroic views, it is argued that no single individual can constitute knowledge: all the individual can do is offer claims, with evidence, arguments, and inducements, to the community for its assessment. Knowledge is the result of the community's evaluations and actions, and it is entrenched through the integration of claims about the world into the community's institutionalized behavior. Since the acts of knowledge-making and knowledge-protecting capture so much of communal life, communities may be effectively described through their economies of truth.[26]

Extending this reasoning, I suggest that Emerson's central role in antebellum American thought was not the establishment of any one particular truth but of a "culture of truth," a moral economy of knowledge created and maintained by those deemed trustworthy. The church lost Emerson's trust by refusing to renegotiate the terms of truth even in the face of the new science, and so science itself, that continual and free negotiation by which the experiences of the many could be grasped in one law, became Emerson's new standard. To be sure, the nature and guarantee of that "freedom" wrapped science and politics together, and the constraints of that "law" were interlaced with cultural definitions; but science then (or now) cannot merely be reduced to a readout of ideology or power. It is far more interesting to consider how elites have sought to make themselves powerful by capturing knowledge to their own ends—or, more accurately, by instilling mutual consent, "commanding through obedience."

Emerson claimed that his quest for truth was grounded in the moral agency of God rather than in collective human action, yet finally he counted not on divine authority from above but on the thought and actions of believing members to generate and sustain his revived community. Words and actions converted principles of belief into forces that energized and constrained the community's members with a power that can only be conceived

[handwritten margin notes: "Plant must have sun, light, soil —", "See Weaver paper", "Tobias Wolf", "Old School"]

as sacred. Thus outer law became inner discipline through "self-culture"—a figure that united Romanticism's "touchstone" metaphor, organicism, with the traditional American and Unitarian conception of the individual as a growing plant. Hence my use of the word *culture* despite the burdens it carries today. It was precisely the word that nineteenth-century intellectuals themselves used to name both the process and the result by which the individual self, through "self-culture" or education, would "grow" and join other selves, collectively constituting the wider "culture" of educated and enlightened humanity. This progressively widening circle of "shared understandings" was "realized, stored, and transmitted" largely in language.

Anthropologists dating back to Johann Gottfried Herder have assumed that all social groups have distinctive "cultures." This relativistic sense of the word applied pressure on Emerson's conviction that all cultures, when freed of artificial distortions, would converge on a single, higher truth—a kind of ecumenical absolutism that was itself a defining characteristic of culture in antebellum New England. A related New England characteristic was the emphasis on culture itself as a process—in Emerson's phrase, "a discovering of human power." Truth was not an end point. That would be the death of the human species, as Carlyle made clear: "could you ever establish a Theory of the Universe that were entire, unimprovable, and which needed only to be got by heart; man then were spiritually defunct, the Species we now name Man had ceased to exist. But the gods, kinder to us than we are to ourselves, have forbidden such suicidal acts. As Phlogisten is displaced by Oxygen," Ptolemy's solar system by Kepler's, and so on down through the continual displacements of old scientific knowledge by new, "the process does not stop. Perfection of Practice, like completeness of Opinion, is always approaching, never arrived; Truth, in the words of Schiller, *immer wird, nie ist;* never is, always is a-being."[27] *[handwritten: Like Derrida — Democracy to come]*

[handwritten margin: Truth not an end point]

[handwritten margin: David Robinson]

The notion of culture as a process of continual growth held together the culture of truth. As David Robinson observes, "self-culture" as a moral ideal was effective because it was ever approached but never reached. Thus the discipline of continual growth, originally the means to the end of static perfection, became an end in itself. One was forever in a state of education, the "self-cultivation" or "educing" of the soul's various hidden faculties and capacities by experience in the outer world. Emerson read in Sampson Reed's *Observations on the Growth of the Mind* that the mind's essential characteristic lay in its "power of acquiring and retaining truth," which was achieved only by active and continual exertion. According to Reed, the human mind was planted in nature like a seed in the earth and grew by feeding on matter, converting it into knowledge "for its own purposes of growth and nutrition." Truth was not "laid up in a mind for which it has no affinity . . . but the latent affections as they expand under proper culture, absolutely require the truth to receive them, and its first use is the very nutriment it affords." Emerson found a similar idea in Joseph de Gérando's

[handwritten margin: Sampson Reed]

Self-Education, which declared that *"the life of man is in reality but one continued education, the end of which is, to make himself perfect."* Emerson had already advised his own congregation to cultivate their minds "because it is necessary to know much in order to think well. . . . All study, all attempts to acquire knowledge . . . are efforts to acquire truth; and truth is the mind's home."[28]

The basis of truth was God, but paradoxically, the pathway to truth lay in "self-trust." The two were, of course, the same: Emerson's congregation learned never to take truth from others. "To reflect—to use and trust your own reason, is to receive truth immediately from God. . . . A trust in yourself is the height not of pride but of piety, an unwillingness to learn of any but God himself." This duty of self-trust required that one "go aside from all manner of society," or, as Emerson would say even more strongly in "The American Scholar," stand in a state of "virtual hostility" to society: "In self-trust, all the virtues are comprehended. Free should the scholar be,—free and brave." At first glance such an ideal might seem to encourage the very atomistic, self-interested individualism the culture of truth hoped to prevent. Quite the opposite, however. The individual isolated from the falseness of society would find a deeper truth: "He then learns, that in going down into the secrets of his own mind, he has descended into the secrets of all minds." Since all minds originated in God, each individual was "not a cause, but a mere effect of some other Cause, and so a mere manifestation of power and wisdom not his own." Compose a whole *society* of such individuals, and it would be not fragmented but unified and coordinated by their willing acquiescence in the greater will of divine power. Self-trust, paradoxically, became the ideal means of social control and cohesion.[29]

The culture of truth thus bound the individual to society by the reciprocal action of self-culture. Each individual would grow and strengthen by assimilating ever more of the social landscape to his own interests, needs, and desires; society would grow and strengthen by assimilating ever more individuals to the collective agreement. Yet not all was harmony. The tension between the high-minded ideals of self-culture and society's materialistic demands split intellectual culture (including both literature and science) away from "materialistic" or "mechanistic" commercial society. One worked through internal power, the other by means of external force—that is, moral, natural, organic "culture" versus immoral, artificial, mechanistic "civilization." This was Coleridge's influential distinction: "But civilization is itself but a mixed good, if not far more a corrupting influence, the hectic of disease, not the bloom of health, and a nation so distinguished more fitly to be called a varnished than a polished people; where this civilization is not grounded in *cultivation,* in the harmonious developement of those qualities and faculties that characterise our *humanity.* We must be men in order to be citizens." This division would in time destroy the intellectual alliance between literature and science, with literature assigned to the humanities and science displaced to a realm beyond humanity. One sees this separation in Thomas De Quincey's memorable distinction between "the literature of *power*" and "the literature of *knowledge*," which he exemplified by contrasting Milton's sublime poetry with a "wretched cookery-book."[30]

For Emerson, however, this separation became a point of hope, not of defeat. In 1837 he complained in his journal that "the high idea of Culture as the end of existence, does not pervade the mind of the thinking people of our community," who failed to understand that "a discovering of human power" rather than trades and occupations and the "motley tissue" of common experience was "the main interest of history." Yet this state of affairs opened an opportunity: Could this human power "be properly taught, I think it must provoke and overmaster the young & ambitious, & yield rich fruits." Emerson continued with an allusion to Coleridge: "Culture in the high sense does not consist in polishing or varnishing but in so presenting the attractions of Nature that the slumbering attributes of man may burst their iron sleep & rush full grown into day. Culture is not the trimming & turfing of gardens, but the showing the true harmony of the unshorn landscape with horrid thickets & bald mountains & the balance of the land & sea." Here, "culture" was not the artistic ornament of the educated elite but the cultivation of knowledge as *power*. So the rules of enculturation were not about "trimming and turfing" the gathered truths of the ages, but about the unrelenting drive for more knowledge, higher truths, deeper power.[31]

Emerson borrowed one of his favorite proverbs from the geologist John Playfair: "A method of discovering truths is more valuable than the truths it has discovered." The burst of energy devoted to methodizing science—beginning with Kant and continuing through John Herschel, Coleridge, and William Whewell—was directed not to finding better boxes for existing knowledge, but to constructing and refining the dynamic engine that had been so dazzlingly effective at making new knowledge. In one of several attempts to renovate Bacon for the Victorian age, Whewell offered to trace "the condition of the progress of knowledge" by looking not to its products but to the *principles* of its production: "But the object of our perambulation in the first place, is not so much to determine the extent of the field, as the sources of its fertility. We would learn by what plan and rules of culture, conspiring with the native forces of the bounteous soil, those rich harvests have been produced which fill our garners." This broad-based metaphor of "culture," loaded with allusions to soil, growth, nutrition, and fertility, lies behind Emerson's declaration in *Nature* that "all the facts in natural history taken by themselves, have no value, but are barren like a single sex. But marry it to human history, and it is full of life." The "ray of relation" that connected humanity to all the beings of nature was not passive but active, a vital penetration of nature by mind whose success could be measured by its fruitfulness. Truth, grown like a crop, was teased from the soil of nature, vital and nourishing as green corn.[32]

The strength of the culture of truth lay in its paradoxical synthesis of skepticism and trust, in which the ideology favoring lived experience covered for the vast amount of knowledge that must actually be taken on faith. "Take any practical action or cultural move in science," Shapin suggests; "then imagine that all trusting social relationships were canceled. Consider the void that would be left." Shapin argues that our understanding of science is deeply paradoxical: it is warranted with individual empirical foundations, by which we assert that we *know* only what we see with our own eyes; yet, strictly held, the notion that we believe only what we see would make scientific knowledge impossible. Nor is scientific knowledge just the aggregate of what individuals know. As Herschel recognized, science is the knowledge of many ordered "so as to become attainable by one." Or, in Shapin's words, "To the aggregate of individuals we need to add the morally textured relations between them. . . . The epistemological paradox can be repaired only by removing solitary knowers from the center of knowledge-making scenes and replacing them with a moral economy."[33]

Emerson worked through this paradox in the early 1830s during his most intense and formative encounter with science. In a sermon of 1831, he reminded his congregation that "to receive religious truth on trust—it is impossible. It cannot be taken on the bare affirmation of any being in the Universe. You don't understand what Truth means. Truth is not crammed down the throat of men, but is something to be understood. Until it is understood, it is not truth to the mind. You may kill a child by thrusting bread into its stomach; it is not food, it is death." The idea remained an important one to Emerson, and as anyone who has faced a classroom knows, it is indeed the key to learning: only the knowledge one owns for oneself is truly *known*, "under-stood." Yet the real impossibility is to refuse to take *any* truth on trust. Emerson was thinking about this problem in 1835 when he observed that

> every man, if he lived long enough, would make all his books for himself. He would write his own Universal history, Natural History, Book of Religion, of Economy, of Taste. For in every man the facts under these topics are only so far efficient as they are arranged after the law of *his* being. But life forbids it & therefore he uses Bossuet, Buffon, Westminster Catechism as better than nothing, at least as memoranda & badges to certify that he belongs to the Universe & not to his own house only and contents himself with arranging some one department of life after his own way.[34]

In 1792, Dugald Stewart had similarly pointed out that to establish the foundation of knowledge in the examination of objects and facts was not to say "that all our knowledge must ultimately rest on our own proper experience." Were this so, human knowledge could barely advance at all, as evidenced by uncivilized peoples. However, in a "cultivated society," children learned *language*, by which they became familiar with classes of objects and

Understanding what we experience

general truths, "and before that time of life at which the savage is possessed of the knowledge necessary for his own preservation, are enabled to appropriate to themselves the accumulated discoveries of ages." The implicit potential for a relativistic view—that truth might vary with language and culture—was resolved in this view by positing a common moral sense. This common sense was, in effect, a natural, inborn or genetic, part of the human mind, a notion that had given its name to Stewart and others of his school: the Scottish "common sense" philosophers.[35]

Emerson adopted their solution for his own. He assured his congregation, for example, that the best evidence for the existence of God and the truth of religion was the fact that such beliefs are so widespread. This proves that "they spring from the constitution of our minds, and so are found to be fundamentally the same, wherever there is a mind at all cultivated." What is "incredible" to us is so because it is inconsistent "with the constitution of our minds. The reason why we know that the Koran is not a Revelation from God, is because some of its laws shock our sense of right and wrong," whereas we receive the Christian religion as "the truth of God . . . because it is agreeable to our moral constitution." The firmest foundation for religious opinions, which otherwise will totter, "is to have them rooted and grounded in the laws of our own mind." In other words, to be educated in the truth is to know that truth to be natural, and thereby the touchstone by which others (like those of the Muslim faith) may be judged.[36]

Cultivation "roots" one to the local soil, but cultivation—the circulation of knowledge through language—also links the local soil to the universal earth. If we understand only what we experience, it is also true that we experience most of what we understand through the language of others. Thus the solitary knower is, as Shapin recommended, removed from the center of knowledge-making and replaced with a moral economy; or, in Emerson's words, by using the books of others we belong to the universe and not to our own house only. Yet since the language of others makes sense to us, or can be appropriated to our own nature, only so far as it can be assimilated to what we already know, our "growth" or "education" ties us ever more deeply into the moral economy that nourishes us, naturalizing the cultural agreements into which we are born, or "planted." Truth is a coevolution of mind and environment, both social and natural, like a river that makes the very banks that confine it.

As the example of the Koran shows, this formation relativizes truth at the deepest level by making it dependent on local circumstances that are then universalized as the law of human nature. However, this relativism is hidden from view by the apparent coherence of one's lived world. The individual looks out upon a deeply harmonious mental landscape cultivated by the culture of truth and illuminated by a single sun. Yet it is not a closed landscape. The key to the strength of the culture of truth is not its exclusiveness but, on the contrary, its evident openness, its flexible ability to incorporate everything such that even falseness, or what shocks, points back to the core of truth rather than beyond it. Thus Emerson could say that every church manifests *some* truth, for "every truth and every error refer to some reality. As every bud on the branch indicates the presence of vegetable life and every excrescence indicates it no less, though distorted by the worm that eats therein." *Every* belief, no matter how diseased or perverted—presumably even the Koran—might be traced "to some fact which gave to its perversions what semblance of truth they had,—to some eternal truth in moral nature."[37]

The gesture of trust made across difference to every human being extends the moral economy even to those beyond its cultural bounds. The greatest incivility of all lies, as Shapin reminds us, in withdrawing our trust in another's access to the world and that other's commitment to speak the truth about it. This is to withdraw even the possibility of *dis*agreement, to close down communication completely—in effect, to deny the other's humanity and cast him beyond the pale. "The great civility, therefore, is granting the conditions in which others can colonize our minds and expecting the conditions which allow us to colonize theirs. It is in this sense that a world-known-in-common is part of the moral fabric of ordinary social interaction." The culture of truth thus has the potential to extend itself democratically to other viewpoints, to accept pluralism as a necessary aspect of civil discourse. Conversely, it also has the potential to shut down the process of culture by denying certain individuals or groups their status as moral equals, effectively denying them their humanity simply by refusing to listen.[38]

Thus the culture of truth can confine or liberate, release or silence, be absolutist or revolutionary. It confines, for instance, when Emerson declares, "I wish to be a true & free man, & therefore would not be a woman, or a king, or a clergyman, each of which classes in the present order of things is a slave." America had eliminated kings, and soon Emerson would free himself from the clergy; but his comment dehumanizes women, and by implication servants and blacks, by excluding them from truth and freedom. On the other hand, as Emerson also discovered, the culture of truth liberates whenever an appeal to "higher" law beyond the "present order of things" opens the moral fabric of society, so that the lives and experiences of women and blacks can be interwoven with the rest by extending to them, too, the civil courtesy of free and truth-seeking discourse. For the battle is not over what counts as truth. Or rather, that battle is softened into the customary texture of life in a society in which truth is internalized by education and exchange rather than dictated from above. The battle is over *who* is allowed to participate in the ongoing process whereby "truth never *is*, always *is a-being*"—who, in short, is capable of "cultivation" and hence inclusion as a citizen of the culture of truth.[39]

METAPHOR: TOWARD A POETICS OF SCIENCE

Few writers were more densely metaphoric than Emerson, yet the paradox is that Emerson himself fought against the metaphoricity of language. What he sought was not linguistic play, but truth, the single reality beyond language. Words, said Emerson, are signs of natural facts, and natural facts signs of spiritual facts. For no word is this more relevant than *truth*. To say that truth is "cultured" or grown responds to the root meaning of the word in Old English: *trēow*, "tree." Truth, like a tree, is firm, solid, steadfast, as is a person whom one "trusts," because his word is his bond, his "troth," and will "endure" over time, or will resolve an argument by declaring a "truce"—truth being, as Emerson said, "the proper peacemaker of the world." Truth is an arboreal construction, a tree well rooted in the material world. Tested by time, it rears its crown toward the eternal heavens, lighted by the sun of reason and watered by the springs of knowledge—all extensions of the landscape metaphor so popular in antebellum writing. After all, "Nature is the symbol of spirit." Men are planted like trees in their native soil, to grow, flourish, flower, fructify, and decay. The organic metaphor for society explicitly links these individual processes to the cultivation of a state, both political and intellectual (or theological), all secured like leaves to branches and branches to trunk by the solidity of a single, shared truth.[40]

Coleridge exploited this metaphor of truth as a tree when he declared theology's precedence over language, history, and philosophy: "the SCIENCE of Theology was the root and the trunk of the knowledges that civilized man, because it gave unity and the circulating sap of life to all other sciences, by virtue of which alone they could be contemplated as forming, collectively, the living tree of knowledge." Under theology were ranged the various aids, instruments, and materials of "NATIONAL EDUCATION," or "the shaping and informing spirit, which *educing, i.e.* eliciting, the latent *man* in all the natives of the soil, *trains them up* to citizens of the country, free subjects of the realm." In short, constitutional states are grown, not made.[41]

This organic metaphor works by dissolving the Cartesian divide between mind and body. As Mark Johnson says, it puts "the body back into the mind"; or, as Emerson said, it projects the mind as an active organizing force into the body of the world. Either way, the concept of knowing as an activity destroys the hierarchical split between fact and meaning, putting in its place a theory of truth in which imagination is central to rationality—or, to repeat Emerson's Coleridgean aphorism, the imagination is "the use which the Reason makes of the material world." The act of imagination that centralizes scattered facts around a principle of order is itself "embodied" in the organic metaphor, as when James Elliot Cabot clinched an argument about the relationship of art and nature by stating that "the interest of

Art lies not in the facts, but in the truth,—that is, in the facts *organized*, shown in their place. It is not that we care more about stocks and stones than [the ancient Greeks] did, but that we hold the key to an arrangement that gives these things a significance they have not of themselves." The metaphor of organicism literally "organized" the world into meaning, giving the mind the key with which to unlock the universe.[42]

So, is such a metaphor linguistic play or truth itself? Since the early 1960s, linguists have contrasted the standard theory of metaphor, which divides literal or "objective" meaning from metaphor's "subjective" ornamentation, with various revisionist theories whereby metaphor is itself fundamentally constitutive of meaning. The revisionist theorists George Lakoff and Mark Johnson declare that "metaphor is pervasive in everyday life, not just in language but in thought and action. Our ordinary conceptual system, in terms of which we both think and act, is fundamentally metaphorical in nature." In language evocative of Emerson, they assert that metaphor allows us to see one kind of thing in terms of another, and thereby "unites reason and imagination." Metaphor, then, is "*imaginative rationality*. Since the categories of our everyday thought are largely metaphorical and our everyday reasoning involves metaphorical entailments and inferences, ordinary rationality is therefore imaginative by its very nature." Furthermore, meaning is always grounded in the acquisition and use of a conceptual system, "such that truth is always given relative to a conceptual system and the metaphors that structure it." This does not mean, they hasten to add, that there are no truths; for our conceptual system is "grounded in, and constantly tested by, our experiences and those of other members of our culture in our daily interactions with other people and with our physical and cultural environments." Truth is grown, then, constituted through the play of language.[43]

Yet not just any play will do. Whereas Lakoff and Johnson emphasize the constitutive role of language, their view is qualified by the cognitive anthropologist Naomi Quinn, who cautions that "metaphors, far from constituting understanding, are ordinarily selected to fit a preexisting and culturally shared model." So although metaphors may structure or "constitute" thought, they do so in interaction within a number of enabling constraints: preexisting cultural models, physical environments, demands for consistency and coherence that prohibit "mixed" metaphors or "unlawful" marriages of mind and matter.[44]

The process is demonstrated by Emerson when he writes:

Every truth that is clearly known to men is a sort of tribunal by which other truth may be tried. For it is a maxim that every truth agrees with all truth. Thus, after all experience has demonstrated the truth of the theory of Gravitation, if any theory should be proposed inconsistent with that law, we should pronounce it false at once. So every acquisition we make in any kind of knowledge is continually furnishing us with a better test for detecting the truth of other opinions. . . . Whenever a more excellent principle is presented, it condemns the less excellent.

The tribunal of truth allows the novel to be tried against the familiar, and accepted or rejected according to the standard of "consistency" or "agreement." Yet this is not simply a conservative standard. As evidenced in science, the same tribunal may accept the novel as "more excellent," condemning the familiar by revealing it to be inconsistent with that higher idea or law which truths are continually approaching. Truth, therefore, is not absolute, but progressive, and "scientific" truth becomes the touchstone by which progress may be tested.[45]

Science writing is not exempted from this debate about metaphor. According to revisionist views, scientific texts do not somehow escape from metaphoricity into a God's-eye view of the world, mirroring reality without mediation, but instead are constituted by their authors "through their intersection with other, multiple discourses"—social, political, religious, cultural. So in using metaphors from science, Emerson was not importing exotic figures from some remote zone beyond literature, but operating in a shared middle ground. A common metaphor such as "polarity," for instance, could be used to think about the technical nature of electromagnetic forces, the generative duality of the universe, or the "bipolarity" of the human mind. Moreover, as Jonathan Bishop observed, the scientist "is not merely providing images for the poet," for the scientist's own use of metaphor is simultaneously "poetic" and "scientific": the scientist who seeks to know the world "puts together his metaphor and thereby constructs the world his metaphor allows him to see." Such metaphors, as James Bono argues, function interactively as both a site and a medium of exchange, "among not only words or phrases, but also theories, frameworks, and, most significantly, discourses." They trade "on the capacity of metaphorical language to shift meaning" and function within—or even create—"an 'ecological' network driven by the tension-fraught need or desire both to 'fix' meanings and to disrupt, generate, and transform them." Scientific discourse, in particular, strives to enforce fixed and stable meanings even as it combats the excess of meaning brought by all words, meaning that escapes authorial control.[46]

For example, nineteenth-century scientists like John Herschel may have wished to confine the word *law* to observed regularities in nature, but the word's association with divine Logos allowed a wide range of writers, both scientific and nonscientific, to insist that the "laws" of science were not merely regularities of nature but moral decrees. Moreover, the ecological interactiveness of a metaphoric network makes it unlikely that any single metaphor can be traced to a definitive origin either within or outside science. Indeed, as Martin Eger notes, because of this interactivity, bits of scientific knowledge do not exert their effect in themselves like chunks of reality in a pudding of poetry, so much as "through the metaphors they support." For instance, Emerson was drawn to the study of magnetism because it offered support for his metaphor of polarity. On the other hand, Emerson's optics remained firmly Newtonian even though the new wave theory of light was attracting attention, for Newton's "corpuscular" theory of light supported a wide range of Emerson's favorite metaphors. Eger adds that through their metaphors "the sciences do indeed have action-orienting power"—itself a metaphor redolent of nineteenth-century polar energies, and suggestive, as Eger points out, of the hermeneutic circle whereby the part explains the whole, which thereby explains the parts. The whole of natural science becomes a hermeneutic for reading the "book of nature"—or, as Bacon put it, for becoming nature's "interpreter," enmeshing the entire enterprise from first to last in the problematics of language and the demands of cultural work.[47]

Even as Emerson threw a net of metaphor across the world, the world itself seemed to escape through its meshes. He used a metaphor from physics to suggest metaphor's frustrating limits: "As Boscovich taught that two particles of matter never touch, so it seems true that nothing can be described as it is. The most accurate picture is only symbols & suggestions of the thing but from the nature of language all remote." As a minister he longed for the moment when the imagery of religious truth would perish before the real thing: "the idea of heaven as a kingdom; of God as a king; of hell as a pit of fire; of Jesus as a judge upon a bench—these words must be pierced for the truth they contain"; for, like the leaves on which they are written, words "are poor and perishable materials, time shall wear them, fire burn them, the wind scatter them, but the truth of which they are the vehicle" would live eternally. Before he entered divinity school, Emerson entertained this as a problem in metaphysics: "Metaphysicians are mortified to find how entirely the whole materials of understanding are derived from sense. . . . A mourner will try in vain to explain the extent of his bereavement better than to say a *chasm* is opened in society. I fear the progress of Metaphysical philosophy may be found to consist in nothing else than the progressive introduction of apposite metaphors."[48]

The point of Emerson's turn to science was precisely to get *beyond* the mere succession of apposite metaphors, beyond the perishable language of men and things, and enter directly the mind of God. Hence his endless metaphorical play. Emerson unceasingly poured the universe through bits of itself, until the point was clearly to arrive at no one triumphant solvent metaphor but at the metaphorical relationship itself. Donald Pease has observed that the transparent-eyeball figure in Emerson's *Nature* is less *a* metaphor than *the* "metaphorizing power": "God himself serves only as the switch point. . . . we find not unity but relations at the origin of Emerson's ontology, and, in a remarkable reversal, all things become metaphors for this original relation." A clue to this urge is found in Emerson's earliest thoughts about God as "the Founder of the Moral law." Before the Creation, Emerson writes, God was solitary, and in that solitude "*relation* did not yet exist*," for there were no objects to which such relations as Creator and creature, or good and bad, could be affixed. From this observation, Emerson concludes that "moral laws are relative," and therefore did not subsist before the Creation made relation possible. So moral laws are the laws that govern not things but the relations between things, and moral knowledge is relational knowledge, subsisting in that space or interaction or convergence *between* subject and object. Metaphor became a way to dissolve the objects of knowledge into relational, or moral, knowledge—the true knowledge of which they are merely the vehicle.[49]

Emerson wished to burn away the object, the perishable body, to arrive at pure truth, a wish notoriously expressed when he imagined himself as a "transparent eye-ball": "I am nothing. I see all. The currents of the Universal Being circulate through me; I am part or particle of God. The name of the nearest friend sounds then foreign and accidental. To be brothers, to be acquaintances,—master or servant, is then a trifle and a disturbance." In this apocalyptic vision, all objects do indeed dissolve and flow before the penetrating eye of God/Emerson. He stands at last not self-willed but "in the will of God," before whom "the universe becomes transparent, and the light of higher laws than its own, shines through it." But as Lee Rust Brown has remarked, a transparent eyeball cannot see. Certainly it cannot see objects, for they, too, are transparent; but neither can it see relations, friends, even brothers. Metaphor, in dissolving objects, has dissolved relationship, even the closest of moral bonds between human beings. This is to become again God before the Creation, self-existent but wholly alone in the absolute in which meaning is meaningless. The ultimate morality is the utter annihilation of moral relation.[50]

In the *Timaeus*, Plato presented the eye as the giver of light to the universe. In a related metaphor, Emerson presented moral science as "the sun in heaven, by whose light we must walk." However, the Platonist within him urged him not to walk by but to *be* that light, the moral center whose radiance lit his universe into meaning and connected all to himself through a "ray of relation." Man was centered as the sun of the system—Kant's Copernican turn. The fully illuminated universe should be completely transparent, offering no resistance to the ray of higher law. But another metaphor intervened: light cannot be seen unless it falls on an object. As David Van Leer notes, "A truly transparent eyeball would have no opaque retina on which to focus the image." Hence, no vision without opacity. Sampson Reed wrote that "it is the tendency of all truth to effect some object. If we look at this object, it will form a distinct and permanent image on the mind; if we look merely at the truth it will vanish away, like rays of light falling into vacancy."[51]

The idea that vision required opacity was a favorite one in Romantic thought, starting with Kant's point that Plato, in leaving the narrow world of the senses to venture "out beyond it on the wings of the ideas, into the empty space of the pure intellect," was like the "light dove, cleaving the air in her free flight," feeling its resistance and imagining that flight would be still easier in empty space. Plato "did not observe that with all his efforts he made no advance—meeting no resistance that might, as it were, serve as a support upon which he could take a stand, to which he could apply his powers, and so set his intellect in motion." Friedrich von Schelling and Coleridge used the same idea to show that nature was the means by which the Absolute "objectified" itself and so made subjective knowledge possible. Thus art acquired its mysterious significance in Romantic thought, and nature itself became the realized art of God. Individual beings were the only site where universals might be grasped: "The universal does not attract us unless housed in an individual," mused Emerson.[52]

If the physical universe is God's way of "housing" the universal, intersecting the light to make truth visible, it is both the stepping-stone out of itself to Platonic ideas *and* the only site where those ideas have any meaning. To step out of the world into the vacuum of absolute truth is not only to leave the universe behind—it is to leave the truth behind. Like the press of wings against the wind, truth is not an essence but a relationship between bodied beings. Emerson's Platonic philosopher wars with his Romantic poet: Is the goal of science to be self-extinction in the final truth, or self-creation under the press of circumstances?

Emerson's refusal to deny either has momentous consequences. For one thing, it makes both options available: we can have our truth "out there," such that the statement of a relationship is really not a statement at all but an objective truth, *and* "in here," such that all statements may be constantly responsive to contingency, coordination, re-vision, and nevertheless be completely "true." If truth becomes an impossible ideal that we approach only asymptotically, our earthly experience of truth can be only the process of striving toward that impossible ideal. The circumstance of being fallen creatures means that truth becomes a *performance*, existent to us only as we "experience" it, performing the actions that might take us closer to it. Truth is not a state of being but a way of living in a universe in which every relationship is a dynamic balance of mutually destabilizing opposites: a tightrope walk with no endpoint in sight.

In conducting this performance, we are constantly using the world to embody our own thought, creating objects so that we may see reflected from them the light of our own eyes. "What we are, that only can we see," Emerson declared. As transformative agents we deploy the world as technology, whether metaphoric or physical, by which we shape it to our own designs, secure in the knowledge that we are divinity walking forth on the Earth, and hence an essential part of the world's design. We can do no real harm.

Nature is fluid

We can dissolve the world into ourselves because we have seen the essence of the world and it is—us. Bacon had said that the understanding must be "thoroughly freed and cleansed" of the idols of the tribe, the cave, the market, and the theater, "the entrance into the kingdom of man, founded on the sciences, being not much other than the entrance into the kingdom of heaven, whereinto none may enter except as a little child."[53]

Emerson echoed Bacon at the end of *Nature,* when he had his Orphic poet prophesy the end and the dawn of vision:

> Nature is not fixed but fluid. Spirit alters, moulds, makes it. The immobility or bruteness of nature, is the absence of spirit; to pure spirit, it is fluid, it is volatile, it is obedient. . . . Build, therefore, your own world. As fast as you conform your life to the pure idea in your mind, that will unfold its great proportions. A correspondent revolution in things will attend the influx of the spirit. . . . The kingdom of man over nature, which cometh not with observation,—a dominion such as now is beyond his dream of God,—he shall enter without more wonder than the blind man feels who is gradually restored to perfect sight.

The world cannot be mastered through force or outward forms. Only bring to the world a principle to see with, and it will coalesce around the power of your perfect vision.[54]

CODA: THE SPHINX

In the vision that concludes *Nature,* the commanding eye learns to see not by "observation" but by its own creative light. This imperial vision virtually wills nature into existence at its own command—an exhilarating metaphor about the power of metaphor to see, to illumine nature from vacancy into being. But metaphor alone is not enough. The poet's imperial dream will build no railroads. Or, as Bacon said in his fable "Sphinx; or Science," it is one thing for the Muses to ponder hard questions, delighting in their very difficulty, for the Muses are free "to wander and expatiate." It is quite another thing when those questions pass from the "Muses" of poetry to the "Sphinx" of science, "from contemplation to practice." Then the questions no longer delight but turn "painful and cruel," "torment and worry the mind, pulling it first this way and then that, and fairly tearing it to pieces. Moreover the riddles of the Sphinx have always a twofold condition attached to them: distraction and laceration of mind, if you fail to solve them; if you succeed, a kingdom." If Emerson's conclusion is consequential to the world, then it is not the Muses but the Sphinx who presides over *Nature.* What does the poet, the man of words, say when the Sphinx crouching at the crossroads demands of him not words but action?[55]

"The Sphinx" is the title both of Bacon's fable about science and of Emerson's most cryptic poem, in which he answers the riddle of Bacon's

Sphinx by creating a dialogue between poetry and science. This poem, originally published in *The Dial* in January 1841, was placed first in Emerson's first volume of poetry, and thus presided over all the rest of Emerson's poems. Later editors found it altogether too enigmatic and moved it further into the volume. Nevertheless, the Sphinx presided over all of Emerson's work—literally: he kept a small figure of a sphinx in his study, and the Old Manse, in which he wrote most of *Nature,* "had a sphinx-head door knocker." Emerson may have had Bacon's figure specifically in mind while he wrote the poem. Not only did he have Bacon much in mind generally, but years later he echoed Bacon's language:

> I have often been asked the meaning of the "Sphinx." It is this,—The perception of identity unites all things and explains one by another, and the most rare and strange is equally facile as the most common. But if the mind live only in particulars, and see only differences (wanting the power to see the whole—all in each), then the world addresses to this mind a question it cannot answer, and each new fact tears it in pieces, and it is vanquished by the distracting variety.

However, the poem itself contradicts Emerson's memory. It is the *poet,* who refuses particulars to live only in the whole, who is given a question he cannot answer.[56]

Bacon's Sphinx is a "monster," for science is "the wonder of the ignorant and unskilful." It is many-shaped, "in allusion to the immense variety of matter with which it deals," and it is female, "in respect of its beauty and facility of utterance. Wings are added because the sciences and the discoveries of science spread and fly abroad in an instant; the communication of knowledge being like that of one candle with another, which lights up at once." Claws signify "the axioms of science [which] penetrate and hold fast the mind, so that it has no means of evasion or escape." Bacon's Sphinx lives high in the mountains, for all knowledge is "a thing sublime and lofty," and it infests the roads, for "at every turn in the journey or pilgrimage of human life, matter and occasion for study assails and encounters us." And those who solve her riddles acquire thereby "a kingdom"—command over nature, the end of natural philosophy—and command over man, "for whoever has a thorough insight into the nature of man may shape his fortune almost as he will, and is born for empire." Finally, "the fable adds very prettily that when the Sphinx was subdued, her body was laid on the back of an ass: for there is nothing so subtle and abstruse, but when it is once thoroughly understood and published to the world, even a dull wit can carry it." Nor does Bacon omit to explain why Oedipus, the one man capable of subduing her, was lame: "for men generally proceed too fast and in too great a hurry to the solution of the Sphinx's riddles," and so she gets the better of them.[57]

Emerson's poem narrates an exchange overheard between the Sphinx

and a poet. A "drowsy" Sphinx with heavy lids and furled wings "broods on the world," wondering who will "tell me my secret." Science, sleepy and quiescent, is not yet called to action, for none have challenged her. Mankind, too, "slumbered and slept" while she awaits the "seer" who can tell her "The fate of the man-child / The meaning of man." Nature's grand cyclical harmony rolls on undisturbed:

> Sea, earth, air, sound, silence,
>   Plant, quadruped, bird,
> By one music enchanted,
>   One deity stirred . . .

Once sung into harmony by the poet Orpheus, Nature waits for the return of the singer, a new Orpheus. Hope turns to the infant, who, still part of nature, lies "bathèd in joy," the "sum of the world / in soft miniature," man, the microcosm of nature. But man, grown up, betrays this hope: he "crouches and blushes, / Absconds and conceals"; he is "An oaf, an accomplice, / He poisons the ground." The Sphinx climaxes her lamentations by repeating the words of "the great mother," nature herself, who, beholding man's infirmity and fear, pleads to know

> Who has drugged my boy's cup?
>   Who has mixed my boy's bread?
> Who, with sadness and madness,
>   Had turned my child's head?

It is at this moment that the poet speaks, responding not to the Sphinx herself but to her invocation of the "great mother": "Say on, sweet Sphinx! thy dirges / Are pleasant songs to me." Surprisingly, the poet offers no solution, for he sees no problem: he supposes that these "pictures of time" will "fade in the light of / Their meaning sublime." The poet enlarges his contemplation, for he rather enjoys the melancholy state of affairs painted by the Sphinx. What drives mankind, he proposes, is actually "love of the Best":

> The Lethe of Nature
>   Can't trance him again,
> Whose soul sees the perfect,
>   Which his eyes seek in vain.

The poet refuses any question the Sphinx is capable of asking. Instead, man's "aye-rolling orb / At no goal will arrive," for his yes-saying eye affirms everything in its orbit, rolling all into a "vision profounder" of the harmonious whole, a new heaven for which he gladly spurns the old. According to the poet, "Pride ruined the angels": *their* search for an answer, for knowledge, led them to challenge God; but shame restores them, for it is sweeter than pride—even pride in the self:

Had I a lover
  Who is noble and free?—
I would he were nobler
  Than to love me.

It is as if the transparent eyeball spoke. It is nobler to love the All than to love any particular individual. Our poet is a transcendentalist who sees himself at the heart of "Eterne alternation," connecting the alternating cycles of pain and pleasure through the pulses of love. He chides the Sphinx for her dull wits and gives her a cheeky prescription: "Rue, myrrh and cummin for the Sphinx, / Her muddy eyes to clear!" The Sphinx of science, according to the poet, has missed the point: the seer she longs for stands before her. She is so blind that she sees him not. The fate of man lies not in the world over which she broods, but beyond it, in a new heaven.

Emerson's Sphinx will have none of it. She repays the vaporings of the self-proclaimed seer with a riddle, the very riddle of the Sphinx: "Who taught thee me to name?" The only possible answer is, You did. Without science to name the particulars of the world—the palm, the elephant, the thrush the poem so lovingly catalogues—there could be no conversation. As she declares, "I am thy spirit, yoke-fellow; / Of thine eye I am eyebeam." Her pronouncement combines Plato and Bacon: the light of the eye is science, for by science the mind shapes nature into meaning, looks sense into the universe; knowledge is the light by which the eye sees. The poet is "yoke-fellow" to science, as the Sphinx goes on to insist:

Thou art the unanswered question;
  Couldst see thy proper eye,
Alway it asketh, asketh;
  And each answer is a lie.

Her real question, the meaning of man, is still unanswered. "Every man's condition is a solution in hieroglyphic to those inquiries he would put," posited Emerson in *Nature*. Similarly, the Sphinx's question cannot be answered by contemplation and escape to another world, but by action, living, in *this* one. Her claws sink deeper: the poet's "*proper*" eye does not *refuse* to ask, but must *always* ask, pressing each answer to the next question, for no answer is absolute; "each answer is a lie" because the only true answer to the riddle of the Sphinx is another question.

So take thy quest through nature,
  It through thousand natures ply;
Ask on, thou clothed eternity;
  Time is the false reply.

No answerable question can be asked of all nature, but only of a thousand natures, each by each, piece by piece: the whole can be grasped only by a particular. "Ask on, thou clothed eternity," commands the Sphinx, bequeathing her role to man; no answer is finite and final, for "Time is the false reply." With that, "Uprose the merry Sphinx," dissolved from flesh or stone to flow everywhere, imbue everything with light and power, and speak her single challenge through a thousand voices: "Who telleth one of my meanings / Is master of all I am." Each part of nature contains the whole, and all parts lead to the same great whole; but the only way to master the whole is to begin with a part, any part.

This brief dialogue between science and the poet asks how to rise from the lower world of particulars, the plain of ignorance, to the "sublime and lofty" realm of the Sphinx. Emerson's poet answers the call of science by rejecting her premises, protesting her earthbound goals, and condemning her dull wits and muddy vision. Emerson, surprisingly, has science answer that it is the poet who cannot see, for by rejecting the light of science he refuses to engage the one question that matters, the meaning of his life on Earth. Without science, the poet will remain in darkness; though without the poet, science cannot aspire to eternity. Each needs, and suggests, the other. Working together as "yoke-fellows," eye and light, they can illumine the universe into meaning and lead humanity into its rightful kingdom. For while we have been entertaining the Muses, the Sphinx has cornered us at the crossroads, demanding that we trade uncertainty and variety for decision and choice—contemplation for action.

CHAPTER TWO
··

Converting the World
KNOWLEDGE, SCIENCE, POWER

THE REPRESENTATIVE SCHOLAR: EMERSON AND BACON
Emerson began *Nature* by inquiring into "the end of nature," but he began
his first major public address by inquiring instead into the character, duties,
and hopes of nature's *interpreter,* "The AMERICAN SCHOLAR." Emerson's
American Scholar was a healer who would fuse the scattered and atomized
individuals of the young republic into a whole, a "nation." For the whole
had been fragmented: man, "this original unit, this fountain of power," had
been spilled into "drops" that "cannot be gathered." Individuals had "suffered amputation from the trunk, and strut about so many walking monsters,—a good finger, a neck, a stomach, an elbow, but never a man." The
young men in his audience were walking brains. They had to be recon-
nected to the trunk, the fountain of power. They should be not books or
brains or parrots of other men's books, but "*Man Thinking.*"[1]

Emerson's vision followed a Romantic tradition: science hands to humanity the tools of thought, but the tools alienate humanity from nature.
Thus modern science launches social progress, but with progress comes
crisis. Materialism destroys the human spirit. Man, fallen from unity
with nature, must achieve a new and higher unity through a new self-understanding: nature is not alien to mind, but *is* mind in its growth and
lawful unfolding; and mind *is* nature, mind bodied forth in matter. Each
individual embodies different potentials, different aspects of the Universal Mind, showing that there are many paths to wholeness; but all paths
share the same destination, the creative union with the creating Spirit of
Nature.

Emerson's own path up to this point—1837, his thirty-fourth year—offers a representative journey from division to integration. From his own
"*divided* or social state"—student, teacher, minister—Emerson's "education" or educe-ment had led to his new role as Man Thinking, freed from
obligations to church or university and hence free to model the ideal of
"self-trust," to be "a man speaking to men." Emerson planned to offer his
Cambridge audience "a theory of the Scholar's office"—a corrective theory,
reminding these bookish gentlemen of books' limited importance. Instead, a

: 33

Scholar "must be able to read in all books that which alone gives value to books—in all to read one, the one incorruptible text of truth."[2]

His performance was successful. Afterward one of his auditors, Charles H. Warren, offered a toast: "I give you *The Spirit of Concord. It makes us all of One Mind.*" In reading the "one incorruptible text of truth," Emerson had forged the many into One—*e pluribus unum,* a consensus, reached not by debate or argument or persuasion but by the power of self-evident truth. Beauty, truth, goodness: each partner in the divine trinity contributed to the unifying creative process which rendered truth self-evident by converting it from the world to the self. As Emerson asked, "Is not, indeed, every man a student, and do not all things exist for the student's behoof?" His address dealt with each partner in turn: nature, the beautiful mirror of the self; books, the transmutation of life into truth; action, the conversion of truth to power. Emerson's scholar would be not truth's carrier but its creative center, converting world into mind, knowledge into power. This scholar would be an author, an authority, not by virtue of his authority as a person but by depersonalizing authority altogether. Once truth was self-evident it was free and open to all, the property of everyone, not a specialized few— although only a few, perhaps, were truly capable of rising to self-evident truth's high disciplinary demands.[3]

The key was Nature. Emerson's American Scholar was above all an "interpreter of nature," in Francis Bacon's sense. Both Bacon and Emerson were drawn to "philosophy" rather than to the practice of science, and Emerson, who had studied Bacon with care from youth onward, proved himself one of Bacon's most acute and sensitive critics. In 1830 Emerson copied into his journal Bacon's vision of philosophy as a great tree with three branches, divine, natural, and human, which joined together "like branches of a tree that meet in a stem" to constitute "one universal science." This great trunk of all knowledge was *"Philosophia Prima,"* the Primary or First Philosophy, and to it belonged observations and axioms that fell within no one specialty "but are more common and of a higher stage." Such knowledge does sound peculiarly Emersonian. For instance, Bacon wondered if axioms of mathematics were not also axioms of "justice." "Is not the precept of a musician, to fall from a discord or harsh accord upon a concord or sweet accord, alike true in affection?" Such resemblances were not mere "similitudes . . . but the same footsteps of nature, treading or printing upon several subjects or matters." It seemed to Bacon that this primary or higher science had not been sufficiently studied. As he observed (Emerson copied out this passage, too), "I see sometimes the profounder sort of wits, in handling some particular argument, will now and then draw a bucket of water out of this well for their present use; but the springhead thereof seemeth to me not to have been visited." Emerson's goal as America's Representative Scholar would be to place himself right at this springhead, whence rose "the great principles that are true in all sciences, in morals, & in mechanics," principles which made possible the "generalizations" that so attracted him in the writings of Edmund Burke, Anne Marie de Staël, and Sampson Reed.[4]

Bacon placed science within this broader vision. The goal of the "interpreter of nature" was to invent not arguments but arts, not things but the principles of things; not to "overcome an opponent in argument" but "to command nature in action." This meant that the scholar first of all had to free his mind from "impediments": the weight of tradition, custom, and received opinions, including—or especially!—those of the highest authorities. Bacon dismissed the endless variety of books as nothing but "endless repetition of the same thing." He dismissed even the vaunted wisdom of the Greeks as immature and childish: "It can talk, but it cannot generate; for it is fruitful of controversies but barren of works." As Emerson said, "The book becomes noxious. The guide is a tyrant." To argue with a bookish authority was to become a "bookworm," valuing books as such, "not as related to nature and the human constitution."[5]

Instead, the true scholar turned from books to nature, for his goal was not to rehearse exhausted arguments but to invent *new* knowledge. The reader might disagree, but Bacon refused to argue. Instead, he issued a challenge. Examine not words but the subtleties of nature. The reader should correct the "depraved and deep-rooted habits of his mind"—and then, when "he has begun to be his own master, let him (if he will) use his own judgment." For what Bacon offered was not any particular theory or body of knowledge but a kind of knowledge-producing device, literally a "machine" that he was confident anybody might operate successfully. Truth, for Bacon, was not something found but something made, intelligence acting in the world.[6]

Bacon's first step was to dismiss received authority, but his goal was not to destroy authority but to depersonalize it by locating it in things rather than in people. In effect, Bacon gave away authority in order to gain it back, "that the mind may exercise over the nature of things the authority which properly belongs to it." To do this, Bacon explicitly ushered himself out of the picture, promising to reappear mainly as a guide to point out the road, an office of "small authority." In this small office he recommended "humility" both in inventing and in teaching: "I have not sought (I say) nor do I seek either to force or ensnare men's judgments, but I lead them to things themselves and the concordances of things, that they may see for themselves what they have." Only in this way, by diverting attention away from the self and toward the world, could the inquiry be left open-ended and capable of continuing expansion and correction: "I so present these things naked and open, that my errors can be marked and set aside before the mass of knowledge be further infected by them; and it will be easy also for others to continue and carry on my labors."[7]

Emerson fully understood the global implications of Bacon's modest

Become one's own inventor of Knowledge — To Invent oneself

little "plan." He noted in his lecture on Bacon that "the whole history of Science since the time of Bacon is a commentary and exposition of his views." By founding and then disappearing from the first philosophy of modern thought, Bacon made himself simultaneously indispensable and invisible—rather like God. Emerson noted elsewhere, "He has so anticipated the progress of thought that his opinions and words are quoted often unconsciously in every professor's chair as the Scriptures in the pulpit." This openness and permeability, however, came at a price: Bacon's own writings lack organic unity. Emerson pointed out that Bacon's works consisted of fragments, in "mechanical" order or strewn along the ground, "the order of a shop and not that of a tree or an animal where perfect assimilation has taken place and all the parts have a perfect unity." Yet this very criticism of Bacon's works revealed their virtue: "So loose a method had this advantage, that it allowed of perpetual amendment and addition," "gradual growth." That is, rather than a polished and accomplished "organic" unity, Bacon offered a *process* of organization. His workshop method draws in not only the scientists, like "Newton, Davy, and Laplace," who continued his work, but every reader. Any reader who wishes to take Bacon at his word must become himself the organizing center of Bacon's words. It is up to the reader to complete Bacon's words in his own mind and actions. Thus to read Bacon well is perforce to *become* Baconian, to become one's own inventor of knowledge—to invent oneself. Emerson found, in the self-deconstructing authority of Bacon, the authority for the American Scholar's primary principle, self-trust.[8]

Of course, to withdraw from the position of authority to let the world "speak" directly to the reader puts one in the most authoritative position of all. Bacon suggests this in his meditation on humility: "And by these means I suppose that I have established for ever a true and lawful marriage between the empirical and the rational faculty, the unkind and ill-starred divorce and separation of which has thrown into confusion all the affairs of the human family." In performing this ceremony of marriage, Bacon places himself in the role of "priest" whose sanction assures that the making of knowledge is "lawful," in accordance with the institutions of church and state, and "productive" not only of "fruit" or works but of a progressive and socially integrated society. Yet it is not the priest but nature who seals this ceremonial union. Nature must assent, and the sign of her assent is the "generation" of "works": "Of all signs there is none more certain or more noble than that taken from fruits. For fruits and works are as it were sponsors and sureties of the truth of philosophies." Bacon's test of truth is, then, entirely pragmatic: truth works. Falsehood is "barren" or, worse, bears the "thorns and briars of dispute and contention," whereas truth's self-evidence places it beyond dispute. No matter what anyone says, it will command assent by its effectiveness in the world: "If it be truth . . . the voice of nature will consent, whether the voice of man do or no."[9]

Nature, then, is the ultimate authority, and all who would claim power must claim it from her. In this Antaeus image, man draws his strength from the earth: systems and sciences will prosper only so far as they are true, "for what is founded on nature grows and increases; while what is founded on opinion varies but increases not." In an early image of organic connection, Bacon writes that whereas false doctrines are "like a plant torn up from its roots," true ones are "attached to the womb of nature" and continue "to draw nourishment from her." The organic image extends outward to become the mighty tree of all knowledge, the sciences its branches and boughs. All the sciences will remain true and progressive only if they remain connected, not "severed and cut off from the stem," the "common fountain." In this sense, truth will be coherent, single, and whole, and, no matter how complex, it will always follow a strict hierarchy from many weak points to fewer strong ones to one common and enduring center.[10]

The catch is the interesting way in which anyone, at least potentially, can climb up the hierarchy. Like Oedipus before the Sphinx, "the lame man who keeps the right road outstrips the runner who takes a wrong one." This race will not go to the brilliant or gifted. Indeed, an excess of talent may lead one astray. Bacon's knowledge-machine accommodates anyone who is willing to follow its rules, for the rules effectively level the playing field: it "places all wits and understandings nearly on a level." This is the purpose of Bacon's machine, not to concentrate power but to distribute it. While only the skilled and practiced few can draw a straight line or perfect circle by hand, anyone can do so "with the aid of rule or compass . . . so is it exactly with my plan." Hence Bacon constructs an authoritarian paradox: whoever submits to the rules may capture the power of God on earth, yet the game is officially open to anyone, absolutely anyone at all. In the strenuous republic of science, even servants might rule, and even kings must serve.[11]

For the aim is to benefit not kings, but humanity. The true ends of knowledge are not pleasure, superiority, profit, fame, or power, "or any of these inferior things; but for the benefit and use of life." Bacon tells us that his business is not "an opinion to be held, but a work to be done," and he assures us he is "labouring to lay the foundation, not of any sect or doctrine, but of human utility and power," "for the common good." For the truth will assuredly be useful. Writing of the difference between the idols of the mind and the ideas of the divine, Bacon states, "The former are nothing more than arbitrary abstractions; the latter are the Creator's own stamp upon creation, impressed and defined in matter by true and exquisite lines. Truth therefore and utility are here the very same things: and works themselves are of greater value as pledges of truth than as contributing to the comforts of life." All along, the source of authority has not been nature at all, but God, whose mind nature bodies forth. The source of the power of

Lame Man out-strips

Republic of

True Knowledge to be used in. life

Cannot command by exerting ones will over another

truth is finally not material but divine, which makes the command of nature not merely physical, but moral.[12]

The culture of truth is profoundly Baconian. One cannot truly command men by exerting force against their will; one commands only by appealing to the knowledge by which men rule their own wills. In Bacon's words, "But yet the commandment of knowledge is yet higher than the commandment over the will; for it is a commandment over the reason, belief, and understanding of men, which is the highest part of the mind, and giveth law to the will itself." The law-giving "commandment of knowledge" can govern only "by force of truth rightly interpreted," the closest fallen humanity can come to the "similitude" of divine rule. To govern by force is to govern by falsehood, for it attacks the will alone, but men who "under-stand" the truth will govern themselves. Instead of being imposed on from outside, they will "own" their own judgment.[13]

Thus *true* power is reciprocal, an exchange ratified by the mutual benefit it brings. In just the same way does man command nature: not by force or imposition but by learning her laws, taking them into "the highest part of the mind" (so that the body "under-stands" them), and so using them lawfully to his advantage. Bacon's great plan thus culminates in perhaps his most famous adage: "Now the empire of man over things depends wholly on the arts and sciences. For we cannot command nature except by obeying her"; or, in its slightly longer form:

Culminat of Bacon great plan

> For man is but the servant and interpreter of nature; what he does and what he knows is only what he has observed of nature's order in fact or in thought; beyond this he knows nothing and can do nothing. For the chain of causes cannot by any force be loosed or broken, nor can nature be commanded except by being obeyed. And so those twin objects, human Knowledge and human Power, do really meet in one; and it is from ignorance of causes that operation fails.

Or, as Emerson glosses for the American Scholar, "So much of nature as he is ignorant of, so much of his own mind does he not yet possess." Knowledge is power, but only if power is obedience. Obedience is the true path to command, submission the only path to freedom.[14]

Knowledge is power

This reciprocal exchange between brute necessity and human power intensifies in Emerson's writings to a pitched battle, even a war; yet this war can be won only by means of knowledge. Science was the most benign, pure, and lawful form of this exchange or "marriage," as Bacon styled it, for it bore the most abundant fruit in theories and things as well as technologies. All of these literally "empowered" humanity, yet in their proliferation and density they took on their own life, the servants turning masters. "Things are in the saddle, / And ride mankind," Emerson warned in 1848. The solution, ever and again, could lie only in reapplying the same dynamic. Emerson reiterated in 1860: "every jet of chaos which threatens to extermi-

nate us, is convertible by intellect into wholesome force. Fate is unpene-
trated causes." Master the oppressor by learning its law; command fate by
the obedience that converts fate into power. "All power is of one kind, a
sharing of the nature of the world. The mind that is parallel with the laws of
nature will be in the current of events, and strong with their strength."[15]

Effective, yes—but command through obedience leaves no room for critique, no way to step outside the creative mirroring by which nature reflects
the self; and the self sees in its reflection not just the affirmation of one
truth, but of the one and only possible truth. Steve Fuller observes that "the
ultimate ground for the 'Knowledge is Power' equation is rhetorical: to wit,
*the thread that connects the history of science from the Greeks to the pres-
ent day is that people came to be convinced that particular forms of knowledge are embodied in the world and are, in that sense, the hidden sources of
power over the world.*" The truth to which nature consents becomes not a
statement about the world, but a truth already in the world, put there by a
mind not our own. Knowledge bears no responsibility for the world it purports only to describe. In this replay of the age-old debate between nominalism and realism, young Emerson was profoundly on the side of realism.
As he worked it out in his journal:

> Man puts things in a row
> Things belong in a row
>   The showing of the true row is Science

Emerson found his realism confirmed and empowered by modern science.
In linking realism and science with such eloquence, enthusiasm, and persist-
ence, he helped to subdue the voices of those who argued, with Baconian
caution, that science could never venture more than statements, and that no
one statement would ever cashier the entire universe.[16]

For Emerson, surrendering the view that science was the recovery of
mind in nature would have meant surrendering everything: his most basic
Christian beliefs, right back to the fall of man and the resulting possibility
of redemption. Bacon was anxious to affirm that the fall of man was caused
not by acquiring natural knowledge but by overreaching moral law: "it was
not that pure and uncorrupted natural knowledge whereby Adam gave
names to the creatures according to their propriety, which gave occasion to
the fall. It was the ambitious and proud desire of moral knowledge to judge
of good and evil, to the end that man may revolt from God and give laws to
himself, which was the form and manner of the temptation." Bacon affirmed that no knowledge whatsoever could restore man to God—only contemplation of God could do that—but knowledge could repair the loss of
Eden, which had featured "that commerce between the mind of man and
the nature of things, which is more precious than anything on earth." This
belief gives the Baconian plan its deeply optimistic character: we can each of
us, in our own lifetime, work to reclaim the lost, heal the broken, and as-

cend back to paradise. "For man by the fall fell at the same time from his state of innocency and from his dominion over creation. Both of these losses however can even in this life be in some part repaired; the former by religion and faith, the latter by arts and sciences." Emerson's work would be very different without this vision of recovery, which assures us that we are not conquering another's territory but reclaiming what is already ours by right of inheritance: man shall enter his kingdom "without more wonder than the blind man feels who is gradually *restored* to perfect sight."[17]

"The American Scholar" was Emerson's attempt to rewrite Bacon into an American idiom. The three "influences" upon the scholar—nature, books, and action—follow the classic Baconian sequence; derive your own authority directly from nature, and you can defy the traditional authorities with their dusty books and acquire power to act directly in the world. Emerson writes that the "school-boy under the bending dome of day" who is open to the influence of nature will see "that nature is the opposite of the soul, answering to it part for part. One is seal, and one is print." Bacon had used the same metaphor: "certain it is that *veritas* [truth] and *bonitas* [goodness] differ but as the seal and the print." The boy who sees the print of God's truth stamped everywhere in nature learns that he, too, is printed from God's seal—and is a "seal" himself, a bit of God's mind delegated to earth.[18]

Such a boy is not recipient but *maker* of knowledge, and in making knowledge he is making himself, taking possession of his own mind by freeing himself from the noxious and tyrannical rule of the past. His "original" relationship with the universe trades the imposition of authorities for the power of authority, the role of authoring his own statements: "The scholar of the first age received into him the world around; brooded thereon; gave it the new arrangement of his own mind, and uttered it again. It came into him—life; it went out from him—truth." The truth is not the world but an utterance about the world, so true authority lies not in the record but in the act—not, that is, in any finished statements but in the creative process of "transmuting life into truth." No one is free from his own local idols, so the process must go on forever, each age for itself, each scholar for himself. "Each age, it is found, must write its own books." Might the process ever be finished? Hardly, Emerson scoffs: "Do we fear lest we should outsee nature and God, and drink truth dry?" Bacon's writings were not complete or organically unified, because they demanded just this process of endless correction and revision. "To make Bacon's works complete, he must live to the end of the world."[19]

By locating truth in nonhuman nature rather than in human authorities, Bacon—and Emerson after him—guaranteed that no human statement would ever capture the whole of truth. The process of making and circulating truth must stay open to continual negotiation. Truth is not a noun but a verb: an action, a performance. Hence action becomes essential to the

Without action -"he is not yet man

scholar: "Without it, he is not yet man. Without it, thought can never ripen into truth." Emerson had praised Bacon as "the first conspicuous example of modern times of the philosopher in action," who "saw things before they were yet out of the doors of their causes." Bacon himself had noted that knowledge is powerful because it operates not on the will but on the "understanding" that commands the will—that is, it is effective *as* knowledge exactly to the degree that it gets *inside* the mind, generating action. For not until knowledge is acted upon is it completed, made real: "Only so much do I know, as I have lived," Emerson states; "the true scholar grudges every opportunity of action past by, as a loss of power."[20]

If truth is action, then truth is useless until it is experienced. As Emerson explained to his congregation, "The virtue and the vice, and knowledge and the power of our fathers" is useful or harmful to us only if it has "entered into our bosoms. . . . What if there is truth in the world? That only is true in *my* world which I perceive. All other, until I perceive it, is as much incapable of helping me as it if were false." As he added a few weeks later: "The knowledge that has been accumulated by others," be it scientific or moral, "is of no use until it has been verified by us. . . . The whole nature and duty and happiness of man have been a thousand times described. . . . yet is the whole theory to be learned again by every one of us in his own proper experience." Only through experience is truth made into *living* truth; truth cannot be found in nature but only in the human act of "verifying" it, realizing it in life.[21]

This convergence of matter and spirit heals the dualism of body and mind. Knowledge as a project continually calls one out into the world, to take the world into the self—like Bacon's bee: whereas ants only "collect and use," and spiders "make cobwebs out of their own substance," the bee "takes a middle course: it gathers its material from the flowers of the garden and of the field, but transforms and digests it by a power of its own. Not unlike this is the true business of philosophy." In Emerson's aphorism, "One must be an inventor to read well"—whether one reads books or nature.[22]

What Emerson's "creative reading" creates is, in the first place, language: "Life is our dictionary." Life and language are continually circulating into each other. Life provides us with words "by which to illustrate and embody our perceptions"; words re-create life as thought, which then must be experienced as life. Emerson borrows his image for this movement, "Undulation," from Newtonian optics: "these 'fits of easy transmission and reflection,' as Newton called them, are the law of nature because they are the law of spirit." Each becomes a resource for the other: "The mind now thinks; now acts; and each fit reproduces the other." And what does this process of undulation create? A trustworthy self. This is the idea toward which Emerson has been building. The ultimate duty of the scholar is "self-trust," gained because in seeking authority in nature the scholar has found it himself, and can face and defy any danger on his path. The scholar strong with

Bacon Philosopher in action

action

Knowledge in action

Truth useless until experienced

Truth to Living Truth

action in animals

Every object printed with the seal of God.

nature's strength—strong with the strength of God—will find that his individual knowledge is not personal after all, but universal: the secrets of his own mind are shared by all minds. The man who is, like all nature, the print of God becomes the seal of truth and can print only goodness in all his actions, and sees, wherever he looks in nature, himself.[23]

This extraordinary technology of self- and world-making is available everywhere, not only in "the sublime and beautiful" but in "the near, the low, the common." Every object, even the most mundane, is printed with the seal of God, and from any object truth may be unfolded. And since every person bears the impress of God, every person can unfold from his self-trusting will a world of his own: "Every thing that tends to insulate the individual,—to surround him with barriers of natural respect, so that each man shall feel the world is his, and man shall treat with man as a sovereign state with a sovereign state;—tends to true union as well as greatness." Self-trust will *strengthen* true union, for each sovereign self shall draw together other sovereign selves, moving in mutual self-trust toward a common source. As Emerson concludes: "A nation of men will for the first time exist, because each believes himself inspired by the Divine Soul which also inspires all men." So ends Emerson's vision of the American Scholar: the spilled drops that cannot be gathered will become seeds that take root and grow. Drinking from a common source, the individual seedlings will grow together into a new whole, individual still but united: a nation of men, the new nation of America, Bacon's New Atlantis realized.[24]

NATURE: THE ART OF GOD

Emerson's doctrine worked so well because it creatively appropriated traditions with deep roots in Anglo-American Protestant culture. Emerson did not, for example, need to argue that nature was lawfully designed by God, for the notion had long been a mainstay of Protestant theology. He needed only to ask, rhetorically, "But what is classification but the perceiving that these objects are not chaotic, and are not foreign, but have a law which is also a law of the human mind?" Such a claim was more than the avowal of a post-Kantian Romantic; it was authorized by the very progress of science. In 1823 Emerson read in John Playfair's history of physical science that when Bacon's contemporaries, Kepler and Galileo, proved that the physical laws that ruled the Earth held true for the heavens also, they smashed the ancient barrier separating terrestrial and celestial systems: "Philosophers had ascended to the knowledge of the affinities which pervade all nature, and which mark so strongly both the wisdom and unity of its author." The barrier between Earth and heaven had also separated the mind of man from the unknowable mind of God, but suddenly God's mind stood revealed as knowable after all—indeed, remarkably human.[25]

The new laws of science struck down the tyranny of the past and put man in an "original" relation with the universe by showing that the creative mind, whether human or divine, was one and the same. Yet in removing one barrier, the new knowledge also erected another, between religion and science, or rather, as Playfair qualified, between "the provinces of faith and philosophy." Everyone knew, Playfair said, that revelation told only of matters beyond human reason, and did so using the language and opinions of the time. Conventional priestly authority had to give way before the new authority of science, which showed "the true system of the world." This true system, however, was not a world divorced from God, but a world designed by God and imbued with his presence. This was proved by the universe's harmonious and lawful structure, which had been triumphantly revealed by Bacon's great successor, Newton. Seventeenth-century science, that is, did not cancel or exclude the authority of religion, but assumed it for itself: "The kind of transcendent truth that Newton believed his science to possess, and that he bequeathed to the Enlightenment, started with a transference. He transferred divine authority to the laws of science."[26]

Newtonian science did not reject religion. Quite the contrary, it made science and religion codependent, for Newton's universe made God a necessary hypothesis. As one historian of science concludes, for Newton matter in itself was "mindless and inchoate." Without God's creative energy, "the world would have languished inert and purposeless." Protestant intellectuals agreed that "rational analysis of nature . . . showed that the world cannot be explained by natural causes alone but must be the work of God. Newtonian science demonstrated the need for a Creator, who graciously sustains a world totally incapable of sustaining itself."[27]

One of Emerson's favorite authors, the Neoplatonist natural philosopher and clergyman Ralph Cudworth, had helped Newton develop the theological implications of his natural system. In *The True Intellectual System of the Universe,* Cudworth attempted to destroy atheism by showing that matter is nothing but a collection of passive, inert atoms, making mind the necessary creative force of the universe. In a phrase that would be recycled endlessly over the following generations (and used against Darwin fully two centuries later), Cudworth put to rest the "prodigious nonsense" that nature could generate itself: "the fortuitous motions of senseless atoms" could never produce "so much as the form or system of one complete animal, with all the organic parts thereof so artificially disposed (each of these being as it were a little world), much less the system of this great world." Nature's imbecility had to be directed by the creative intelligence of God.[28]

According to Perry Miller, the English Puritans imported this belief to the New World, adopting science as a part of faith, and accepting the new physics readily once they were assured by propagandists for the newly formed Royal Society of London that the new science did not undermine piety. On the contrary, "to study nature is to study God in the act of willing, to marvel at the ingenuity of His plan and the minuteness of His care."

Bacon → Newton
A World cannot be explained by natural causes alone
Newton did not reject religion
God
Royal Society forming

Nature is not at all a self-subsisting independent agent but "ordered as the wheels of a watch are ordered . . . arranged in hierarchies of value and form" according to necessary rules. In short, nature is "the 'art' of God." Miller is referring to the American Puritans, but Emerson also encountered the idea in Cudworth, who announced that "what is called nature is really the Divine art," "a living stamp or signature of Divine wisdom; which, though it act exactly according to its archetype, yet it doth not at all comprehend nor understand the reason of what itself doth."[29]

Nature, then, is art exactly in the sense that it cannot create itself but must be created by another toward purposes and designs outside itself and beyond its comprehension: the clock does not know the time. From the Latin Middle Ages onward, nature had been imagined as a book full of meaning, and a book can hardly write itself. In an early version of the monkeys-at-a-typewriter argument, Cudworth noted that the fortuitous motion of dead matter could no more produce order than an "illiterate" man, by taking up a pen and "making all manner of scrawls with ink upon paper," could write "philosophic sense." If nature is a book, it obviously has an Author. And so the biblical metaphor of God's Two Books, the book of Revelation and the book of Nature, was appropriated and redefined by the scientific revolution of Bacon, Kepler, Galileo, and Newton. Since science had shown that the book of Revelation had been written by man, not God, the Bible could no longer overrule the one *true* book, the one that had without question been authored by God himself: nature. Once nature was seen as the embodiment of the Word of God, the divine Logos made flesh, or stone, or vegetable tissue, the study of nature became an enormously important way of relating to God.[30]

So religion and science were still partners, but their relationship had changed in one crucial respect. Ever since the first century A.D., science had been viewed as the "handmaid" to theology, meaning that science was not to be pursued for its own sake but for the light it could shed on "the interpretation of Holy Scripture": "The glorification of God was the ultimate goal of the scientific study of nature." Natural knowledge served scriptural knowledge. There was no doubt which, in cases of conflict, must give way before the other: things in nature must be seen to conform with the words of scripture. But this was not at all what Bacon declared, even though he agreed that "natural philosophy . . . is rightly given to religion as her most faithful handmaid." In throwing off the authority of books, Bacon reversed the ancient relationship. Reason could not perform its proper service if it had to conform to scripture: "For our Saviour saith, *You err, not knowing the Scriptures, nor the power of God;* laying before us two books or volumes to study, if we will be secured from error; first the Scriptures, revealing the will of God, and then the creatures expressing his power; whereof the latter is a key unto the former."[31]

Following Bacon, the Two Books stood in the relationship not of master and servant but of lock and key. It was the truth of nature that unlocked the "true sense" of the scriptures. Thus, in cases of conflict it was *scripture* that had to be made to conform. James R. Moore calls this crucial step "the Baconian compromise" and notes that this compromise "became a convention in English-speaking scholarship for more than two hundred years: the basis of congenial relations between naturalists and exegetes, and a chief sanction for the growth in volume and expertise of physical research." Emerson participated in this Baconian compromise when he wrote, for example, "The world,—this shadow of the soul, or *other me,* lies wide around. Its attractions are the keys which unlock my thoughts and make me acquainted with myself." As Bacon said, the book of nature is "a key" that unlocks the will or mind of God; by conflating divine Mind with human mind, Emerson took the next logical step, reading in nature the key to the self. He who answers the riddle of the Sphinx unlocks the mind of God and inherits an empire.[32]

Emerson leaving Church

Emerson did not drift passively into his acceptance of the Baconian compromise. It was a position he worked hard to earn over his years as a minister, when he had to conform the teachings of his own reason to the scriptural doctrines of the Unitarian church. Not until he was unwilling to conform his thoughts and actions to the words of the church could he accept resignation as an option, and it was the Baconian compromise that made that option acceptable. It assured him that he would not be surrendering religion, only its husk, the dying forms of outgrown convention, which could be revivified by turning to the true "fountain" of established teaching, "the Moral Nature, that Law of laws, whose revelations introduce greatness,—yea, God himself, into the open soul. Christianity could have nothing to fear from knowledge. Surely the progress of learning, which showed the historical nature of Jesus and the Bible, could not threaten the truth of the scriptures; that could happen only if one assumed that "the age of inspiration is past, that the Bible is closed"; why, the very fear of the knowledge that Jesus was a man showed "the falsehood of our theology." Applying his idea that truth can be known only by experience, Emerson told the students of the Harvard Divinity School that truth is "guarded by one stern condition; this namely; It is an intuition. It cannot be received at second hand." The teacher cannot instruct, only provoke: "What he announces, I must find true in me, or wholly reject. . . . be he who he may, I can accept nothing." The minister who fails to understand this can only "clatter and echo unchallenged," like the ridiculous specter Emerson holds up in the Divinity School address as an object lesson for the students—and for himself.[33]

Truth

The alternative is to accept a new faith, a *true* faith, whose text should be "its power to charm and command the soul, as the laws of nature control the activity of the hands,—so commanding that we find pleasure and honor in obeying." This transference of Bacon's famous dictum locates us as obe-

Religion commands by Obedience to Nature

dient servants not of nature but of the true faith, which shall be natural law in its moral aspect. False theology closes the Bible against knowledge; true theology reopens the Bible to the influences of the book of nature, which will rewrite it perpetually anew. The choice is dire: accept the errors of the church, which cause "that calamity of a decaying church and a wasting unbelief"; or else "go alone . . . refuse the good models, even those most sacred in the imagination of men, and dare to love God without mediator or veil." The divinity student, like the American Scholar, must spurn society to recover trust in the inner soul, which shall be the Soul of all, its integrity at last recovered complete—which, shockingly, meant rejecting even the divinity of Jesus. At the end of his Divinity School address Emerson triumphantly prophesies the new dawn of a religion that commands by obedience to science: "I look for the new Teacher, that shall follow so far those shining laws, that he shall see them come full circle; shall see their rounding complete grace; shall see the world to be the mirror of the soul; shall see the identity of the law of gravitation with purity of heart; and shall show that the Ought, that Duty, is one thing with Science, with Beauty, and with Joy." Emerson's new Teacher might be a theologian—but he must be a scientist first.[34]

For all its shock value, Emerson's prophecy nevertheless put him firmly in the aged and many-sided tradition of natural theology, which argued that the harmonious and lawful design of nature proves the existence of a divine designer. This tradition was central to Anglo-America, where it had been burnished by the high prestige of Newtonian science. In Emerson's day, natural theology reigned. Nearly every book of science he opened declared that that particular science affirmed "the conviction of the existence, wisdom, power, goodness, immutability, and superintendency of the Supreme Being! So that without an hyperbole, 'An undevout astronomer is mad.' " So announces the florid opening of *Ferguson's Astronomy*. Humphry Davy, introducing his *Elements of Chemical Philosophy*, proclaims that the goals of chemistry are to increase human comforts *and* to demonstrate "the order, harmony, and intelligent design of the system of the earth." John Herschel opens his magisterial *Preliminary Discourse on the Study of Natural Philosophy* by reiterating that the study of science leads one "to the conception of a Power and an Intelligence superior to his own," bewildering him in admiration until "his intellect falls back on itself in utter hopelessness of arriving at an end." Charles Lyell reminds his readers in an 1826 article that "it is therefore clear to demonstration, that all, at whatever distance of time created, are parts of one connected plan. They have all proceeded from the same Author, and bear indelibly impressed upon them the marks of having been designed by One Mind."[35]

William Paley's *Natural Theology* (still a standard text at Harvard well into the 1850s) opens with a famous scene in which a man happens across a watch lying on the open heath, and knows instantly that he has found the

Watch must have a maker

product of intelligence: "the watch must have had a maker"; for "there cannot be," Paley pronounces, "design without a designer; contrivance, without a contriver; order, without choice. . . . Arrangement, disposition of parts, subserviency of means to an end, relation of instruments to a use, imply the presence of intelligence and mind." Paley came under criticism for his artificial and mechanistic approach to nature, but Emerson's exceedingly mechanistic images in an early sermon show his indebtedness to Paley: "Now the tree, the vegetable, may be properly considered as a machine by which the nutritious matter is separated from other elements. . . . The little seed of the apple does not contain the large tree that shall spring from it; it is merely an assimilating engine which has the power to take from the ground whatever particles of water or manure it needs, and turn them to its own substance and give them its own arrangement." Emerson continues, echoing Cudworth, that the seed, that little "engine," can hardly know what it does or how glorious a show it puts on. "Can the wheat admire its own tasselled top? or the oak in autumn its crimson foliage, or the rose and the lily their embroidery? If there were no mind in the Universe, to what purpose this profusion of design? It is adapted to give pleasure to us." Design points not just to a designer, but also to us, designers ourselves, whose own creative powers are drawn out by the superadded beauty of God's design, beauty that turns machinery into art.[36]

Thus truth and beauty are twinned; but where is the moral link? To those who searched for a defense against skepticism, Paley's fault was his unrelenting attention to the evidence of the senses. His bodies and organizations and contrivances carried no moral meaning beyond utilitarian benefit and sensual delight. As Coleridge complains, all the merely physical evidence of God everywhere in the universe would not prepare man to recognize his moral creator and governor: "Hence, I more than fear, the prevailing taste for Books of Natural Theology, Physico-Theology . . . &c. &c. *Evidences* of Christianity! I am weary of the word." Only rouse man to *self*-knowledge, pleads Coleridge, and he will find his own evidence. Paley's mechanistic, sensual natural theology opened a moral gap between man and God that would be closed by making man not God's applauding spectator, but the creative center of the universe.[37]

The idea was hardly a new one. Bacon had laid the groundwork by insisting that man must not passively adopt the opinions of others, but actively seek out the truth of nature by his own best sights. Bacon added that God had framed the human mind as a kind of "mirror or glass capable of the image of the universal world, and joyful to receive the impression thereof, as the eye joyeth to receive light." Moreover, God had framed a *thinking* mirror, which both "receives" images on its surface and actively reaches out beyond its surface to "find" and "discern" the principles of order. Emerson alludes to Bacon in *Nature* when he declares: "Parts of speech are metaphors because the whole of nature is a metaphor of the

Bacon

human mind. The laws of moral nature answer to those of matter as face to face in a glass. 'The visible world and the relation of its parts, is the dial plate of the invisible.' The axioms of physics translate the laws of ethics." This densely allusive passage combines the reference to Bacon, a quotation from Swedenborg, and a borrowing from de Staël's observation that the thought of the Creator had been "transfused into two different languages, and capable of reciprocal interpretation. . . . Almost all the axioms of physics correspond with the maxims of morals."[38]

The thread that unites these diverse sources is the principle of "correspondence," which healed the moral gap by making mind and nature aspects of each other. Through correspondence, the visible world points back, like the gnomon on the dial plate of a sundial, to the reality of the invisible. Without this principle of correspondence the mind would be set adrift, unable to make sense of the "ordinances and decrees" which govern it. In other words, the mind can arrange all it wants; what really excites Emerson is the possibility that the arrangements of the mind can achieve the consent of nature, and thereby, as he says, "be strong with its strength."

> There is in nature a parallel unity, which corresponds to this unity in the mind, and makes it available. . . . This design following after, finds with joy that like design went before. Not only man puts things in a row, but things belong in a row. The immense variety of objects is really composed of a few elements. The world is the fulfilment of a few laws.
>
> Hence, the possibility of Science.[39]

As Emerson adds in "The American Scholar," "science is nothing but the finding of analogy, identity in the most remote parts." Science can follow a metaphor, like "design," out into the world and find that world captured by the metaphor, confirming that the scientist's act of creative vision recapitulates the original creative act of God. Given that this is the nature of science, one can find in the writings of science a trove of moral insights, as when Emerson draws a lesson from chemistry:

> Read Chemistry a little, & you will quickly see that its laws & experiments will furnish an alphabet or vocabulary for all your moral observations. Thus very few substances are found pure in nature. There are metals like potassium & sodium that to be kept pure must be kept under naphtha. Such are the decided talents which a culminating civilization produces in illuminated theatres, or royal chambers; but those souls that can bear in open day the rough & tumble of the world must be of that mixed earthy & average structure such as iron, & salt, atmospheric air, & water.

Science is the creative appropriation of truth in nature, and may itself be creatively appropriated to moral reasoning and aesthetic insight: "Truth, and goodness, and beauty, are but different faces of the same All." For

Emerson, correspondence underwrites the possibility of science. Take nature into the mind as idea, then cycle that idea back out into nature: the
more it captures, the truer it is, and the higher on the scale ascending from
many facts to one law.[40]

The correspondence between moral and natural law was, according to
Robert Spiller, "the core doctrine of the Transcendental movement—a suspended dualism of natural and moral laws in parallel, with which one could
live only if one found a way to accept apparent contradictions and to forego
final answers." Spiller observes that Emerson inherited the doctrine "from
the easy assumptions of these earliest teachers"—his father, the Reverend
William Emerson, and the Reverend William Ellery Channing—"when he
was still a boy and was merely questioned, doubted, tested, and reaffirmed
as he grew older and wiser." The doctrine of correspondence could with-
stand such extended and intensive scrutiny because it centered all inquiry in
the deep embrace of a divinely designed universe which excluded the unknowable by fencing in certainty with the guarantee of coherence, and
which structured inquiry by providing a theoretical framework that was,
until Darwin, capacious enough to appear invisible.[41]

It was also flexible enough to contain such exotic variants as Swedenbor-
gian idealism, Coleridgean metaphysics, and German *Naturphilosophie,* all
domesticated under the overarching cultural metaphor of design. Thus
Emerson was prepared to accept Swedenborg by none other than the Scottish rationalist Dugald Stewart, who stated confidently that "the number-
less references and dependencies between the material and the moral
worlds, exhibited within the narrow sphere of our observation on this
globe, encourage, and even authorize us to conclude, that they both form
parts of one and the same plan;—a conclusion congenial to the best and noblest principles of our nature, and which all the discoveries of genuine science unite in confirming." As he added shortly, "The presumption unques-
tionably is, that there is one great *moral system,* corresponding to the
material system," with the connections we trace among the objects of sense
intimating "some vast scheme" comprehending human intelligence. To
clinch the point, Stewart offered Newton, whose conjectures that gravity
extends from Earth to the heavens were, according to Stewart, of exactly
the same sort as conjectures that extend the law of gravity from nature to
morality. They differed only in Newton's ability to appeal to physical evidence for proof; but since the principle was sound, such proof was no more
than an accidental bonus. Newton and the moral philosopher alike succeeded in marrying Earth and heaven, uniting by a creative act of mind the
dualism imposed by the original act of Creation.[42]

Yet there is a problem here, as Bacon understood profoundly well. The
mirror of the mind is hardly "a clear and equal glass" reflecting reality accurately: "nay, it is rather like an enchanted glass, full of superstition and
imposture, if it be not delivered and reduced." The rigor of Bacon's self-dis-

ciplinary program was designed precisely to "deliver" the mind from its en-
chantments and "reduce" it to purity. Bacon feared that phantoms created
by the mind could be taken for truths of nature; hence the mind must clear
itself of Bacon's four famous idols, the tribe or human nature, the cave or
individual self, the marketplace or the "alliances of words and names," and
the theatre or the "received systems," which "are but so many stage-plays,
representing worlds of their own creation after an unreal and scenic fashion." Human nature being what it is, man is prone to see only what he
wants to see, to create the world in his own image. To counter this limitation, Bacon strove to create a "machine" that might eliminate from science
everything but the laws of matter.[43]

four idols

This leaves a whole second realm free: poesy, imagination, "which, being
not tied to the laws of matter, may at pleasure join that which nature hath
severed, and sever that which nature hath joined, and so make unlawful
matches and divorces of things." The narrower science is, the freer is poetry,
approaching in its freedom something like the divine capacity for creation.
Emerson read and marked Bacon's key passage on poetry: "it was ever
thought to have some participation in divineness, because it doth raise and
erect the mind, by submitting the shews of things to the desires of the mind;
whereas reason doth buckle and bow the mind unto the nature of things."
In *Nature*, Emerson dramatizes Bacon's assertion: the poet "unfixes the
land and the sea, makes them revolve around the axis of his primary
thought, and disposes them anew. Possessed himself by a heroic passion, he
uses matter as symbols of it. The sensual man conforms thoughts to things;
the poet conforms things to his thoughts." Scientist and poet are thus driven
apart, into radically opposed positions. The scientist records and conforms;
the poet creates and makes conform.[44]

Emerson reads Bacon

Yet Emerson, unlike Bacon, opposes to the poet not the man of science
but the "sensual man." His entire project depends on elevating science to a
creative art—not just marrying science and poetry, but merging the two in a
new, prophetic power: "The soul active sees absolute truth; and utters
truth, or creates. In this action, it is genius. . . . Genius creates. To create,—
to create,—is the proof of a divine presence." The distance from Bacon to
Emerson is not so great as it might seem. The disciplined scientist who
purges himself of the idols intrinsic to the human condition does approach,
through the truth of nature, the mind of God. Yet for Bacon it is still only
an approach: humility prohibits the Baconian scientist from assuming for
himself any of the attributes of the divine.[45]

Elevate Science to Art

Part of the distance between Bacon and Emerson was closed by Newton's
friend Ralph Cudworth. As we have already seen, Cudworth defended
against atheism by showing that matter, lacking life and purpose, had to be
continuously shaped and moved by "Divine law and command." Yet how?
Inanimate objects could hardly hear a verbal command, and God could
hardly intervene to manage every detail. So Cudworth proposed an inter-

mediary, "a plastic nature . . . which, as an inferior and subordinate instrument, doth drudgingly execute that part of his providence, which consists in the regular and orderly motion of matter." It's as if "the art of the shipwright, were in the timber itself," "acting immediately on the matter as an inward principle." This makes for two kinds of art in the world: human art, which acts "from without and at a distance" and with the clatter and bang of tools; and nature, "another kind of art, which, insinuating itself immediately into things themselves, and there acting more commandingly upon the matter as an inward principle, does its work easily, cleverly, and silently. Nature is art as it were incorporated and embodied in matter, which doth not act upon it from without mechanically, but from within vitally and magically . . . as an inward and living soul, or law in it."[46]

Well over a century later, Coleridge would take up Cudworth's idea as his own, making it famous as the principle of organic form. Emerson copied into his journal Cudworth's passage advising that perfect art—the art of nature—"doth never consult or deliberate . . . as unresolved what to do, but is always readily prompted." Thus *true* art springs naturally and spontaneously from its own inner law, which is, as Cudworth continues, the "omniscient art of the Divine understanding, which is the very law and rule of what is simply the best in every thing." Nature *is* God's art, then, "reason immersed and plunged into matter"; and if nature is reason embodied, the art of God, then human art is human imagination given body by the work of reason in the world. Whether it takes form as poetry, art, or science, the creative act is the same. When Emerson defines imagination as "the use which Reason makes of the material world," his immediate subject is Shakespeare, but he soon broadens the scope: "Whilst thus the poet delights us by animating nature like a creator, with his own thoughts, he differs from the philosopher only herein, that the one proposes Beauty as his main end; the other Truth. But, the philosopher, not less than the poet, postpones the apparent order and relations of things to the empire of thought." All that has had to happen between Cudworth and Emerson is the seizure of law by human agency. Man steps from the margin, where he creates from outside noisily and mechanically in a feeble imitation of God, to the center, where, like God, he creates by an act of reason. The poet makes free with the world, which is beautiful; but science *makes* that world.[47]

Dugald Stewart, Emerson's Common Sense philosopher, would have rejected this as nonsense, just as he dismissed Cudworth's "plastic nature" as a "wild hypothesis." Yet Stewart praised Cudworth's idea when it was transferred to the plastic power of the human mind. Without mind, we are like monkeys trying to read typescript. The senses see in the book of nature nothing but "figures and colours," and in a written book nothing but "so many inky scrawls." But the human mind, that chip of divinity, will see beauty and knowledge and will also "clearly read the divine wisdom and goodness in every page of this great volume, as it were written in large and

Science must turn to imagination

legible characters." That is, if the mind is a passive register of sensual impressions it will see nothing but unmeaning scrawls. Seeing is an act of creation, the composition of chaos into sense by the active mind. Thus it is wrong, Stewart declares, to say that knowledge is stored in books. All we actually *see* is "a multitude of *black strokes drawn upon white paper*." It is our "acquired habits" that communicate meaning and significance to those strokes: "The knowledge which we conceive to be preserved in books, like the fragrance of a rose, or the gilding of the clouds, depends, for its existence, on the *relation* between the object and the percipient mind; and the only difference between the two cases is, that in the one, this relation is the local and temporary effect of conventional habits; in the other, it is the universal and the unchangeable work of nature." Others had celebrated Kant's so-called Copernican revolution for refuting the skepticism of Hume by putting the creative human mind at the center of the universe. Stewart, by contrast, rather coolly remarks that Kant's leading idea in the *Critique of Pure Reason* had already been much better expressed by Cudworth, from whom, Stewart suggests, Kant liberally borrowed. Although parts of Kant's system may be new to Germany, "it certainly could have no claim to the praise of originality, in the estimation of those at all acquainted with English literature."[48]

Reading Stewart's *Dissertation* in 1821, Emerson saw Kant's Copernican revolution domesticated as a British home truth. Reading de Staël a year later, he saw the same pattern, although her domesticating agent was not Cudworth but Bacon. Science must turn toward imagination, and imagination "must support herself upon the accurate knowledge of nature"; and both must turn toward a general philosophy that, like Bacon's, unites sensibility, imagination, and reason all together. "The soul is a fire that darts its rays through all the senses; it is in this fire that existence consists; all the observations and all the efforts of philosophers ought to turn towards this ME, the centre and the moving power of our sentiments and our ideas." De Staël's Kant becomes the healer who can rescue the soul from modern materialism, which "gave up the human understanding to the empire of external objects, and morals to personal interest; and reduced the beautiful to the agreeable." By his Copernican turn, Kant established that we can perceive nothing except through the laws of our reasoning, laws that are therefore "in ourselves, and not out of us." According to de Staël, Kant has thus shown that the fracturing universe can and must be made whole again through the activity of the mind:

> The new German philosophy is necessarily more favorable than any other to the extension of the mind; for, referring every thing to the focus of the soul, and considering the world itself as governed by laws, the type of which is in ourselves, it does not admit the prejudice which destines every

Must know the whole
Whole

man exclusively to such or such a branch of study. The idealists believe that an art, a science, or any other subject, cannot be understood without universal knowledge, and that from the smallest phenomenon up to the greatest, nothing can be wisely examined, or poetically depicted, without that elevation of mind which sees the whole, while it is describing the parts.

Must be able to see the whole while examining the parts

The Baconian thesis has reemerged in its fully Romantic formulation: the laws by obedience to which we achieve command of nature are the laws of our own mind.[49]

As Emerson read, the degree of consensus he encountered was remarkable. On every hand—in histories of science, scientific treatises, works by avowed natural theologians and rationalist philosophers and Romantic radicals—he was told that nature was a creation of mind. Such deep power demanded careful disciplining to keep the creative mind from dangerous excesses. Stewart, for example, was especially pleased with Bacon's discussion of education as "the *Georgics of the Mind*,"

> identifying, by a happy and impressive metaphor, the two proudest functions entrusted to the legislator—the encouragement of agricultural industry and the care of national instruction. In both instances, the legislator exerts a power which is literally *productive* or *creative*; compelling, in the one case, the unprofitable desert to pour forth its latent riches; and in the other, vivifying the dormant seeds of genius and virtue, and redeeming, from the neglected wastes of human intellect, a new and unexpected accession to the common inheritance of mankind.

Minds are not born

Minds, then, are not born or manufactured but grown, "cultivated," as Coleridge would say, into a state of "culture" through the productive or generative power of law, whose virtue extends equally to nature or the classroom or the legislative chambers of government. Emerson found the metaphor literalized by Sampson Reed: "It is not sufficient that the letter of the Bible is in the world. . . . The book must he eat, and constitute the living flesh." The mind grows by chewing the world, consuming it, digesting it, and assimilating its substance to the mind's own nature. This process is as greedy and dynamic as the metaphor suggests. Truth remains truth, *living* truth, only through action. Memory is not a passive accumulation but an active possession. Time is not a train that carries us passively forward; *we* generate time by the changes we make, and we generate matter by the power of the thoughts we think. Joseph de Gérando's *Self-Education* explained to Emerson how the world calls out the mind, setting it on a lifelong course of self-education by creative submission to moral/natural law.[50]

A similar idea motivated the English philosopher and historian of science William Whewell, who was developing his ideas of science about this time

Whewell

Whewell's

in a book favorably reviewed by Theodore Parker in *The Dial*. Whewell's first three aphorisms on science are:

I. Man is the Interpreter of Nature, Science the right interpretation.
II. The *Senses* place before us the *Characters* of the Book of Nature; but these convey no knowledge to us, till we have discovered the Alphabet by which they are to be read.
III. The *Alphabet,* by means of which we interpret Phenomena, consists of the *Ideas* existing in our own minds; for these give to the phenomena that coherence and significance which is not an object of sense.

Such thinking was not confined to the sophisticated theorists of science. William B. O. Peabody assured his American readers that the beauty and grandeur of nature excite little interest in us because "we are wanting to ourselves. We expect to find these associations already existing, while it depends on us to form them. Our eyes do not see,—our imagination must create them," and with the right spirit, "we can easily and almost without effort create 'tongues in the trees' . . . 'books in the running brooks' . . . and 'sermons in stones.' . . . This is the work of poetry,—that is, *creative*, where this quickening power exists, the desert need not be barren; and there is an inviting field for its action in a land like ours." Peabody suggested that the American had a special creative role in a land of trees, brooks, and stones.[51]

Nor was Peabody the first to link science and Shakespeare: in his *Preliminary Discourse,* John Herschel, too, extolled the virtues of creative perception. One of "the great sources of delight" imparted by the study of natural science is the ability it gives its votaries to walk ever "in the midst of wonders":

Herschel

A mind which has once imbibed a taste for scientific enquiry . . . has within itself an inexhaustible source of pure and exciting contemplations:—one would think that Shakspeare had such a mind in view when he describes a contemplative man as finding

"Tongues in trees—books in the running brooks—
Sermons in stones—and good in every thing."

The cultivated mind—the mind that has learned to draw nourishment from its surroundings—has thereby learned that the true source of nourishment is "within itself," an inexhaustible fountain of creative power to see and connect in seeing every object with principle, with instruction, impressing him, as Herschel said, "with a sense of harmony and order" and engaging his mind in "constant exercise."[52]

Emerson was reading, from the 1820s and into the 1830s, on every hand that the mind was the creative center of the universe, and that true science was not the dull accumulation of facts but the dynamic process by which mind assimilated nature to itself, making nature's strength its own. By the early 1800s, the view that science was a function of the creative imagination was widespread. Yet instead of settling the debate on the nature of science, this view precipitated new arguments. If nature is a text open to "creative reading," does nature exist beyond the act of reading? Barbara Packer reminds us that texts "also need us," since a book "exists" only as it is being read. This alters the balance of power between text and reader: the text becomes "less like a landed estate that we may, if we are lucky, inherit than like a lump of capital we can, if we are enterprising, learn to manipulate." The enterprise of science was exploring just how manipulable the text of nature might be, and so Emerson found himself in the midst of a debate not only over the nature of "science," but more deeply about the nature of mind. Could the creative human mind ever hope to see beyond itself? Did any world that really mattered exist beyond the mind's creative vision? Was nature the art of God—or the art of man? Or was it no one's art at all?[53]

BOOKS: READING SCIENCE

By the time Emerson embarked on the series of lectures on science that led up to the composition of *Nature,* he had steeped himself in a wide range of readings on science, becoming familiar with the thoughts of the age's key theorists and practitioners: Bacon, of course; Kant, if at second hand; Herder, Goethe, and Alexander von Humboldt; Georges Cuvier and Alphonse de Candolle; Stewart, Playfair, Humphry Davy, Coleridge (of prime importance to Emerson), Herschel, Whewell, Lyell, Mary Somerville, and others. What Emerson learned from them all is distilled and transformed in *Nature,* the product of Emerson's own "creative reading" of science. Thus to understand the key role that science played in his thought, both in and beyond *Nature,* it is necessary first to glance at those science writers from whom he learned the most, and at his first attempts to frame his learning into his own statements about the nature and purpose of science, in his journal and his early lectures.

The legacy of Bacon was central. Despite their fundamental differences, each of England's major theorists of science—Herschel, Whewell, and Coleridge—claimed Bacon as a guide and forefather. Emerson took his foundational idea of the goal of science from Bacon: "Believing that every object in nature had its correlative in some truth in the mind he conceived it possible by a research into all nature to make the mind a second Nature, a second Universe. The perfect law of Inquiry after truth, he said, was, that nothing should be in the globe of matter, which should not be likewise in the globe of crystal; i.e. nothing take place *as event,* in the world, which did not exist *as truth,* in the mind." Emerson condensed this global analogy to arrive at his own most succinct definition of science: "the reconstruction of nature in the mind." The muddy Earth would be reconstructed as a crystal globe. The problem, then, was method. How was this global project of reconstruction to be carried out? Here, as Emerson's sources chimed in agreement, was Bacon's greatest contribution.[54]

: 55

Metaphysic
Physic
Natural facts

Bacon's method

What was the Baconian method? As Emerson put it, Bacon's new method, or "novum organum," was "a slow Induction which should begin by accumulating observations and experiments and should deduce a rule from many observations, that we should like children learn of nature and not dictate to her. His favorite maxim was 'Command Nature by obeying her.'" Bacon proposed for science a distinctive shape: the pyramid. The broad base would be the facts of natural history; "Physic" would explore their "connexion"; crowning them was "Metaphysic," or union into "a perpetual and uniform law." This structure presents science as cumulative and progressive, from many facts to one law, but how exactly does one ascend this pyramid? There are, says Bacon, two and only two ways, one fashionable and one practical. Both ways begin with "the senses and particulars" and end in generalities, "but the difference between them is infinite." The currently fashionable method flies at once to "abstract and useless generalities," whereas his new and as-yet-untried method "rises by gradual steps" which, tortoiselike, will beat the hare every time. Lest we revolve forever in the circles of our own imaginings, "We must begin anew from the very foundations" and build a structure that will last—as long, one gathers, as the pyramids of Egypt.[55]

Bacon seems to say that the mind ascends by slow and patient steps, its vision chained to the ground, never venturing to the perilous heights of speculation. This was (and remains) a common misconception, one against which Bacon's nineteenth-century followers spent much time defending him. Even Bacon acknowledged, if cautiously, that the true road "does not lie on a level, but ascends and descends; first ascending to axioms, then descending to works." Yet in his world the real battle was against not pedestrian plodding but unconnected flights of fancy. "The understanding must not therefore be supplied with wings, but rather hung with weights, to keep it from leaping and flying. Now this has never yet been done; when it is done, we may entertain better hopes of the sciences."[56]

Hope - give science wings

By the nineteenth century, these hopes had been triumphantly realized. The successes of science had proven that the method of science was the one sure road to truth. Kant, for one, had begun with the premise that the only way to combat the devastating skepticism of Hume was to establish metaphysics as a science, winning for it at last the "universal and lasting applause" and ceaseless progress characteristic of other sciences. Playfair, Stewart, Herschel, and Whewell had all fully justified Emerson's remark that science itself started with Bacon. Yet ironically, as Whewell observed, Bacon himself failed when he tried to apply his own method. In the abstract, it was perfect; in practice, unworkable. It was not clever theorizing about science that corrected the flaws in Bacon's scheme, however, but actual practice over succeeding generations, "till the actual progress of science had made men somewhat familiar with the kinds of steps which it in-

Kant

Bacon could not apply his own method

cluded." Science was making itself up as it went along, filling the gaps in Bacon's scheme with actual experience.[57]

It was time, then, to update Bacon. In Whewell's eyes, Bacon's greatest flaw was that he had slighted the role of ideas, and so it fell to Whewell to correct this by establishing "that *Ideas* are no less indispensable than facts themselves; and that except these be duly unfolded and applied, facts are collected in vain." Stewart also noted that Bacon had underrated the value of hypotheses, which frequently put philosophy on "the road of discovery." Although knowledge of facts came first, "yet a hypothetical theory is generally our best guide to the knowledge of useful facts." What emerged from these nineteenth-century theorists was a picture of the proper scientific method as a circle from facts, to ideas, back to facts, a carefully managed sequence combining induction and deduction. As Herschel concluded, the circle of science proceeded by linking facts to truth and truths to facts, creating "the grand and only chain for the linking together of physical truths, and the eventual key to every discovery and every application." Science was chain and key: the chain that linked the universe into a growing network of interconnections; a key that unlocked the secrets of the Sphinx.[58]

Emerson took a great deal of his understanding of science from Herschel, whom he read and admired from 1831 onward. Herschel's *Preliminary Discourse* would in fact become "a widely read conduct manual for proper scientific practice," the classic treatment establishing the true and proper scientific method as conned by generations of students ever since. An anonymous reviewer in the London *Quarterly Review* (which Emerson read steadily) recommended Herschel's book to "all English readers whom the intellectual progress of the present age has affected." Emerson did not know that the reviewer was William Whewell, the pioneer philosopher of science who was himself about to attempt his own revision of Bacon. Whewell's "review" amounted to an outline of his own ideas on science, which not only reinforced the pronouncements given by Herschel but did so under the veil of anonymity, giving Emerson the voice not of a particular writer but of Science itself. "If," said the Voice of Science,

we conceive the facts of external nature to lie before us like a heap of pearls of various forms and sizes, mere Observation takes up an indiscriminate handful of them; Induction seizes some thread on which a portion of the heap are strung, and binds such threads together; while Deduction, the purely reasoning faculty, employs herself in measuring spaces on the naked thread, and in devising patterns which can only be of use so far as the jewels will fit the places thus assigned them.

Induction, then, binds truths into higher truths, so that "a vast structure of science" is "built up," comprised only of phenomena as its material, "but in which phenomena are cemented by these successive modes of connected representation, and form a series of laws each supporting other laws more general; so that we obtain, combined with all the physical reality of the external world, all the permanence and symmetry of mathematical relations." Bacon's terrene globe would be fully interwoven with the crystal globe of ideas: someday in the future, the rules of method would string the beauties of the external world like pearls on the threads of mathematical reasoning.[59]

"Temperament is the iron wire on which the beads are strung." Thus wrote Emerson in "Experience." How did Whewell's lustrous and open-ended vision turn into so haunting an image of limitation, of Emersonian necessity? One clue lies in the slightly altered version of scientific method that emerged in Coleridge's writings, which stand above all others in their influence on Emerson. Bacon was frequently contrasted with Plato, in a contest pitting empirical study of nature against idealist conceptions of the universal. Coleridge, however, folded the two together as opposite poles of a fundamentally "bi-polar" philosophy. According to Coleridge, "Plato treats principally of the truth, as it manifests itself at the *ideal* pole, as the science of intellect," whereas Bacon treated "the same truth, as it is manifested at the other, or material pole, as the science of nature." While Plato directed himself to the objective truths of the intellect (or the "subjective" pole), Bacon directed himself to truths whose signatures are in nature (at the "objective" pole). While Plato called ideas "LIVING LAWS," Bacon called "the laws of nature, *ideas*." Whereas Plato's method was "inductive throughout," Bacon's scheme was "Platonic throughout"—and so forth.[60]

This interfolding by which opposites are reconciled as the two poles of the same power came about by way of Coleridge's own "Method," which reconstituted science as a function of his redefined "Reason." Coleridge hoped to render eighteenth-century empirical reason obsolete by redefining it as Reason, the "spiritual organ" that subordinates experience "to ABSOLUTE PRINCIPLES or necessary laws." As Barbara Packer points out, perfect knowledge of such laws could belong only to the Supreme Being, but Coleridge was willing to extend Reason in a secondary sense to science, which he defined as the study of the "properties of things by means of the Laws that constitute them." It was capital-R Reason, then, that closed the divide between body and mind by looking straight through the material world to the preexisting causal "idea" of God. Since nature was the art of God, the cause of nature had to be a single self-subsisting idea in the mind of God, an "antecedent method, or self-organizing PURPOSE," which, moving through matter rather like Cudworth's "plastic nature," organized it into physical form. To comprehend this one generative idea was to comprehend how all the scattered pieces are actually united parts of one whole.[61]

The name for this generative idea was "LAW," which predetermined the relationship of parts to each other and to the whole "by a truth originating in the *mind*," as in the physical sciences or mathematics. Clearly, however, some sciences had not yet reached this higher insight. Thus there was also a second and lesser ordering principle, "THEORY," in which observation or experiment suggested "a given arrangement of many [objects] under one point of view . . . for the purposes of understanding, and in most instances of controlling, them." Botany and chemistry, for example, were still little more than huge catalogues of names, for they had not yet received from philosophers the higher idea "matured into *laws* of organic nature." To Coleridge, this was only a sign of their immaturity: these sciences, too, would someday be "methodized" when their one creative idea germinated in some one great mind.[62]

The true connective principle, when sent forth into nature, would gather all the wild diversity of its forms into one united whole, in the same way that "water and flame, the diamond, the charcoal, and the mantling champagne, with its ebullient sparkles, are convoked and fraternized by the theory of the chemist." Coleridge reconciled poetry and science as polar manifestations of the same creative insight: just as in Shakespeare "we find nature idealized into poetry," so in great men of science like Sir Humphry Davy, "we find poetry . . . substantiated and realized in nature; yea, nature itself disclosed to us . . . as at once the poet and the poem!" Both Plato, "the Athenian Verulam," and Bacon, "the British Plato," understood this, and followed a "religious instinct" that bade us, once again quoting Shakespeare, to "find tongues in trees; books in the running streams; sermons in stones: and good . . . in every thing." Such correlations from nature naturally fostered a belief that the productive power that acts in nature "is essentially one . . . with the intelligence, which is in the human mind above nature." Coleridge's scientific method comprehended nature by replacing the creative mind of God with the creative mind of man. When the creative principle of connection in the mind was true, it was sanctioned by "the correspondency of nature," and then we would *truly* see nature—we would see it as God sees it, as the crystal globe of the mind realized on the instant eternity.[63]

This would seem to open the possibility that those "sermons in stones" were written not by God after all, but by the human intelligence, which purports to find what it has actually made itself. The mind, that is, may not *re*-construct the world according to God's blueprints so much as construct that world which most satisfactorily realizes blueprints of its own. If so, the sanctions offered by a "corresponding nature" are less a final lock on "real" reality than a provisional agreement on a possible reality. Coleridge's method opened both possibilities, for it encouraged experimental constructions of truth while affirming that, if they held, they were not constructions at all, but discoveries of preexisting truth. This approach built both flexibility and certainty into the production of knowledge, making it possible to be simultaneously "Baconian" in cautious inductive empiricism and "Platonic" in speculative idealism. Systems that seem like errant fantasies to twentieth-century sophisticates—Richard Owen's archetypes, Arnold Guyot's geography, Coleridge's own theory of life—were thus created and justified in their own day as models of inductive reason.

Herschel

The strength of Herschel's scientific method lay in the way it chained together small loops of trial and error according to one big directing idea, which thereby captured more and more of nature to itself. By contrast, Coleridge's method placed more emphasis on the big directing idea that grasps everything at once, that "unfixes the land and the sea, makes them revolve around the axis of his primary thought, and disposes them anew." Despite this distinction, Emerson did not see the methods of Herschel and Coleridge as radically different. To Emerson, both were at heart an imaginative grasp of reality by which "the solid seeming block of matter has been pervaded and dissolved by a thought." Emerson had set out to find a definitive refutation of Hume, and no refutation was so definitive as that offered by science: it worked. It gave the law by which "phenomena" were successfully "predicted," and taught—"impressively"—both "the presence and antecedence of Spirit" and that that Spirit was human, "that nature proceeds from a mind analogous to our own." Such was the payoff of the Copernican turn that made the creative mind the sun of the universe: poet and man of science were both alike channels for a higher power. The poet "turns the world to glass, and shows us all things in their right series and procession"; or, in other words, the poet "uses forms according to the life, and not according to the form. This is true science." And the scientist? Like the poet, in his view "the universe becomes transparent, and the light of higher laws than its own, shines through it." Neither poet nor scientist was the actual author of what he spoke; both were prophets who spoke with the voice of God.[64]

"A profound thought, anywhere, classifies all things. A profound thought will lift Olympus." A profound thought will make the individual larger than the universe. This is part of what Emerson meant when he spoke of "the infinitude of the private man": according to the way the universe speaks through you alone, you alone will uniquely construct a "representative" world. As he mused in 1836: "A man is a method; a progressive arrangement; a selecting principle gathering his like to him wherever he goes. 'Half is more than the whole.' Yes let the man of taste be the selector & Half is a good deal better than the whole or an infinitesimal part becomes a just representative of the Infinite. A man of taste sent into Italy shall bring me a few objects that shall give me more lively & permanent pleasure than galleries, cities, & mountain chains. A man is a choice." In this way, method links all individuals as unique manifestations of universal law, parts of the determining whole by whose grace they exist as parts at all. At this point in Emerson's thought, the organizing metaphor of "method" is still directed outward, toward infinitude domesticated by sympathy, toward choices that are gains rather than losses. By 1844 and his essay "Experience," the same organizing method points inward, to the "iron wire" of temperament that has clamped choice in the vice of necessity. Method becomes a selecting principle that confines rather than enlarges, threatens to turn the world that mirrors the infinitude of the self into a world that reproduces the limitations of the self.[65]

Meanwhile, the "profound thought" that "classifies all things" does so by relating them to the self, "humanizing" the world by connecting it to man: "All is naught without the Idea which is its nucleus & soul: for this reason no natural fact interests us until connected with man." To "build science upon ideas" is to guarantee that science will be, as it should be, "humanly studied": "When science shall be studied with piety; when in a soul alive with moral sentiments, the antecedence of spirit is presupposed, then humanity advances, step by step with the opening of the intellect and its command over nature." Yet, as he reminds himself, there is work to be done: "The savant is unpoetic, the poet is unscientific." In this admission lies Emerson's own mission: to correct the poet who "leaps to a conclusion which is false," and the men of detail who "cling to the cadaverous fact until Science becomes a dead catalogue, and arbitrary classification." Truth lies in their convergence, for, as he would show in his poem "The Sphinx," each corrects the other: "I fully believe in both, in the poetry and in the dissection."[66]

In this belief lies a new hope. Stewart wondered, in a low moment, whether human invention was limited, "like a barrel-organ, to a specific number of tunes." Yet he offered instead the fairer inference, that pure imagination was narrow "when compared with the regions opened by truth and nature to our powers of observation and reasoning." Empirical science could open the eyes, too. Emerson speaks to this truth with the voice of a convert, a voice Thoreauvian in its intensity. Poetry gives us "an oratorio of praises of nature;—flowers, birds, mountains, sun, and moon,—yet the naturalist of this hour finds that he knows nothing, by all their poems, of any of these fine things." Poets have been repeating idly a few glimpses in their song,

> But go into the forest, you shall find all new and undescribed. The honking of the wild geese flying by night; the thin note of the companionable titmouse, in the winter day; the fall of swarms of flies, in autumn, from combats high in the air, pattering down on the leaves like rain . . . any and all, are alike unattempted. The man who stands on the seashore, or who rambles in the woods, seems to be the first man that ever stood on the shore, or entered a grove, his sensations and his world are so novel and strange.

The poets tell Emerson that there is nothing new to be said; but nature, "an alien world; a world not yet subdued by the thought," calls him out of himself, "takes down the narrow walls of my soul, and extends its life and pulsation to the very horizon. *That* is morning, to cease for a bright hour to be a prisoner of this sickly body, and to become as large as nature." At a time when poetry seemed choked with repetition and religion dying from convention, to bear "an original relationship with the universe" meant to open one's eyes to the dawning worlds of natural science.[67]

This, then, was the mood in which Emerson approached his reading in science: creative reading, to be sure, not to absorb facts but to take down the narrow walls of self—if also, in apparent contradiction, for assurance that the self reflected in nature was indeed familiar and known. Emerson pursued his reading of science at a time when science was expanding and changing rapidly, too rapidly for anyone to keep up. Fields were emerging and separating and mutating in a wild stew of energy. Emerson "tracked himself," as Thoreau said, through a wide range of scientific texts. His most intensive reading in science occurred from 1830 to 1834, the years leading up to and following his resignation from the ministry and his first trip to Europe. However, Emerson continued to read widely in science right into the 1870s, including the latest books by prominent natural and physical scientists: Charles Darwin, Richard Owen, John Tyndall, T. H. Huxley, Ludwig Büchner, John Herschel, Max Müller.[68]

The changes in science across Emerson's lifetime can be tracked along two axes. First is the perceived continuum between "mature" sciences such as astronomy and physics and "immature" sciences such as botany and chemistry. As Emerson expressed the difference, "Science immature is arbitrary classification. Science perfect is classification through an Idea." Second is the increasing professionalization of science, with the concomitant increase in scientific specialization. Although this is not the place for discussion of the professionalization of science, it can be observed here that "amateurs" continued to play an important role in astronomy and natural history to the end of the century, as indeed they still do. In the United States, science can be said to have emerged as a profession rather than an avocation at midcentury, with the establishment of the Lawrence Scientific School at Harvard and the formation of the American Association for the Advancement of Science, both efforts in which Emerson's friend Louis Agassiz played a central role. This opened up science as a career pathway (too late for Emerson or Thoreau, but just in time for William James), a new alternative to the traditional choices available to the middle-class American gentleman: divinity, law, medicine, business.[69]

Until science became a viable, paying profession, individuals interested in science who were not independently wealthy tended to be clergymen, either ministers or professors. The very word scientist as designating a specific type of intellectual was not invented until 1833, when at a meeting of the British Association for the Advancement of Science Coleridge himself stood up and forbade the roomful of learned gentlemen to dignify themselves with the term they had been using, philosopher. As an alternative, William Whewell spoke up to propose "that, by analogy with artist, they might form scientist"—but as Whewell added ruefully in the 1834 article where he first tried to introduce this novel term to the public, "this was not gener-ally palatable." Richard Yeo notes that influential scientists like Michael Faraday and T.H. Huxley rejected the term *scientist* because they "pre-ferred to think of their work as part of broader philosophical, theological, and moral concerns." Nevertheless, no other adequate replacement for *philosopher* was found. By the end of the century, English-speaking scien-tists had accepted the neologism, and Coleridge, as usual, had prevailed.[70]

Thus Emerson knew no "scientists." He knew instead "natural philoso-phers" (or just "philosophers"), "naturalists," "savants," or "men of sci-ence," terms that, as Coleridge suspected, tended to blur the distinction between knowledge of nature and knowledge more generally. As profes-sionalization developed and the distinctions sharpened, a new class of writ-ers emerged, beginning in the 1870s, to fill the widening gap between the knowledge produced by practicing professionals and the reading public: the popularizer or science journalist. Thus what Emerson read was written not by self-conscious translators of specialist knowledge to a nonspecialist public, but by practicing scientists, such as Davy, Cuvier, Lyell, Herschel, Whewell, Somerville, Agassiz, Asa Gray, and Darwin, who assumed that their audience included the literate public; or by intellectuals who assumed that any educated and interested intellectual could fairly pronounce on the wider meaning of scientific knowledge: Goethe, de Staël, Sampson Reed, and Emerson himself.

A complete survey of Emerson's reading in science would fill many vol-umes, for a variety of individual scientists had an impact on his thought that could usefully be traced through any number of his journal volumes, ser-mons, essays, lectures, and poems: Goethe, Schelling, Lorenz Oken, Georges Cuvier, Jean-Baptiste Fourier, Laplace, Adolphe Quetelet, Davy, Faraday, Lyell, Herschel, Alexander von Humboldt, Agassiz, Darwin, and Tyndall, among others. Beyond such individual stars, Emerson's reading in science fell into several categories. First, his Harvard education trained him in the Anglican tradition of natural theology, which embraced science as the knowledge of God's beneficent design and therefore of God himself. William Paley's *Natural Theology* was, as has already been noted, the nine-teenth century's classic text; but Emerson also knew such earlier classics as Ralph Cudworth's *True Intellectual System of the Universe* (which de-fended against atheism by adopting Platonism into Christian theology) and William Wollaston's *Religion of Nature Delineated* (which demonstrated that moral distinctions have their basis not in feeling or sentiment but in reason, from which it followed that true religion was founded on nature).[71]

Natural theology assumed that the universe was purposeful and each of its myriad details meaningful, a legacy that carried natural science deep into nineteenth-century culture, both British and American, as generations of natural historians pursued truth into every corner. The classics in this tradi-tion which Emerson read included Gilbert White's *Natural History and An-*

tiquities of Selborne, William Bartram's *Travels,* and James L. Drummond's didactic essay *Letters to a Young Naturalist.* Some books in this tradition opened up the overlooked marvels of God's creation, such as William Kirby and William Spence's *Introduction to Entomology;* others carried the high seriousness of the increasingly outdated Paley into the mid-nineteenth century, notably the so-called Bridgewater Treatises of the 1830s, eight works deeded by the Earl of Bridgewater and commissioned by the Royal Society on "the Power, Wisdom and Goodness of God, as manifested in the Creation." Emerson read at least three: *The Hand,* by the Scottish physiologist Sir Charles Bell; *On the Adaptation of External Nature to the Physical Condition of Man,* by Oxford professor of medicine John Kidd; and *Animal and Vegetable Physiology, Considered with Reference to Natural Theology,* by Peter Mark Roget—the secretary of the Royal Society, but better known for his famous *Thesaurus,* a product of his system for organizing all knowledge. Natural history thus developed in close cooperation with natural theology, and indeed the life sciences carried the "argument from design" well into the final third of the century.[72]

While natural history studied and appreciated the plenitude of God's creation, natural philosophy concentrated on the underlying principles of matter in motion, generally setting aside arguments about final causes the better to understand secondary or efficient causes. Here the model was Newton, who had made the universe a true uni-verse by finally joining the terrestrial and celestial spheres under one mathematical rule—the kind of all-comprehending general truth that made natural philosophy "philosophical." Emerson's Harvard reading included William Enfield's *Institutes of Natural Philosophy,* a textbook covering astronomy and physics, which he supplemented with such volumes as James Ferguson's *Astronomy Explained upon Newtonian Principles* and John Playfair's gracefully written survey, *Dissertation Second: Exhibiting a General View of the Progress of Mathematical and Physical Science.* In 1832 he looked into the French mineralogist René-Just Haüy's somewhat outdated *Elementary Treatise on Natural Philosophy;* he brought himself up to date by reading such books as Herschel's *Preliminary Discourse* and *Treatise on Astronomy,* Dionysius Lardner's *Treatise on Hydrostatics and Pneumatics,* and David Brewster's *Life of Sir Isaac Newton*—not to mention such periodicals as the (London) *Quarterly Review* and the *North American Review,* which faithfully reported the latest scientific insights in long and meaty articles written by participating men of science.[73]

Newton had set the standard not just for the physical sciences but for all science. It was widely assumed that the same ideal, the reduction of many phenomena to a few determinative laws, would in time order and control the bewildering chaos of particulars in natural history, making the "history" or description of nature into a truly philosophical study, or "science." This was the implicit assumption in Emerson's distinction between mature

reduce all nature to physics

and immature science, which he may have derived from Coleridge. Herein lay another controversy, which Emerson also tended to elide: determinative, mathematical laws were doing a magnificent job of predicting phenomena of physical nature, but there was a real question whether organic nature could also be reduced to predictive laws. Kant had considered the question and answered no: this would reduce all nature to a special case of physics. Yet, as Kant pointed out, *nature* really had two meanings: first, "the primal, internal principle of everything"; and second, "the sum total of all things insofar as they can be objects of our senses and hence objects of experience" (or what Emerson, borrowing a medieval topos, would call *natura naturans,* or nature "the quick cause," and *natura naturata,* nature "passive" or created). Kant made it clear that *science* could refer only to the former, for only the "rational cognition of the coherence of things," their derivation from an internal principle, could be necessarily true; "cognition that can contain merely empirical certainty is only improperly called science."[74]

Thus Kant laid down the rule: "A rational doctrine of nature, then, deserves the name of natural science only when the natural laws that underlie it are cognized a priori and are not mere laws of experience." True science, in short, was Newtonian science, and if natural history was to become a true natural *science,* those "merely empirical" certainties had to be reduced to necessary truths. In response to this challenge, traditional natural history rapidly diversified into a number of new and flourishing law-seeking natural sciences: chemistry, physiology, comparative anatomy, biology, geology, physical geography, botany, zoology, and, late in the nineteenth century, ecology. Each science hoped for its own "Newton": for instance, Kant's student Herder looked ahead to the Newton of "geogony" or earth science; Cuvier opened his popular *Discourse on the Revolutions of the Surface of the Globe* by wondering, "Why should not natural history one day boast also of her Newton?"[75]

Why not indeed? Kant's distinction became a challenge, for to admit that one's own field of study could not be rationalized according to a priori laws was to surrender the prestige of Newton, of science itself, and to admit to being a mere stamp collector. The challenge was taken up on a wide range of fronts, and Emerson made himself acquainted with virtually all of them. In chemistry, he read Davy's *Elements of Chemical Philosophy* and *Elements of Agricultural Chemistry,* which showed that chemistry, too, was lawful, and applied those laws to the growth and cultivation of plants (and nations). His reading of Cuvier's *Discourse* and Charles Lyell's 1826 review article opened up the lawful development of Earth through the "immense antiquity" of geologically deep time. In botany, de Candolle's *Elements of the Philosophy of Plants* began the hard task of replacing Linnaeus's artificial system of classification with one based on the formative law of natural design.[76]

Emerson's reading in these works, in the late 1820s and early 1830s, laid the basis for his later reading in natural science, which included the comparative anatomists Louis Agassiz and Richard Owen; Lyell's *Principles of Geology* and Edward Forbes on glaciers; and the evolutionary theories of Robert Chambers and, later, of Charles Darwin. He also derived much of his understanding of science from the more aggressively metaphysical theories of the Germans, which, as Jonathan Bishop notes, gave him the insight he needed to reinterpret the old argument from design not as "a proof of the existence of an anthropomorphic divinity, but evidence of the human mind in its natural action of finding patterns in any scene it comprehends." Emerson first encountered the German theories through Coleridge (whose works were constantly at his elbow from 1819 onward), de Staël, Carlyle (another constant companion, from 1829 on), Frederick Henry Hedge, and James Marsh's seminal "Preliminary Essay" to Coleridge's *Aids to Reflection,* which introduced America to Coleridge. He responded immediately to Goethe's theories of plant and animal morphology, and to the more holistic aspects of Humboldt's thought. All of them helped to feed the Emersonian synthesis, and led him in the 1840s to look into some of the primary texts: Schelling's *Ideas for a Philosophy of Nature,* Johann Bernhard Stallo's *General Principles of the Philosophy of Nature,* Johann Gottlieb Fichte, Oken, and others. Given this background, Robert Chambers's notorious evolutionary theories, published anonymously in 1844, struck Emerson as rather tame, while Agassiz's and Owen's archetypes and the "arrested and progressive development" Emerson attributed to John Hunter were normal science, and even Charles Darwin's evolutionary theories seemed perfectly unsurprising.[77]

Finally, the successes of the physical and natural sciences inevitably suggested that a similar success could be won if such philosophical rigor were applied to the study of the human mind. In 1822 Emerson read Joseph Priestley's *Lectures on History and General Policy,* in which Priestley observed that human history, too, evidenced the ways of God; and seven years later he looked into Johann Gottfried Herder's *Outlines of the Philosophy of the History of Man,* where Emerson found the creative life force, "the finger of God," in the developing embryo, in the course of a human life, through many lives in the course of a society, and through many societies in the course of human history—life at every level forming itself in the context of its local environment, using energy to assimilate matter to its own generative idea.[78]

Dugald Stewart also turned to science for a model, seeking to lay down the rules of investigation into "the original and universal principles and laws of human nature" as distinguished "from the adventitious effects of local situations." Herder would have quarreled with Stewart on the grounds that mind could not possibly be understood apart from its "local situation," since mind created itself in interaction with environment; and Stewart would have rejected Herder as unscientific for not teasing apart what Co-

Science - model + method for

leridge would call the objective and subjective poles of investigation. Nevertheless, Herder and Stewart—and Coleridge, too—all turned to "science" as a model and method for studying not the world apart from man, the perceiver, but man as the very agent of perception. Emerson went on to read widely in the nascent sciences of man, from Johann Casper Lavater's *Physiognomy,* to Sampson Reed's *Observations on the Growth of the Mind,* to Fourier the social engineer, Quetelet the statistical theorist, James Cowles Prichard's *Natural History of Man,* Henry Charles Casey's *Principles of Social Science,* Max Müller on language and religion, and Hegel's *Lectures on the Philosophy of History.*[79]

Moreover, Emerson strove to place himself in this tradition, from his 1821 Bowdoin Prize essay (heavily indebted to Stewart), to his repeated and unsatisfying attempts, from the 1850s on, to complete a "Natural History of the Intellect" in which the laws of nature, reflected from mind to nature and back to mind, would help to explain ourselves to ourselves. By then, the sciences of man were no longer nascent but had crystallized into the developing new fields of sociology, anthropology, psychology, and linguistics, all of which demanded training and expertise a generation or two beyond Emerson's early days at Harvard. The sturdy branches of the mighty tree of scientific knowledge had so multiplied themselves and grafted one onto the other that the clean solid stem of science itself, Bacon's "First Philosophy," was no longer truth's anchoring parent but yet one more territory in the tangled web of knowledges.[80]

CHAPTER THREE

..

Gnomic Science
THE BODY AND THE LAW

THE LIGHT OF THE EYE: EMERSON'S EARLY CAREER

One day sometime in the middle of 1824, the twenty-one-year-old Emerson sat down and drafted a "Letter to Plato," in which he wished he could seek from that philosopher the answer to a pressing question: "How could those parts of the social machine whose consistency & just action depend entirely upon the morality & religion sown & grown in the community, how could these be kept in safe & efficient arrangement under a system which besides being frivolous was the butt of vulgar ridicule?" It is a utopian question: How could a community be bound into a true organic whole, a smoothly running "social machine," when the traditional means, the Christian church, was both frivolous and ridiculous? The question went right to the heart of the problem that drove Emerson. A young American society, "aloof from the contagion" that had brought the rotten state of Europe into decline, "hath ample interval to lay deep & solid foundations for the greatness of the New World." Yet it seemed that America, too, was infested with the European disease of skepticism, materialism, and self-interest. In his occasional dark moods, Emerson outlined the consequences: denying God would leave man alone "in a Universe exposed to the convulsions of disorder and the wrecks of systems where man is an atom unable to avert his peril . . . destitute of friends who *are* able to control the order of Nature." Denial of God would abandon humanity to a loveless, "dire dominion of Chance"; it was to "put out the Eye of the Universe."[1]

Emerson spent much of his life writing to Plato, whose *Timaeus* offered a theory of vision that eventually provided the answer Emerson needed. Timaeus offers his auditors a detailed creation myth, including the origin of the human organs of sense. Of these organs, the gods "first contrived the eyes to give light"—not to receive, but to *give* light—as follows:

> So much of fire as would not burn, but gave a gentle light, they formed into a substance akin to the light of everyday life, and the pure fire which is within us and related thereto they made to flow through the eyes in a stream smooth and dense. . . . When the light of day surrounds the stream of vision, then like falls upon like, and they coalesce, and one body is

formed by natural affinity in the line of vision, wherever the light that falls from within meets with an external object.

The stream of vision diffuses "the motions of what it touches or what touches it over the whole body, until they reach the soul, causing that perception which we call sight." The stream of vision flows out through our eyes from the fire within us and meets its like in the light of day. From affinity comes sight. What we see we half create, and what we create we will not see unless it meets its like in the world; but when the eye creates, and when nature consents, vision and light coelesce into a body, and truth enters the soul.[2]

Hume had said there was no causal relationship between what the eye saw and what really existed. Kant had said it was Hume who spurred him to drive out the infection of skepticism. Young Emerson set himself to "combat" Hume, too, just as the Neoplatonist Cudworth had combatted Hobbes, the Hume of his own day. It was a veritable battle: the scholar of Emerson's day had been born, Emerson wrote, "in a time of *war*. A thousand religions are in arms. Systems of Education are contesting. Literature, Politics, Morals, & Physics, are each engaged in loud civil broil. A chaos of doubts besets him from his outset."[3]

Curiously, though, the solution to this crisis lay partly in Hume, who had said one fine thing: the abstract objects presented by natural religion "cannot long actuate the mind." They needed to affect the senses and the imagination as well. There needed to be, as Plato had said, some "affinity" between the soul within and the world without, some mediating element that brought the two together and made them one. In a word, light; or, in another word, religion, from the Latin *religare,* to tie or bind fast, the "invisible connection" between heaven and Earth. Division had made possible connection; connection brought beings into a relation, and all relationships were governed by "moral laws" that had come into being the instant God created objects and beings. Emerson, even before he turned twenty, understood the nature of the moral relationship between mind and world: knowledge, the "discipline" whereby the mind was "exercised" and brought out of itself into relationship with the world, by bringing the world into relationship with the soul. Without knowledge, we are shut up in the dark closet of the self; with it, the self becomes a transparent window opening the soul to the sunshine: "The Oyster complains that the Universe is a very dark hole. The office which men of science & philosophy do the mind is not to create or sing the beauty of the world but to take down the thick walls of our house & put glass windows instead. The sun and the Universe were shining there before we saw them." Once science has opened "the thick walls of our house," the universe becomes our oyster: "the world is a school for the education of the mind."[4]

Emerson repeated this message, in various ways, so long as he was able to write anything at all: "Education is the drawing out of the soul"; the "perishable *material universe*" is an "organ of our trial or education"; in a world in which moral law did not govern, "the chaos of thought would make life an unsupportable Curse. . . . Rend away the darkness, and restore to man the knowledge of this principle, and you have lit the sun over the world & solved the riddle of life." Once the riddle is solved, the world becomes the food of the soul: the mind "feeds with unsated appetite upon moral & material nature" and is "perpetually growing wiser & mightier by digesting this immortal food." Knowledge feeds us and makes us strong not by any inherent virtue but by virtue of the power that it gives the mind: "the great end of abstruse study is not the knowledge it procures but the cultivation of the intellect which necessarily results." Bacon was right: every event feeds the student, increases his "knowledge and in consequence his power"; the kingdom of heaven begins "in thus trying your powers & bringing out each one in order until the whole moral man lives & acts & governs the animal man."[5]

This very process, the exercise of knowledge that gives power, refuted Hume, for only in a world bound by moral "relation" could this be so. The myth of Timaeus is about not sight but *in*sight, knowledge. It is the light of our minds meeting the light of the Eye of the Universe that kindles the world into mind, into knowledge, and that this happens at all proves the affinity of like with like, our mind with God's mind. In 1830 Emerson copied quotations from Pythagoras into his journal: "As light is perceived by the eye, sound by the ear because of affinity between object & organ so the universality of nature by reason because of a consanguinity betwixt them." The basis of man's "consanguinity" with nature was, then, not body but reason. Emerson's developing system assumed that the universe was, *must* be, rational. What, then, could explain the existence of pain and evil? "If God is good," wondered Emerson, "why are any of his creatures unhappy?" One possible explanation was chance, but in the end Emerson found the idea of chance too repulsive to take seriously. This rejection of chance solidified his lifelong commitment to the argument from design, even with its troubling limitation on the freedom of the individual: "No Chance or Fortune sits as the blind lord of the changes that take place but that an Order is appointed, a System is proceeding. [The mind] can perceive a harmonious whole, combined & overruled by a sublime Necessity, which embraces in its mighty circle the freedom of the individuals, & without subtracting from any, directs all to their appropriate ends." There was, then, a purpose to pain, a "great & primal Necessity" that "may make Impossible an Universe without evil." Even pain and evil could be rationalized as the introduction of lawfulness into the human universe.[6]

From the first Emerson was building into his universe a dynamic polarity, a balance of contending forces. His insight that God does not strike us personally for blasphemy, but appeals to us impersonally through our reason, laid the basis for an early essay on the moral law, or nature's "equal & pervading Compensations." Thus, commit an outrage upon the moral law, and in all cases "pain and misery" will follow. The "bold misdoer" has "disordered a part of the moral machinery of the Universe and he is in peril of being crushed by the mischief he has caused." Conversely, Emerson comforts himself for his own deep pain—the eye disease that nearly blinded him—by reminding himself that "Compensation has been woven to want, loss to gain," and if every misdeed brings down its own punishment, so "every suffering is rewarded." A personal God might seek revenge for misdeeds or bestow arbitrary rewards, but in a rational universe governed by impersonal law, one's own actions will command the consequences. Our life, our own free choice, calls the world around us into being, and in this sense we are free after all, free in this world of law, free to create our own fate not by fighting but by soliciting nature's "irreversible decree." All living is a form of prayer, Emerson explained in his first sermon: there is a "pre-existent harmony" in the universe by which prayers become effects, by which our life in this moment realizes our whole future: " 'Every man,' said the Roman maxim, 'is the architect of his own fortune.' "[7]

By his twenty-first birthday, then, Emerson had already woven a tight-knit theory of the universe in which moral law was activated in every decision, every moment of living, every thought. His "rational" universe was acutely alive to the possibilities of mind because it established mind and matter as reciprocal agents of power. Knowledge put in play a moral relationship, a practice of self-making by which mind brought itself into being by "relating" to the material world. Thus Emerson's early dualism consisted not of the epistemological gap between mind and body—the problem of how the immaterial mind can know a material universe—but of the moral gap between chance and certainty, chaos and hierarchy, unmeaning and self-realization through informed submission to order.[8]

Indeed, it was by rejecting chance that Emerson forged mind and body into reciprocal halves of one whole and teleological order. "This is the amount of all our insight into nature, the discovery of the purpose," he sums up. All objects in nature are parts of a plan, which the advancement of knowledge realizes. Mind, as it advances, expands and unites things long severed, restoring order to chaos. The name of this process is science, and the goal of science is to discover the purpose of nature; otherwise fact is "monstrous." Deny this purpose, this goal of science, and, Emerson rhetorically concludes, when change and destruction obey their mighty lord, "when Thought is gathered through all its infinite channels to its Divine Fountain, & Goodness to its reward, & Matter is dissolved—then can his will bridle the ministers of the Universe, and stop the almighty operation?" Whoever casts himself on chance and denies moral design sets himself "adrift upon wild and unknown seas."[9]

This tightly wound reciprocity between mind and the material universe puts tremendous pressure on the individual in his relationship with society. Growing up in society, men learn "to view mankind *as a society,* acting freely for itself & to distinct & finite ends—not as the machinery of another Being concealing secret agencies of another World." The latter view is, however, the correct one. Emerson follows it to its logical extreme: "Extinguish the Sun. Annihilate this solid fabric of earth. Forget the forms of life & beauty which adorned it. What is any worth?" If society is indeed "the machinery of another Being," intended as our "organ" or tool for self-education, it is *we* who make social relationships, as well as natural ones, to *our* will: man should make institutions from the bottom up, not institutions the men who inhabit them.[10]

Yet entrance into those institutions exacts its toll. Young Emerson notes without irony that "men pay a price for admission to the civilization of society." He means this literally: he and his mother pay, he estimates, twenty-six hundred dollars a year to get the "higher seats," and the calculation makes him sound like quite the social conservative: "Every man who values this bargain which he drives so zealously must give the whole weight of his support to the public, civil, religious, literary institutions which make it worth his toil. Keep the moral fountains pure. Open schools. Guard the Sabbath." Who is making whom here?[11]

A few pages later Emerson asks the same question: do "institutions affect a people or a people modify their institutions"? The question's relevance soon becomes clear: Emerson has been weighing his own institutional commitments, and, surprisingly for those who know only the Emerson of "Self-Reliance," he does not come down on the side of the individual. It already seems to him that the successful author follows public feeling rather than leading it, that he "catches the tune of the times and relies on this implicit obedience to its will to ingratiate himself in the world." The author makes himself by obedience to the social will; great men, the "Newtons, Bacons, and Lockes," are *not* self-made, or they would be "as often bred in shops & stables as in colleges—it would indeed be deeply discouraging to the cause of Education. But the fact is that *all* genius has owed its development to literary establishments." Well then, he will surrender himself to the religious establishment. After expressing his unease with the decision, and declaring even greater unease with the alternatives—law, or medicine, or teaching— he submits: "But in Divinity I hope to thrive." He will make "an entire conquest of himself" for the sake of achieving private influence over other men (for he confidently expects success in public preaching); he will let the institution make him to its mold. "My trust is that my profession shall be my regeneration of mind, manners, inward & outward estate."[12]

It was shortly after this momentous decision to enter the ministry that Emerson drafted the letter to Plato in which he wondered how to bond together a moral community when religion was the butt of ridicule. He already knew his answer, but he continued slowly to work it through: by instilling a love of knowledge, the "cultivation of the intellect." The truths of science would not do for all, however: not everyone was able to "comprehend Newton's Principia." By contrast, the truths of Revelation were simple and open to all; everyone was made "to receive the belief of a God. . . . Hence the difference between Religious and Scientific thought, that the one is to be *felt,* & the other to [be] elaborated." That said, the primary model was still to be science. Emerson added on the next page that feeling, love, "warms the mind into a ferment. . . . Then, embryo powers . . . are nursed into the manhood of mind. A powerful motive is to the character what a skillful hypothesis is to the progress of Science; it affords facility & room for the arrangement of the growing principles of our nature." The formulation would become a classic one in Emerson's thought: love or beauty warms the heart and opens the way to truth, to knowledge, to science. Conversely, only knowledge based on love, truth experienced and hence true to the heart, is truly one's own knowledge; and only beauty or poetry based in the truth of nature is connected to the fountain of power.[13]

Progress of Science

Emerson was beginning to imagine a certain role for himself, a role that the ministry might enable but that reached well beyond the limits of a local congregation. Yes, he would enter the church, but what Emerson really wanted to be was a writer. Most books, he had noticed, only "record the progress of science or exhibit only successive forms of taste in Poetry, letters & fiction." But now and then books of another sort appeared, books "which collect & embody the wisdom of their times & so mark the stages of human improvement." Such had been the works of Solomon, of Montaigne, and—"eminently"—the essays of Bacon, and such would be, perhaps, someday, the essays of Emerson: "I should like to add another volume to this valuable work." A few pages later he expanded on this thought: someone needed to write serious books not for the learned, and not for children and the childhood of the mind (not, that is, "Novels & Romances"), but for "the third class of men" who make up "the great body of society" and conduct its business. Why not himself? "I shall therefore attempt in a series of papers to discuss in a popular manner, some of those practical questions of daily recurrence, moral, political, and literary, which best deserve the attention of my countrymen." He continued to entertain the idea over the following months, concluding that he would deserve the most thanks of the knowing reader "if I shall shew him the colour, orbit, & composition of my particular star." His ambition as a writer and intellectual would be not universal knowledge or systematic science but the truth of the immediate, local, anecdotal—his "particular" star. By demonstrating the "ray of relation" from every object of interest to *himself,* he would persuade not by argument but by example, not build systems of knowledge for others to live in but show a mind in the living and the building.[14]

Mark stages of Human improvement

Emerson wrote

He would be, then, a peculiarly private kind of public intellectual. His topic would be the power of knowledge—not knowledge accumulated, nor

Private individual

even knowledge applied, but knowledge in the act, on the fly, shooting the gulf, at the moment of drawing the mind into the world and the world into the mind. Although he conceived this project as early as 1824, it would be some years before he perfected the voice that would allow him to achieve it. A first attempt, roughed out in the journal he kept during this period, shows the distance Emerson had yet to travel: he self-consciously rehearsed the topic "Civilization Moral," on the theme that "the great expansion of knowledge which has been the fruit of printing and Commerce in modern times has indeed added worlds to the dominion of the mind. . . . & inasmuch as knowledge is power it has increased the ability to do mischief or good." Knowledge as social power was potentially dangerous if not contained by knowledge as personal power, the disciplined practice of the self that would make knowledge a truly moral relationship.[15]

It was this practice that most interested Emerson. He had been impressed with Bacon's "fixed resolution at eighteen years old—to reform Science"; little older than eighteen himself, he used the examples of Bacon, Milton, Luther, and Newton, whose "perseverence, thro' all obstacles" had made them great, to encourage his own ambition. The practice of self-education was serious work, but even before he read Sampson Reed's *Observations on the Growth of the Mind* in 1826, he understood that the work and uncertainty of the process was precisely what made the mind grow, made it "great": God had "sowed truth in the world" but had let men "arrive at it by the slow instrumentality of human research." Such was human fate: "It is a striking feature in our condition that we so hardly arrive at truth." What seemed to be certainty was often but "bigoted faith." Thus humans were immersed in probabilities, breeding the skepticism that science fought with the weapon of truth, and religion with the weapon of conscience; "The final cause of this is no doubt found in the doctrine that we were not sent into this world for the discovery of truth but for the education of our minds."[16]

Skepticism in this view becomes not the enemy to vanquish but the necessary challenge that goads us into taking up the project of self-culture. As Hume was converted from enemy to ally, skepticism is converted from opponent to resource, a tool of power. All this would come at a cost, of course; Emerson noted grimly that "the Mind is enlightened by Misery. If knowledge be power, it is also Pain." The cost, for example, of freeing the slaves would be the downfall of innocent men and the loss of properties and lives. The original challenge of skepticism had been to find a moral ground from which evils such as slavery could be condemned, and which would justify sacrifice for a greater good. The moment Emerson arrived at this ground was the moment he conceived his ambition to speak, from his private and particular experience, to the world: his own "fixed resolution" would center on the necessary work of the single self, alone with the truth, standing in the eye of the public. Social knowledge began as personal knowledge, and finally all true knowledge could be only personal, since the only person one could hope to educate was oneself. Similarly, one fought to free the slaves less for their sake than for one's own. Only in the convergence of knowledge as personal power, with many individuals working toward one truth, could one entrust knowledge as social power. In a rational universe, no other alternative was possible; and Emerson refused to consider a universe that was not at least as rational as he.[17]

Even as he was centering his fixed resolution on the practice of the self, Emerson in his twenties was committing himself to the institutional practices of the church. As a professional minister, he was embedded in an institution whose ultimate authority was Gospel, the book of Revelation. The institution was not, however, remaking him, but pulling him apart. Publicly he could avow to his congregation what appeared to be standard natural theology: "Now the discoveries of science—not the instructions of revealed religion but the discoveries of science,—have compelled men to believe, that nothing is made without a purpose; no limb, no bone, no antennae, no hair, no feather, without a distinct purpose that is disclosed as our knowledge is increased." Science, not Gospel but science, Emerson drove home, proved the existence of God and provided the basis for religious belief.[18]

So he said publicly, but privately he explored the other half of the argument: the Gospel was at best a dubious source of authority. Sampson Reed, he wrote, came to him as "a revelation," although Reed told Emerson nothing he had not already worked out for himself; but Reed did effectively give Emerson permission to cut loose, as he did, immediately. The evidences of natural theology, he ventured, lay not so much in the thousand particular facts as in "a new evidence discovered only by uniting them together." By contrast, "[The testimony of crowds and of ages has been found on scrutiny almost worthless & the ingenuity of scholars has thrown into irrecoverable doubt parts of history which had been reckoned the best authenticated. But moral evidence, the evidence of final causes when it can be procured is unerring & eternal.]" "What is the authority of this imposing religion?" the young minister asked: the "dismal history of opinions" threw doubt on all—even on the truth of the Gospel itself. Skepticism has drawn young Emerson close to the agnostic's uncertainty. Yet, he reasons, in a universe without God, one could not "enforce a law of obedience in this *dumb* magnificence, this speechless universe." In a speechless universe, the voice of authority must lie in the "moral evidence" provided by man's ability to make nature speak.[19]

Where would such evidence be found? Immediately Emerson pointed to a recent "Geological article" in the *Quarterly Review* for "some prodigiously fine remarks." Its anonymous author was Charles Lyell, announcing the progressive plan of nature in an Earth of unimaginably deep antiquity. Moreover, Emerson ventured a progressive new concept of his own, going against his previous conviction that the knowledge of moral truth, unlike scientific knowledge, could not advance. On the contrary, he now asserted: "Understand now, morals do not change but the *science* of morals does advance; men discover truth & relations of which they were before ignorant; therefore, there are discoveries in morals." Public worship should, then, be adapted to changes in society, should make way for novel and progressive views of moral relations—such as the views he was about to propose.[20]

On the model of science, then, Emerson, too, could be an authority, not merely obedient to the preexisting authority of Gospel and church. This path led him to agnosticism's very brink: "But now it must be admitted I am not certain that any of these things are true. The nature of God may be different from what he is represented. I never beheld him. I do not know that he exists." On April 17, 1827, Emerson felt he was on the very edge of society: "There is a pleasure in the thought that the particular tone of my mind at this moment may be new in the Universe. . . . I commence a career of thought & action which is expanding before me into a distant & dazzling infinity. Strange thoughts start up like angels in my way & beckon me onward." Nor was he wrong. The authority of the book of Revelation was merely human, and no authority over him. The source of authority lay not in the contentious and warring factions of human society but in the nonhuman realm of nature—or, more exactly, in the mind's ability to make nature speak the truth. In other words, the new authority was science. Working completely from scratch, Emerson had begun to transform traditional natural theology into a new formation, one that in two generations would pervade intellectual culture: scientific naturalism.[21]

The most immediate effect of his insight was a still tighter version of compensation in which the law was not just circumstantial and eventual but internal and immediate. The force of authority, once it is no longer external but intrinsic to the self's relation with the world, cannot be resisted. The sinner may think he has escaped:

> But God has not so poorly framed the economy of his administration. He devised no fallible police, no contingent compensations. He secured the execution of his everlasting laws by committing to every moral being the supervision of its own character, by making every moral being the unrelenting inexorable punisher of its own delinquency.
>
> In the hour when he sinned that hour his own fate avenged on him the majesty of the laws he had broken.

The force of law polices each individual internally, makes each delinquent self its own self-punishing policeman. This puts every individual at the controlling center of his circumstances, and so both independent of them and completely formed, even trapped, by them.[22]

Very quickly Emerson arrives at a new image for this subtending force: magnetism. "I have seen a skilful experimenter lay a magnet among filings of steel & the force of that subtle fluid entering into each fragment arranged

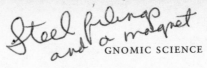

them all in mathematical lines & each metallic atom became in its turn a magnet communicating all the force it received of the loadstone." Individual and society are related like steel filings and magnet: each individual is aligned by the force of the magnet, which arranges every one, by accordance with law; but individuals are aligned not by passive acquiescence but by active energy, for each little filing, each atom, in responding to the lodestone becomes itself a miniature magnet, aligned north and south because it, too, has made of itself a north and a south pole. Not everyone is capable of joining this field of force. Years later, Emerson joked, "Did ye ever hear of a magnet who thought he had lost his virtue because he had fallen into a heap of shavings?" In magnetism, to capture is to *be* captured, in a reciprocal alignment of axis to axis, generating a common field. This became clear as Emerson reworked the image a few months later, late in 1827:

> The soul has a divine power of assimilating all its acquisitions to its own nature. If it is weak & little, events over bear it, and give it their own hue & complexion. If it be great it converts them with instantaneous magic to its own predominant character. It converts calamity to knowledge; knowledge to power; hope to happiness. . . .
>
> Facts become knowledge; events become discipline—15 December.

Nothing was new here, nothing that Emerson had not said a hundred times before, yet everything was new: the certainty of the voice, the collusion of natural science and moral truth, backed by the electric specificity of fact; the discipline capable of converting all to itself, the orientation of self to society; the axis of things and the axis of vision aligned along the conjoint and mutually generated axis of power.[23]

Physics gave Emerson other images for the byplay between center and margin, particular and universal; images in which every particular was, like the magnetized steel filing, a general truth and in which every object was, as Newton's theory of gravity had shown, the center of its own gravitational field: "Think of God as a being whose Providence is particular because it is universal," Emerson had advised himself early in 1828. A bit later he rose to the thought: "I am backed by the Universe of beings. I lean on omnipotence." Yet he immediately broke off and cautioned himself against entertaining "this overweening conceit," reminding himself that God uses our puny strength for *our* education, not for any purpose of *his*. He played with other physical parallels: "Praise & blame are the two forces that keep the moral spheres true; centripetal & centrifugal." Or, "like the hydrostatic paradox as naturalists call it; the Ocean against a hair line of water, God against a human soul." It would be another three years before Emerson commented on de Staël's statement that the axioms of physics are also moral rules, but he was already testing the physical universe for the moral truths it could yield to the inquiring mind.[24]

Meanwhile, though, this man who was rewriting the relationship be-

[margin note: Emerson + Ellen Tucker]

tween man, God, and nature was being claimed by social institutions that he could not rewrite. On December 17, 1828, he became engaged to Ellen Tucker; on January 11, 1829, almost exactly two years after he had hovered on the brink of agnosticism, he was called by Boston's Second Church to become its pastor. He recognized in these events "the hand of my heavenly Father," with happiness and "a certain awe." In anticipating his role as a pastor, Emerson arrived at a solution that moderated his recent theorizing, while holding on to the core of his new beliefs: "The world to the skeptical eye is without form & void. The gospel gives a firm clue to the plan of it. It shows God. Find God, & order & glory & hope & happiness begin." The office of the priest was "to see the creation with a new eye, to behold what he thought unorganized, crystallize into form, to see the stupendous temple uplift its awful form, towers on towers into infinite space, echoing all with rapturous hymns." Gospel provided the key to nature, God; and that key unlocked nature's purpose and meaning, which would "crystallize into form" following not the skeptical, difficult path of inductive science but the affirmative and far more conventional path of Coleridge's method.[25]

[margin note: No just physical science]

Emerson was indeed reading a great deal of Coleridge at this time, both *Aids to Reflection* and *The Friend,* including Coleridge's "Essay on Method." He was also reading Herder, and both helped him supplement with more organic imagery the train of thought he had been developing though the physical sciences. Nature was not the push and pull of forces but generative, "seminal": "We live among eggs, embryos, & seminal principles & the wisest is the most prophetic eye." Herder's epigenetic theories seeped further into Emerson's thinking, with a boost from Laplace: "The government of God is not on a plan—that would be Destiny; it is extempore. . . . The omniscient Eye makes each new move from a survey of all the present state of the game." Some months later Emerson added: "We do not grow up like a plant according to the conformation of a seed. On the contrary it is the privilege of our nature over that of flowers & brutes that we are our own law." Newtonian optics and the shape of the globe reiterated the message: every man was by God's arrangement "absolutely imperially free. When I look at the rainbow I find myself the centre of its arch. But so are you; & so is the man that sees it a mile from both of us. So also the globe is round, & every man therefore stands on the top.—King George, & the Chimney sweep no less."[26] JmN

[margin note: Don't grow like a plant. Newtonian every man free]

Such truths came not from scholarly or social authority but directly from God, which meant that one's personal authority did not come from oneself at all: "I do not affect or pretend to instruct—oh no—it is God working in you that instructs both you & me. I can only tell how I have striven & climbed, & what I have seen." Emerson generalized this observation: "That man will always speak with authority who speaks his own convictions," not what he has heard or read, but what "he hath perceived with his inward eye—which therefore is true to him"—although it may be distorted by some "disease of the soul."[27]

Emerson was moving closer to the language of "Self-Reliance," still ten years away: "Then it seems to be true that the more exclusively idiosyncratic a man is, the more general & infinite he is. . . . In listening more intently to our own reason, we are not becoming in the ordinary sense more selfish, but are departing more from what is small, & falling back on truth itself & God." Authority is personal and individual, but it is so by way of the very impersonality and universality of God: increasingly, the particular is universal, the subjective is objective. Borrowing Coleridge directly, Emerson made this explicit: by a series of approximations, what is truth for you will become truth for all: "You lend every arrow its point. . . . The character of each man shall form his Imagination. The Beings of the imagination shall become objects of unshaken faith, that is, to his mind, Realities. As the man becomes wiser these subjective deities & demons approximate those of the good mind—that is, truth—as the man becomes sinful & ignorant they separate farther & multiply from the Eternal Standard mind." In effect, each subjective mind creates its own reality, but the convergence of minds that are connected through the advance of "wisdom," or education, guarantees that they will all separately arrive at one truth. "Men know truth as quick as they see it," Emerson repeated. His own deep personal honesty would coelesce with the honesty of others into the body of shared truth.[28]

It is hard not to read, in all this, premonitions of the coming conflict between Emerson and the church. For all his relentless intellectual honesty, Emerson was entertaining truths that if spoken aloud would have horrified his listeners. He was stretching institutional forms to the breaking point under the assumption that truth was consensual, but since that consensus was overseen and ratified by God, he was fundamentally secure in pursuing truth idiosyncratically. Truth would take care of itself, and of him. But the consensual model he was developing was in fact a social one, reliant on the gradual accumulation of authority, a kind of cultural capital, which was centered in the very institutions and traditions Emerson was about to defy. "The power of the individual depends upon the power of society," he observed in November 1829. Emerson had been certain that truth as consensual—ratified by God acting in society—would converge with truth as individual—God acting in his sincerest heart. When these two came into conflict, negating both personal and social authority, what would be left? Emerson had the answer ready, prepared when the conflict was still only implicit: nature, the only term remaining to him.[29]

The terms of the conflict were set when, on September 30, 1829, Waldo (as he preferred to be called) and Ellen were married. Only sixteen when they had met two years before, Ellen was, in Waldo's phrase, "delicate and noble," beautiful, intelligent, witty, and already frail from tuberculosis. They were deeply in love, and, having recovered from tubercular weakness himself, Waldo perhaps hoped to rescue her. Their life together would be bitterly short. That fall after the wedding, rest, exercise, and fresh air seemed to improve Ellen's health, but it worsened again in the winter. All during 1830 moments of hope gave way to spells of weakness and depression as she coughed up more blood. In the winter of 1831 Ellen continued to ride out into the cold nearly every day in the mistaken belief—supported by the best medical practice—that fresh air would clear her lungs. Her death that February was swift. By the end her suffering was so intense that death came as a relief.[30]

Throughout 1830, as Ellen's health worsened and his public and private minds diverged, Emerson engaged in his journal in a great deal of reflection about truth. "Truth," he quoted Byron as saying, "is stranger than fiction." Truth was "practical"; truth was "idiosyncratic"; men who consulted their own thoughts spoke with the force of "immortal truth"; when a man perceived truth, "the sword of the spirit" would not join but sever spirit from flesh, God from his works; truth brought with it its own authority; and, most poignantly, "Every science is the record or account of the dissolution of the objects it considers. . . . We are driven to Truth by the decays of the Universe." Truth had been the universal in the particular, the spirit embodied; now the body did not so much embody as betray truth. The shift is startling: "all eloquence is uniform, one. Every thing bad is individual, idiosyncratic. Every thing good is universal nature. Wrong is particular. Right is universal." The particular, idiosyncratic individual was no longer the leader of a new consensus but was cut off from good social sense: "The moral sentiment of a multitude is I believe more correct than of an individual." *Vox populi, vox dei*, as both Coleridge and Stewart had said: the voice of the people was the voice of God, when the republic was enlightened and God was the unity of every nation.[31]

The isolated self was not the channel to truth but its obstruction, which had to dissolve and give way. Society was not many men, but "One Man," incorporated by the truth: "When the earth shall be Christianized every hand will act for God, every tongue speak for him. Now it is diseased; it meets with obstructions of self every where; then, the blood from the mighty heart of Humanity will roll with one pulse from the centre to the extremities. It will make one nation far more than laws or languages." Truth, which in his earlier thinking had been individual and republican, now became universal and imperial. Early in 1831, as Ellen lay dying, Emerson's vision clarified and consolidated: "The greatest man is he that is not man at all but merges his human will in the divine & is merely an image of God." Truth lay not in the body, so fragile, so traitorous, but in the soul, the principle and law of which now not only generated bodies but annihilated them, gave "the welcome stroke / That severs forever this fleshly yoke." So wrote Emerson in memory of Ellen Tucker Emerson, who had died days before, "8th February. Tuesday morning 9 o'clock."[32]

Emerson had indeed been "driven to Truth by the decays of the universe." One of his first entries after Ellen's death proposed to collect "necessary" truths, such as " 'Design proves a designer,' 'Like must know like,' " and so on. "It would be well for every mind to collect with care every truth of this kind he may meet, & make a catalogue of 'necessary truths.' They are scanned & approved by the Reason far above the understanding. They are the last facts by which we approximate metaphysically to God." For necessary truths were weapons in the arsenal of power: "Every word of truth . . . is from God. Every thought that is true is from God. . . . There is but one source of power—that is God." Not a star rolls, not a pulse beats, not a bird drops, not an atom moves "but is bound in the chains of his Omnipotent thought—not a lawless particle. . . . The soul rules over matter. Matter may pass away. . . . But the Soul is the kingdom of God."[33]

It is startling, in this newly militant context, to turn the page and see: "The Religion that is afraid of science dishonours God & commits suicide. It acknowledges that it is not equal to the whole of truth, that it legislates, tyrannizes over a village of God's empire but is not the immutable universal law." God is on the side of science, and there is no longer any question that religion must not lead but follow. "Again man is greater by leaning on the greatest. Nature is commanded by obeying her. God lends his strength to the good." Such obedience is a form of passionate surrender, which arranges all knowledge with the force of a principle, "melts away all resistance," and "turns every thing like fire to its own nature." The command is like a drumbeat: "Natura non imperature nisi parendo," Emerson repeats; "*Imperat parendo.*" Nature is not commanded except by obedience. Command by obeying. Truth is all he has left: if there is no logic by which his thoughts cohere, then "the mind itself uttering necessary truth must be their vinculum." "Obedience is the eye which reads the laws of the Universe." Yet even in the midst of the Baconian drumbeat of knowledge and power, command and obedience, this new Emerson turns Bacon inside out. "Make known the law & you can dispense with collecting the particular instances." The law has burned away the very facts and particulars which Bacon had commanded, and which Emerson—an eon ago, in 1824—had praised, saying, "I apprehend every thinking man's experience attests the accordance to Nature of the Baconian maxim, of not building our theories except upon the slow & patient accumulation of a sufficiency of experiments." But by 1831 Emerson had fully assimilated Baconian induction to Coleridgean method. It was no longer the myriad world of facts that led patiently to the law, but the law that threw open the world to view: "The world becomes transparent to Wisdom. Every thing reveals its reason within itself. The threads of innumerable relations are seen running from part to part & joining remotest points of time & space." After returning from his trip to Europe, he would refine the same idea to its final form: "To an instructed eye the universe is transparent. The light of higher laws than its own shines through it."[34]

Less than a month after Ellen's death, Emerson had declared that the religion was dead which could not embrace the whole of scientific truth. It would take more than a year for the conflict to work itself to a resolution, but he had already, at some level, made the decision to resign his ministry. In answering the question "What Is Enlightenment?" Kant had strictly separated the public and private realms. Enlightenment required "the freedom for man to make public use of his reason in all matters." For Kant, public use meant "before the entire reading public," whereas private use meant "in a civic post or office." Such use was necessarily not free: a clergyman, for instance, had no latitude for personal preferences but "teaches according to his office as one authorized by the church." In such a position one "is not permitted to argue; one must obey." The priest could even teach doctrine he did not believe, so long as it did not violate "inner religion"; but

> should he believe that the latter was not the case he could not administer his office in good conscience; he would have to resign it. . . . As a priest (a member of an organization) he is not free and ought not to be, since he is executing someone else's mandate. On the other hand, the scholar speaking through his writings to the true public which is the world, like the clergyman making public use of his reason, enjoys an unlimited freedom to employ his own reason and to speak in his own person.[35]

Emerson had begun his ministerial career with the ambition to be, in Kant's sense, a public scholar. As a minister, he spoke with all the authority, and all the constraint, of the church; his speech was "private" even as his goal was public, to be, as Kant suggested, a man speaking to men. The American Scholar would never be beholden to a church or a doctrine, and when doctrine conflicted with self-trust, the American Scholar would have to resign—as, indeed, he did. Emerson was worried about the power he was losing—now that he knew just how much there was to lose—in conforming himself to others, to the institution that had indeed regenerated him but that he was clearly, in his own eyes, outgrowing. In a moment of candor he admitted to himself, "The difficulty is that we do not make a world of our own but fall into institutions already made & have to accommodate ourselves to them to be useful at all." The accommodation was beginning to come apart. He could live on "a hundred dollars a year"; why was he toiling "for 20 times as much?" He took courage from Brewster's inspiring biography of Newton: " 'Nature alone is the master of true genius.' Adhere to nature, never to accepted opinion."[36]

On March 29, 1832, Emerson visited Ellen's tomb, opened her coffin, and looked in. What prompting of nature drew him so very far from accepted opinion? He had to see for himself, as Robert Richardson suggests. At this juncture in his life Emerson had to make certain that his own newly forming faith was strong enough to accept the truth—*all* the truth; otherwise he had no right to abandon the old one. "Indeed is truth stranger than fiction," he reminded himself yet one more time. Even the truth revealed within Ellen's coffin must not have deterred him, for soon after he is marveling: "For what has imagination created to compare with the science of Astronomy? What is there in Paradise Lost to elevate & astonish like Herschel or Somerville? . . . God has opened this knowledge to us to correct our theology & educate the mind." "Astronomy hath excellent uses," he continued a few days later; "it irresistibly modifies all theology."[37]

Just after accepting both marriage and a pastorate, Emerson had written: "Truth is irresistible." Now, on the verge of resigning, it was still: "The irresistible effect of Copernican Astronomy has been to make the great scheme for the salvation of man absolutely incredible. . . . Thus astronomy proves theism but disproves dogmatic theology." To be a good minister, he would have to leave the ministry. But first he should tell his congregation that "religion in the mind is not credulity & in the practice is not form. It is a life." He argued with himself to the end, but finally yielded to the irresistible force of truth, resolving that for all the good it might do others, he could not live what he did not believe. "But this ordinance is esteemed the most sacred of religious institutions & I cannot go habitually to an institution which they esteem holiest with indifference & dislike."[38]

After a long autumn of negotiations, the Second Church of Boston accepted his resignation on October 28. On December 22, 1832, Emerson sent a farewell letter to his former congregation in which he explained: "To me, as one disciple, is the ministry of truth, as far as I can discern and declare it, committed, and I desire to live no where and no longer than that grace of God is imparted to me—the liberty to seek and the liberty to utter it." Although he had continued preaching through the summer of 1832 and would return to the pulpit on many occasions for several more years, he had left formal Christianity behind like an outgrown shell: "Is not then all objective theology a discipline, an aid to the immature intellect until it is equal to the truth, & can poise itself?" One doesn't, he added some weeks later, "get a candle to see the sun rise." The light of a new dawn was (as Plato had promised) joining and affirming the light of his own eyes. Sure of the consent of all nature, Emerson set off to explore his new world: he knew the principles, but truly to put them to the test required him to turn his back on America and sail toward the sunrise, to explore his new world by returning first to the Old.[39]

Before embarking for Europe, Emerson borrowed a saying from Walter Savage Landor to keep in mind: " 'The true philosophy is the only true prophet.' He that hath insight into principles alone hath commanding prospect of remotest results." By the time he returned from Europe he was ready to be that true prophet—no less than America's philosopher of truth:

"I talk of these powers of perceiving & communicating truth, as my powers. I look for respect as the possessor of them." It was a daunting task, and Emerson rebuked himself for exercising those powers for such short and irregular periods; but he *had* exercised then, and *did* know himself their possessor, for he had learned the one key truth of all: "All the mistakes I make arise from forsaking my own station & trying to see the object from another person's point of view. I read so resolute a self-thinker as Carlyle & am convinced of the riches of wisdom that ever belong to the man who utters his own thought with a divine confidence that it must be true if he heard it there." All the authority in the world had thus far taught him one thing: the only authority that mattered was his own. Now he knew the principles; would not the rest take care of itself?[40]

THE FUTURE AT THE BOTTOM OF THE HEART:
THE EARLY SCIENCE LECTURES AND *NATURE*

Emerson set sail for Europe on Christmas Day 1832. He arrived by the back door—not through one of the great metropolitan centers, but through Malta, proceeding from Italy to France, ending in England, and arriving back in Boston on October 7, 1833. His mode of travel allowed him to regenerate himself according to no particular institution; he visited them all—churches, art museums, and the great literary minds: Wordsworth, Coleridge, Carlyle. As he traveled his interest in science grew, and he took care to visit the great scientific sites of Europe as well, from Italy to the Sorbonne, the Collège Royale de France, and the Muséum d'Histoire Naturelle in Paris; in London, the Zoological Gardens, the Gallery of Practical Science, and the Royal College's Hunterian Museum (just before it was closed for remodeling in 1834). He made sure to meet some of the age's great scientific minds, too—the botanist Adrien L.H. de Jussieu, the astronomer Jean-Baptiste Biot, the physicist François Arago—and to hear lectures on philosophy at the Sorbonne by Théodore Jouffroy, and on chemistry at the Jardin des Plantes by Louis-Jacques Thénard and Joseph-Louis Gay-Lussac. Yet all this would have been but a wearying pile of details without a moment in Paris when it all crystallized into a vision of such beauty and power that Emerson's path was redirected forever.[41]

— lacking a plan

It happened on July 13, 1833, after three desultory weeks in Paris, not long before he was to leave for England. Emerson was strolling among the cabinets filled with specimens and the outdoor plantings at the Muséum d'Histoire Naturelle. "How much finer things are in composition than alone," he recorded in his notebook. Nature at the Muséum had been extracted from the wild and recomposed according to the latest categories of classification, as worked out by the anatomist Cuvier, the botanist Antoine Laurent de Jussieu, and the mineralogist René Haüy, among others. What staggered Emerson was the coherence of the vast assembly, gathered from all parts of the world. His first reaction became a key statement:

> You are impressed with the inexhaustible gigantic riches of nature. The Universe is a more amazing puzzle than ever, as you look along this bewildering series of animated forms—the hazy butterflies, the carved shells, the birds, beasts, insects, fishes, snakes, & the upheaving principle of life every where incipient in the very rock aping organized forms. Not a form so grotesque, so savage, nor so beautiful but is an expression of some thing in man the observer. An occult relation between the very scorpions & man. I am moved by strange sympathies. I say continually, "I will be a naturalist."

The coherence evidenced by this "bewildering" series was "the upheaving principle of life every where incipient." Of course, life was rampant everywhere in wild nature, but what impressed Emerson here was the *principle* of life, made visible by the method of organizing single specimens such that each one bespoke its necessary place in an ascending hierarchy of categories—species, genus, family, class, order—ending only with the highest possible category of all, uttermost totality. So what Emerson saw was not an impossibly huge collection of separate objects, but the organizing idea which had created them, and which skillful preservation and arrangement had re-created. Ironically, it took a carefully edited collection of dead things to teach Emerson the principle of life. As Lee Rust Brown notes in his definitive treatment of this moment, "Emerson realized in the Muséum that nature *was* natural history"; here was, in Jonathan Bishop's words, "a living metaphor that was literally true." Emerson's private vow, repeated in public almost immediately after his return to the United States, located himself not as a spectator of this wonder but as one of its organizing sources: "I will be a naturalist."[42]

From this vow, this declaration of solidarity with natural science, Emerson never swerved. Of course he did not become one of those laborers who actually gathered, identified, prepared, or labeled specimens; such work was best left to those with technical expertise. His own goal was higher, "philosophical" in Coleridge's sense, or "scientific" before science became a profession. Ironically, it was the very year Emerson was in Europe and England that Coleridge would stand up at a meeting of the British Association for the Advancement of Science and tell the assembled members they could no longer apply to themselves the term *philosopher,* prompting Whewell to offer his initially unpopular substitute, *scientist.* Emerson would never acknowledge the resulting bifurcation, which violated everything Cuvier's Muséum had taught him. Whewell's neologism never appeared in Emerson's writings.

Emerson arrived in Paris well primed to take the lesson the Muséum stood ready to offer him. His reading in science had already been extensive,

Scientist born

Coleridge told Scientist could not be philosophers any more

and Bacon, Coleridge, and John Herschel had already excited him with the power of scientific method; his allegiance to scientific principle had already allowed him to leave the church and define for himself a new role, that of the seeker of natural truth. One of the books he carried through Europe was Goethe's *Italienische Reise,* in which Goethe detailed the goal of his own journey, to find the primal or ur-plant, the primitive form that, unfolded, would exhibit the "complete chain" of life. What the Muséum did for Emerson was not to present him with the *idea* of nature's creative order, for that had long since been familiar to him; rather, the Muséum gave him the *reality* of the idea, the complete chain, offering in material form the keystone that locked the arch of reasoning into place. The effect was catalytic. As Lee Rust Brown details, Emerson could see how to compose not just nature but his own life's experience and its fragmenting written records into a stable but endlessly expansive whole.[43]

Throughout this period Emerson had been working out a metaphorical sequence that shows both the larger implications of his turn toward science and the way in which he domesticated these European insights into a familiar and homely image. While still in the throes of his decision to resign from the church, Emerson had misquoted a line from Landor's *Imaginary Conversation* with Newton: "Philosophy is the only true prophet." A few days later he repeated the line, with some elaboration: " 'The true philosophy is the only true prophet.' He that hath insight into principles alone hath commanding prospect of remotest results." Shortly thereafter Emerson quoted Landor's epigram a third time, adding a note of his own that connected "principles" with the body's innermost sanctum: "May I not add *the whole Future is in the bottom of the heart.*" Through this restatement Emerson made Landor's thought his own, and he reentered his new creation into his *Encyclopedia* under "My Proverbs." The idea continued to grow: aboard ship to Boston from England, Emerson took up the image again, linking it this time with the passionate declaration of love for "moral perfection" that he declared had separated him from men, watered his pillow, and driven him from sleep: "It is always the glory that shall be revealed; it is the 'open secret' of the universe; & it is only the feebleness & dust of the observer that makes it future, the whole *is* now potentially in the bottom of his heart."[44]

On the heels of his revelation at the Paris Muséum, Emerson began to link futurity and the idea of "the whole" to the present instant and the organ of feeling. That this potential wholeness was simultaneously moral and scientific becomes clear when he returned to the idea a few months later, in April 1834: "Could it be made apparent what is really true that the whole future is in the bottom of the heart, that, in proportion as your life is spent within,—in that measure are you invulnerable. In proportion as you penetrate facts for the law, & events for the cause, in that measure is your

Science an affair of the heart

knowledge real, your condition gradually conformed to a stable idea, & the future foreseen."[45]

Emerson's resignation from the church had been made possible by a crisis of belief in which his "heart" had told him he could no longer participate in a sacred ritual that he regarded with "indifference and dislike." But it was made possible at all because he had come to accept the correspondence of moral and physical law, and of science as the expositor of both. One might expect science to be an object of the intellect, but as Emerson here makes clear, science was also an affair of the heart. That is, knowledge of the law might give power—but only love of the law could give invulnerability. To "penetrate" facts for law was to see them, as he had seen the objects at the Muséum, in the light of their idea, which was always there but normally hidden by everyday, unscientific and superficial sight. To open the "iron lid" of the eye was to open the heart and reveal the "open secret," as Goethe had put it, that had been there all along, the moral cause from which all events unfolded. At times of crisis, therefore, one's proper appeal was not to the conflicted human understanding, but beyond, to the nonhuman, to nature—that is, not the collectivity of natural beings but the principle of life itself, the "real" knowledge and "stable idea" behind a world in flux and turmoil.[46]

inward / outward

In his 1838 lecture "Holiness," Emerson used the metaphor in its religious sense to link truth, trust in God, and self-trust, that withdrawal to the "inner" life that builds invulnerability to outer cares. To the "heart" of the aspirant, all nature would reveal that such withdrawal was more truly a turn outward, to the universal and thus public mind. As Emerson concluded, one who has learned this will calmly "look forward to the future in the negligency of that trust which carries God with it, and so hath already the whole future in the bottom of the heart." In his "Divinity School Address" some months later, Emerson concluded by calling for the new teacher who would "see the identity of the law of gravitation with purity of heart," particularizing the image with a reference not just to causal law but to Newtonian physics, carrying the identification of ethical purity and physical law a step further.[47]

But perhaps the most interesting use of the "future in the heart" metaphor comes in "The Over-Soul," where the metaphor developing since 1832 finally reaches publication as the climax of one of his greatest essays. Once we are free of tradition and of rhetoric—those classic Baconian idols—God "will fire the heart with his presence," in a "doubling of the heart itself" with a power of growth that can reach to infinity, inspiring in man "an infallible trust" that dismisses "all particular uncertainties and fears." Emerson then moves the metaphor toward physics: "The things that are really for thee, gravitate to thee," he counsels, making it unnecessary to run toward those you desire. They are sure to find you, for their heart and yours are shared: "the heart in thee is the heart of all; not a valve, not a wall, not *folding science in the heart & heart in America*

Bottom of the Heart

an intersection is there anywhere in nature, but one blood rolls uninterruptedly, an endless circulation through all men, as the water of the globe is all one sea, and, truly seen, its tide is one." By paragraph's end, gravity has pulled the Earth itself into one great body, beating with the common heart of humanity. The concluding paragraph recycles the conclusion of "Holiness," with a new emphasis on the lesson taught by the Muséum: "the universe is represented in an atom, in a moment of time. [Man] will weave no longer a spotted life of shreds and patches, but he will live with a divine unity." He will cease pursuing the frivolous and "calmly front the morrow" in negligent trust, knowing, one last time, that he "hath already the whole future in the bottom of the heart."[48]

Emerson's man achieves this "divine unity," against all odds, through the resolution of part and whole: how can a man, a part, a mere fragment, partake in the universal whole? In the manner of the heart, which, as Emerson says of the soul, is "not an organ, but animates and exercises all the organs." The metaphor of "heart" folds together physics and physiology, Sun and Earth, light and water, with the lifeblood that circulates uniting periphery and center. All things telescope into the "heart" from which all flows out again, but now ordered and interwebbed through all space and unfolding into all futurity.[49]

To withdraw into his "heart" was simultaneously to become less private and "more truly public and human," and this was precisely the trajectory upon which Emerson had set himself. In his prescriptive appeal to the non-human—to the law at the bottom of the heart, whose unfolding prophesies the future—Emerson was completing his self-creation as a thoroughly "modern" man, one who definitively set man and nature at odds so that they might, and indeed must, be "married" in acts of creative vision, compulsive iterations of the act of seeing wholes in fragments and fragments as potential wholes. In his long career Emerson worked to modernize America, and nowhere more so than when he adjured his readers to identify the truth of the heart with the "self-evident" law of nature, producing America as the generative heart of empire, and nature as simultaneously transcendent over and subservient to humanity. Emerson thus helped to reify science as the carrier of absolute truth, for he imagined that science alone escaped the human. One of the lessons of Europe was that there was much work to be done, and he was the first, in either hemisphere, to be doing it. What Emerson brought back from Europe in the bottom of his heart *was* the future; and once he stepped ashore in Boston he set about making his own private heart public, folding science into the heart of American—of modern—culture as the new form of prophetic vision.[50]

Emerson's ideals for America

Of all that had impressed Emerson in Europe, the first he sought to communicate to the American public was the revelation at the Paris Muséum d'Histoire Naturelle. Hardly had he stepped off the dock in early October

Bruno latour - weve never been modern

when he accepted an invitation to speak before the Boston Society of Natural History. "The Uses of Natural History," delivered on November 5, 1833, was in effect, as Bishop notes, a "first public draft of *Nature*," and the embryo from which would unfold his new, his public, career. The lecture establishes a hierarchy of five "advantages" Emerson expects "to accrue from the greater cultivation of Natural Science": health, service, delight, moral improvement, and, finally, the explanation of man's true place in the universe. Emerson opens modestly, with an apology for his lack of specialized knowledge or reading, a lack that he immediately belies both by showing that natural history is the common province of all men, not just of specialists; and by offering a wealth of examples drawn from a surprisingly wide range of the sciences: botany, zoology, entomology, comparative anatomy, physiology, astronomy, geology, physics, chemistry, meteorology. As he says, we all have a presentiment of our relations to external nature that "outruns the limits of actual science," so scientific knowledge is less an end than a beginning.[51]

In a parallel to the "transparent eyeball" passage that opens *Nature*, Emerson begins with his epiphany at "the Garden of Plants in Paris." This time it is not he but nature that is rendered transparent, revealing in its sheer plenitude the creative principle of life. Emerson recounts the experience with the passion of a convert, ending with a stirring declaration of solidarity with his audience of natural historians: "I say I will listen to this invitation. I will be a naturalist." What has moved him so is "the richest collection in the world of natural curiosities arranged for the most imposing effect"; for this display the entire Earth has been "ransacked" to render an account of nature "to the keen insatiable eye of French science." If indeed, as Emerson says, "The earth is a museum," then the Paris garden is its quintessence, distilled and redacted from every region of the Earth. Emerson follows with excited descriptions of the contents of the Muséum's building and grounds, starting with the living wonders of the zoological gardens—the camelopard "nearly twenty feet high," the lions, the elephants, "our own countrymen, the buffalo and the bear"—and moving next to the wonders of the botanical cabinet, where he reads de Candolle in *living* pages and revels in "this natural alphabet, this green and yellow and crimson dictionary." Next he strolls into the building itself, whose marvels he spends the next two pages extolling: the chambers of stuffed birds, "a finer picture gallery than the Louvre"; then the "stuffed beasts," insects, reptiles, fishes, minerals, and the apartments of comparative anatomy, where Emerson found skeletons set out in "a perfect series" from the whale "to the upright form and highly developed skull of the Caucasian race of man." It is here, standing in the stone halls, that Emerson is most deeply impressed. "The universe is a more amazing puzzle than ever, as you look along this bewildering series of animated forms, the hazy butterflies, the carved shells . . . ," and so he repeats the paragraph he jotted in his working notes of the day.[52]

: 89

"I hate museums," wrote Thoreau. "They are catacombs of Nature. They are preserved death." The contrast is revealing. Whereas the unnaturalness of the museum offended Thoreau, to Emerson it was in a sense more natural even than nature itself, more real than reality, since unlike in undisciplined wild nature, here every object was a "specimen," a "proof-type," purified to the *idea* of itself. The "occult relation" Emerson feels between himself and "the very worm" originates in this human ability to see even the lowly and despised as necessary stages in the unfolding of the greatest idea of all, life.[53]

Following the account of this inspiration, Emerson shows the advantages of natural history accruing steadily, unfolding in an upward ascent. Health comes first, for man, like Antaeus, "is invigorated by touching his mother earth," guaranteeing the recreational and educational uses of botany, ornithology, and geology. Second, the study of natural history communicates "useful economical information," not so much through the individual gadgets—water pumps, chimneys, refined sugar, metals, glass, cloth, paints, and dyes that surround and fill the parlor—as through the manner in which such inventions collectively permeate and alter all life. Or rather, all civil life, for the advances in science are, it becomes clear, synonymous with the advancement of civilization: science is not merely the product of civilization, but its producer. The advantage of any one specific invention fades before the great advantage offered by the invention of science itself, namely progress, with no limit in sight: "To the powers of science no limit can be assigned. All that has been is only an accumulated force to act upon the future."[54]

Yet better still is delight, the third advantage. Emerson has said in opening that "the beauty of the world is a perpetual invitation to the study of the world." Now he adds the next step: study leads to delight, a value in itself, but also valuable for its role in the cultivation of the self, for delight leads to yet further study in a spiral of pleasure and knowledge. Every fact we learn enlarges the world we are building by educing that much more of our inner power: "Every fact that is disclosed to us in natural history removes one scale more from the eye; makes the face of nature around us so much more significant." The more we know, the more nature signifies; and the more nature signifies, the more alive we are to the world. Ultimately even the very stones will "speak" to us, and knowledge will "clothe with grace the meanest weed," taking away deformity and giving even the monstrous its own beauty. In this sense truth will be beautiful, and ugliness is merely the illusion wrought by ignorance.[55]

The soul that can see truth will be a beautiful one, and accordingly Emerson focuses next on moral improvement. Observation in itself will teach us to see the world's beauty, even sublimity, but better still is the state of mind the observer must attain to make such observations. Thus science becomes a *moral* discipline that perfects the self: "the state of mind which nature makes indispensable to all such as inquire of her secrets is the best disci-

Science produces civilization

Early Emerson

Science a moral discipline that perfects the self

pline. For she yields no answer to petulance, or dogmatism, or affectation; only to patient, docile observation." He who would learn from nature "must go in the spirit of a little child. The naturalist commands nature by obeying her." From the intellect this spirit of submission will pass to the affections, making "the whole character amiable and true"; and the affections, which led the intellect to study, will be retrained by the intellect, and the spiral of pleasure and knowledge will end in the education of the self. For all his "honour for science," Emerson concludes, he esteems "this development of character" to be worth all of nature itself: "to be worth all the stars they have found, all the bugs or crystals or zoophytes they have described, all the laws how sublime soever, which they have deduced and divulged to mankind."[56]

Yet there is still one more advantage, the greatest of all: the study of natural science will "explain man to himself." Such knowledge has already corrected "many wild errors" of politics, philosophy, and theology; ultimately, "The knowledge of all the facts of all the laws of nature will give man his true place in the system of being." Such knowledge will, that is, at last answer the riddle of the Sphinx: "Who taught thee me to name?" It is precisely the linguistic problem of naming on which the answer will turn—or, as Emerson says, "the power of *expression*" by which external nature names our "inward world of thoughts and emotions" through "correspondence," by which nature "is suited to represent what we think." It is this correspondence which teaches us our place in the system of being: namely, that we are the "keystone," for it is the mind of man that "marries the visible to the Invisible by uniting thought to Animal Organization." It is thus through language that the marriage is consummated: by virtue of language, things become words and nature a "book" that only man can read. The transcendent value of the Paris Muséum, then, is that it presents both the vocabulary and the generative grammatical rules for the language of nature, a kind of Rosetta Stone by which the known can be leveraged into the unknown.[57]

Here at the end of this lecture Emerson initiates his career-long discussion of the nature of language. Speech tells us there exists a nonmaterial world, yet our only language for it is borrowed from the material. We are, then, both bound and liberated by metaphor, by which we use the body to express the mind. For as Emerson argues, in addition to our more obviously metaphoric expressions, even our "most literal and direct modes of speech—as right and wrong, form and substance, honest and dishonest etc., are, when hunted up to their original signification, found to be metaphors also. And this, because the whole of Nature is a metaphor or image of the human Mind. The laws of moral nature answer to those of matter as face to face in a glass." As he will reiterate in *Nature*, "that the axioms of geometry and of mechanics only translate the laws of ethics" reveals a universal "undersong" of "perfect harmony" or "analogy" that is strengthened by every law revealed by "Davy or Cuvier or Laplace." Nature, then, is a living language, the nourishing matrix for that dead language encased in dictionaries printed and bound. It is the vital language of God, and in nature "is writ by the Creator his own history."[58]

The value of a book does not lie in paper and ink. Emerson's lecture ends with the testimony that to study nature is to know "that there is a meaning therein before whose truth and beauty all external grace must vanish, as it may be, all this outward universe shall one day disappear, when its whole sense hath been comprehended and engraved forever in the eternal thoughts of the human mind." The upshot of Emerson's reasoning is that Baconian method, command over nature by obedience to her laws, leads to the ultimate Platonic payoff: the earthly globe will dissolve into the crystal globe of the mind, rather as the limitations of specific metaphors dissolve into the global flow of metaphoricity itself. Ironically, this means that the ultimate goal of natural science is to make nature disappear—that is, the highest imaginable "use" of natural history is to eliminate nature altogether: in Barbara Packer's phrase, "to make nature publish *and* perish." However, that is not how Emerson would have seen it. Rather, natural history teaches us how to "use" nature, as raw material for our civil development; how to go beyond nature, intellectually; how to rise above it, morally; how, in sum, to penetrate the flux of facts for the permanence of principle. In this first lecture Emerson introduces two of his key concepts: discipline, a reframing of his old standby, compensation; and correspondence, which charges nature with moral significance. Both are very early and deep layers in Emerson's thought, which his romance with science will bring to fruition.[59]

Emerson's next lecture, delivered in Boston a few weeks later, is his purest and most complete presentation of the classic natural theological argument from design—virtually a miniature Bridgewater Treatise, showing many traces of Emerson's recent reading in Charles Bell's *The Hand* and John Kidd's *On the Adaptations of External Nature to the Physical Condition of Man,* as well as Cuvier and Lyell. In "The Relation of Man to the Globe," Emerson moves from the uses of natural history to its fruits, of which the sweetest is the knowledge that the globe was designed, over "a thousand thousand ages," for us. The wonder that captures Emerson's imagination is that our globe was not materialized by some fiat of divine pronouncement, but that it made itself over unimaginable stretches of deep time, presenting us today with a "monument" on which its history is inscribed "in gigantic letters." What those letters spell out is only now being read by diligent and patient observers, who have brought the most surprising news: man is no "upstart" but has been prophesied since the beginning of the Earth, which has labored through the ages to "produce" him. Nature's glory, the very flower of deep time, is man.[60]

Emerson's evidence for this startling assertion is of two kinds. First is the long history of the planet, which shows "progressive preparation" for us. Earliest was the deposition of the bedrock granite; then came the successive formation of soils (which were "*cooked* for the nourishment of animal life" by the first vegetables), then the gradual changes wrought by successive races of animals, punctuated by the "repeated great convulsions" that shaped mountains and continents, until finally, when our home was ready for us—"when the house was built and the lands were drained and the house was ventilated and the chimneys of the volcano opened as a safety valve and the cellar stocked—the creature man was formed and put into his habitation."[61]

Once the labor of the ages has completed the preparation of a planet so perfectly adapted to our needs, the very existence of those fine-tuned adaptations constitutes Emerson's second body of evidence. Everywhere he sees "that a proportion is faithfully kept" between the forces of nature and the powers of man. The atmosphere contains just the ratio of oxygen to hydrogen that best sustains us; the Earth is just solid enough to bear our weight, but not so solid that we cannot till it; coal has been providentially lifted just within reach of man's "little hands" to support our "comfort and civilization." This is a view confident to the point of smugness. Emerson's universe in this lecture is so benign, so utterly in and for our possession, that it cannot seriously threaten us with real danger. Having recently completed two sea voyages without mishap, Emerson is exhilarated by the power of man to sail in the teeth of danger and yet always come out safe—not by chance, but by the exquisite adjustment of the strength of "spars and rigging to the ordinary forces of wind and water." "These chances are all counted and weighed and measured," contends Emerson. Although danger is incessant, so is the seaman's attention, and so "the real risk is small" and fatal accidents rare. The reader who knows that circumstances will lead Emerson to write, following the death of Margaret Fuller by shipwreck, that nature "will not mind drowning a man or a woman" finds this a painful passage—a little like watching teenagers turn a fast car onto an open road, with a new license in hand and immortality at heart. One wants to warn him, but can't.[62]

This finely calibrated balance between human power and natural force is not intended merely to make man a good housekeeper. "This symmetry of parts is his equipment for the conquest of nature." Our wants acquaint us with our powers and lead us on to "the possession of the globe" through global commerce, the servitude of the lesser races of animals, and the repairing of such parts of the globe as are insufficiently commodious to man in their original state. But these imperial uses are merely utilitarian; following Emerson's usual pattern, the real goal is to ascend from lower to higher. So, from the service nature renders to man, Emerson rises to the love man feels for nature, love that opens our senses to the beauty of nature and our intellect to the beauties of the natural sciences. It is the pursuit of science that opens the newest and noblest view of man's relation to the globe, for "with the progress of the cultivation of the species the globe itself both in

the mass and in its minutest part, becomes to man a school of science."
Now is the globe truly opened to us, in whole and in least detail: to the astronomer, it becomes a *"moveable observatory"* sailing through space; to
the geologist, a "register of periods of time" otherwise lost to us; the
anatomist classifies worms too small to see; the chemist resolves all created
things into gas; the physicist approaches the elemental secrets of nature in
the principles of polarity; and soon the physiologist will lay bare the laws of
life. From the pinnacle of this ascent Emerson can only exclaim in wonder,
"I am not impressed by solitary marks of designing wisdom; I am thrilled
with delight by the choral harmony of the whole. Design! It is all design. It
is all beauty. It is all astonishment."[63]

Twenty-two years later, in *Benito Cereno*, Herman Melville would por-
tray the type of the American in the figure of Captain Amasa Delano, master of his universe, a universe that is sublimely benign because *he* is benign
and he rules it. Melville's great novella eviscerates this view with damning
precision. Every detail in Delano's field of vision serves his own interest. His
oblivion plays into near catastrophe, but even afterward blindness keeps his
world self-serving to the end; to his blighted Old World counterpart, Captain Benito Cereno, Delano offers the beneficence of nature as redemption,
a redemption that Cereno finds pointless. Emerson most closely approaches
Melville's parody in this 1834 lecture, which encapsulates the ideology of
expansionist, liberal America and shows the basis of that ideology in science. As Melville suggests, not only does this self-protecting and self-stabi-
lizing view operate at the expense of those taken to serve it, but also it is
desperately precarious. To sustain it, enormous cultural and intellectual
work would be needed over the next several generations. How much of that
work would be performed by Emerson?

One answer came less than two weeks later, when Emerson delivered his
third science lecture, "Water," to the Boston Mechanic's Institution. Read
one way, this is Emerson's most scientific essay, his single attempt at formal
"science writing." Read another way, this lecture offers a fable showing the
virtue of service, a demonstration in detail of the general assertions made in
"The Relation of Man to the Globe." That relation is, in a word, possession; and water is both the means of possession and the symbol of the good
servant. To say that this essay is Emerson's most "scientific," that is, most
"factual," is to say that this is his most *made* essay. Most of the facts it con-
veys were manufactured in a laboratory, and, in the mirroring of mind with
Mind, water itself is "the laboratory of Nature" in which "the great
Chemist" himself manufactures continents, using water "as a receiver, a
cover, a solvent" just as our own chemists do, and its "immense force of
pressure" just like our own "mechanicks." Indeed, the chemist, as if he himself were God in his laboratory, can manufacture not just facts about water
but water itself: put eight parts oxygen to one of hydrogen in a glass receiver, pass through it an electric spark, and lo! "they burn, and water is

immediately seen trickling down the sides of the vessel." When weighed it is "found exactly equal to the weight of the original gases."[64]

Water, visible and invisible, is found everywhere on Earth, from our washbasin to the oceans, from the veins in our body to the vapor in the upper atmosphere. And everywhere it is an "obedient, useful, and indispensable . . . household servant," "a friend who sends us favors unsuspected." This is so from the single combined "particle" of hydrogen and oxygen to the great mass that connects and so largely composes the globe itself, meaning that water unites all the sciences and demonstrates the power of each. Chemistry leads, for it first demonstrated that water, one thing, is yet composed of two, which can be recomposed and decomposed infinitely. That matter with the solidity of ice or the force of steam can yet be resolved into gas suggests the insubstantiality of all matter, that matter itself is only phenomenal—although Emerson does not pursue the implications here.[65]

Instead, he reverses scale from micro- to macroscopic to consider water in its global mass, breaking up solid rock into soil, washing mountains into the sea, yet replenishing the lost soil through the plant life it supports and re-forging mountains by compressing deep-sea sediment, heated by the furnace of the Earth's core, into the materials of the mountains of the future. Hence water's agency is cyclic: "The same power that destroys in different circumstances is made to reproduce." Water acts as a thermostat, maintaining the equilibrium of Earth's surface temperature: thawing ice absorbs the surplus heat of the summer, while vapor forming into snow releases heat into the winter cold; "As long as ice remains to thaw, or water to freeze, the temperature of the atmosphere can never vary beyond certain limits." Nor does water just support life. It is the vehicle of life, giving plants their structure and carrying to them their food—and it is equally essential to animals.[66]

The ocean and its rivers "subserve the world" as well by composing one of the great kingdoms of nature, giving us marine life of all kinds, from animalcules to fish to whales (in the 1830s whaling was the oil industry of America). But we are all "bathed in an invisible ocean overhead, and around us"; water circulates through all of nature, from ocean to vapor to clouds, to rain to rivers back to ocean, aligning meteorology with physiology: "The circulation of the water in the globe is no less beautiful a law than the circulation of blood in the body." Finally, in the strangest providence of all, water defies the law of all other liquids, to expand when it freezes. If it were not for this, seas and ponds that are protected in winter by a thick coat of ice would instead be frozen solid, never to be thawed, and pleasant climates would instead be "uninhabitable domains of frost."[67]

All these are the properties of water, in itself, as described by science. But Emerson does not end here. Instead, he closes with the ways in which man has captured water's mechanical power. By harnessing hydraulic pressure, man can "cut through a thick bar of iron as easily as through a sheet of paper." Capillary attraction will split millstones: simply insert a dry wedge of wood, and wet it. And the only limit to the force of the steam engine is the strength of our materials: "It is convertible to the aid of every action of man precisely in the degree of his knowledge." Once again, the natural economy of nature serves the industrial economy of man: given the providential supply of coal, water, heated by that fuel, now becomes the agent of explosive power: "Its might and flexibleness seem to annihilate the obstinate properties of matter, to make the hard soft and the distant near. Such virtue lies in a little water."[68]

There is a double irony in that last sentence, which stands as a single paragraph. Clearly, the wonder is that there is so much in so little: so many virtues in such a simple and common substance, of course, but also so much power in such small compass. Emerson writes that "the greatest power may be exerted by a very small quantity of water"—if, that is, it is properly distributed by the agency of human machinery. Give me a lever, said Archimedes, and I will move the world. Or, as Emerson says, a drop of water may be made to balance the ocean: that is, not water alone, but water leveraged through human machinery, which will create a power that no material can resist, not the very Earth itself. And finally, as that word *virtue* suggests, there is a moral dimension here, which Emerson would make explicit in "New England Reformers": "The familiar experiment, called the hydrostatic paradox, in which a capillary column of water balances the ocean, is a symbol of the relation of one man to the whole family of men." Similarly, in "Montaigne," the "moral sentiment" becomes "the drop which balances the sea." In "Water," by contrast, Emerson withholds the moral analogy, encouraging the hearer to draw it for himself: we perceive the boundless resources of the Creator when we learn that a bucket of water contains force enough "to counterbalance mountains, or to rend the planet," and when we trace the "manifold offices" performed by a particle of water "in the pulse, in the brain, in the eye, in a plant, in mist, in crystal, in a volcano, and it may exalt our highest sentiments to see the same particle in every step of this ceaseless revolution serving the life, the order, the happiness of the Universe."[69]

By this point, the lecture's conclusion, that infinitesimal particle of water has acquired not just physical but moral force, and the lecture, factual as it is, becomes also a fable about relating the one and the many. It is the collective force of multitudes that can wear down and rebuild mountains, and it is the terrific power of the one that can counterbalance the masses. Yet that immense power is strictly subordinated to service of the whole, to which it obediently "ministers" aid, comfort, and delight—rather like the lecturer Emerson himself, who, having composed a particle into the key that unlocks the universe, removes himself from the essay, so that it appears not artful or made but "scientific," even as the object of the essay, water, is removed from the laboratory of the chemist to the laboratory of God, whom it will serve as agent and minister. And so, as Emerson liked to say, "The world globes itself in a drop of dew."[70]

Science worthless if not connected to men — general studies connected to other studies (handwritten)

Why did Emerson write no more science essays? As this lecture shows, he was certainly capable of it. One likely reason was the need he evidently felt in such writing to subordinate the moral dimension. Another was the need to research and collate the writings of others instead of originating his own. A third reason is suggested by the final lecture in this sequence, "The Naturalist," presented at the annual meeting of the Boston Society of Natural History in May 1834 (a more formal occasion than the one for the lecture delivered the previous November).

Faced once again with an audience of naturalists, Emerson writes what is in effect an apologia and farewell in which he explains why he will not, despite his earlier vow, become one of them. Yet he wishes still to come "here to school" and continue to read their general results and understand their discoveries. From that perspective, he has a crucial role to play: he will remind them always to connect their studies of nature to the study of man, for "Man is the only object of interest to Man," and he stands in "a central connexion" with all nature. *goes back to Harvard* (handwritten) Thus the focus of interest for Emerson here is "the place of Natural History in Education"—that is, in educing knowledge of ourselves that would otherwise be hidden, latent. As the American Scholar might say, the naturalist should be not a disembodied eye, "a mere thinker," but *Man Thinking.*" The value in studying nature, Emerson reminds his audience, is that in so doing we are studying perfection. Nothing in nature is "false or unsuccessful." "A willow or an apple is a perfect being; so is a bee or a thrush." However, by contrast, "The best poem or statue or picture is not," for whereas the willow or thrush has "its completeness within itself," a work of art is completed only outside of itself, in the mind of the artist. Therefore, as Goethe said, "no man should be admitted into his Republic, who was not versed in Natural History." By studying the work of the great Artist, man can continue and extend the work of nature without changing it in any fundamental way.[71]

The primary lesson learned, then, in the study of nature is "Composition." God's artistry composes by creating much from simple materials, like the French cook with forty recipes for macaroni. No single object, however elegant in itself, is as beautiful alone as it is in composition or context, as Emerson argues in the poem "Each and All": "All are needed by each one; / Nothing is fair or good alone." Ironically, then, despite his earlier assertion that objects of nature are complete in themselves, it is more accurate to say they are not: the willow or thrush is perfect only within the total context of its natural environment. This, too, is a key concept, for the tension it creates—are individuals best read as complete in themselves, or as elements in the larger whole they compose?—increasingly dominates Emerson's writing.[72]

As in all his writings on nature, Emerson highlights certain specific uses or advantages accrued by the study of nature. This time he focuses on the tendency of such study to discourage "imitation," by encouraging creation from facts and truth rather than by "servile copying"; and on the training given the mind in "discrimination," by teaching the student how to classify according to the real distinctions of nature rather than by arbitrary convenience. However, Emerson's real interest in this lecture lies elsewhere. He wants to issue a warning: "We are not only to have the aids of Science but we are to recur to Nature to guard us from the evils of Science."[73]

Initiating a critique that will return with redoubled force in the 1850s, Emerson worries that men of science too often mistake means for ends, losing themselves in the aids "of nomenclature, of minute physiological research, of the retort, the scalpel, and the scales"—all invaluable, to be sure, but none of them the true end of science. Too often the botanist knows the name of a plant, but not its properties, a loss Emerson seeks to repair. His cause is crucial, for natural history gives us the key to the universe: "Natural History seeks more directly that which all sciences, arts, and trades seek mediately—knowledge of the world we live in. Here it touches directly the highest question of philosophy, Why and How any thing is?" Since nature is a hieroglyphic language, it is natural history that can provide the "key or dictionary": since nature is a language, the natural historian must be a poet as much as a scientist, even as the poet must also study natural history. For want of the "marriage" of theory and fact, the poet "loses himself in imaginations and for want of accuracy is a mere fabulist," whereas the "savant on the other hand losing sight of the end of his inquiries in the perfection of his manipulations becomes an apothecary, a pedant." Emerson roundly rejects the division that would ultimately, despite his protests, lead to the parting of the "Two Cultures": "I fully believe in both, in the poetry and in the dissection." This is what it means "to make the Naturalist subordinate to the Man"—to search out the "proximate atoms," but then to integrate them again "as in nature they are integrated" and to be alive to the beauty and the moral "impressions" they convey. In his trinity of values, truth must ever be alive to beauty and goodness.[74]

In closing, Emerson turns again to the image of the cabinet and his awakening at the Paris Muséum: there he found his thoughts expressed not in words but in something better than words. The greatest thought of all is "Method," or "Theory," the law that reconnects all the "various and innumerable works" of nature into one. And so the office of the natural historian is a high one indeed: "The eternal beauty which led the early Greeks to call the globe κόσμος [Cosmos] or Beauty pleads ever with us, shines from the stars, glows in the flower, moves in the animal, crystallizes in the stone. No truth can be more self evident than that the highest state of man, physical, intellectual, and moral, can only coexist with a perfect Theory of Animated Nature."[75]

The close of this lecture points directly to the opening of Emerson's book *Nature* (published two years later, in 1836), in which he states: "All science has one end, namely, to find a theory of nature." By now Emerson is fully secure in his own role as the interpreter of nature, putting himself precisely in the position he advocated in "The Naturalist": he is the poet "versed" in natural history, an American Goethe throwing open the gates of his new "Republic." *Nature* was Emerson's first major publication, the culmination of all his long preparations. It opened the public phase of his career with the annunciation of his own "Theory of Animated Nature," his own organizing method, his solution to the riddle of the Sphinx and to Bacon's challenge that to regain Eden, we must reclaim Nature.[76]

The first edition of *Nature* contains an epigraph attributed to Plotinus but actually derived from Newton's Neoplatonist friend Ralph Cudworth: "Nature is but an image or imitation of wisdom, the last thing of the soul; nature being a thing which only do, but not know." What Cudworth sought to establish, and what Emerson presumably approves, is the doctrine that nature in and of itself is essentially dead. It has no self-subsisting or self-organizing powers; all activity in nature is directly or indirectly generated by God. This division underlies Emerson's crucial opening move, in which he accepts the separation of the "ME," or Soul, from the "NOT ME," or Nature, which encompasses "both nature and art, all other men and my own body." Of course, as he adds, he will also use the "common sense" definition of nature as "essences unchanged by man; space, the air, the river, the leaf," as distinct from art, which is "the mixture of his will with the same things, as in a house, a canal, a statue, a picture." Although both definitions rely on division—nature from art, or soul from nature *and* art—both sets of division create the opportunity for exciting mixtures or hybrids. It might seem that even in Emerson's time, with the Industrial Revolution well under way, the evidence of such mixing suggested that nature could not remain pure or "unchanged" for long. After all, at the far side of this revolution, intellectuals would speak not just of change in nature but of its "end," meaning not "goal" (as when Emerson asked, a few lines earlier, "to what *end* is nature?"), but "terminus." But Emerson here is interested less in actual, physical nature than in its "essence," which by definition cannot be altered by physical operations. All human operations taken together, then, "are so insignificant, a little chipping, baking, patching, and washing, that . . . they do not vary the result." The conclusion of the paragraph leaves us where we began, with mind confronting world, the two in constant interaction but ultimately devoid of real contact.[77]

Given Emerson's reputation for pantheism and celebration of the spiritual wholeness of nature, it might seem surprising that his defining move is to reinstate the Cartesian divide between mind and matter. Yet only by vacating "Soul" from nature, emptying everything outside the "ME" of self-subsisting powers, can he prepare for the "transparency" that will follow immediately in his most notorious passage: "I become a transparent eyeball. I am nothing. I see all. The currents of the Universal Being circulate through me; I am part or particle of God. The name of the nearest friend sounds then foreign and accidental." Emptying nature clears it to become the unimpeded channel for those "currents of universal being," enabling Emerson's powerful extended metaphor of fluid, flow, circulation, hydraulic pressure, and organic growth. Yet this image of emptying dramatizes the very difficulties created by forcing the Cartesian divide to such a logical extreme: a truly transparent eyeball cannot see, just as light intersected by no particular object will fall away into vacancy. To see all is to see nothing in particular, just as a universe rendered "transparent," such that "the light of higher laws than its own, shines through it," cannot be seen in any of *its* particularity. As Emerson adds a few lines later, "There sits the Sphinx at the roadside, and from age to age, as each prophet comes by, he tries his fortune at reading her riddle." How can one see, how can one know both the universe in its fullness and the higher law that shines through it? How can Mind in its billowing totality be resolved into the single instance? How can nature be named?[78]

In *Nature*, Emerson's answer is: "A Fact is the end or last issue of spirit. The visible creation is the terminus or the circumference of the invisible world," making material objects the "scoriae," or excrement of the Creator—as Kenneth Burke says, "nothing other than God's *offal*." Or, to follow instead the light metaphor, the object that makes the light visible also terminates it. Thus the "end" of nature is "terminal" after all—for if facts are the "terminus" of spirit, to return to spirit is to annihilate nature, to dissolve the object back into light. But such idealism leaves the spirit "in the splendid labyrinth of my perceptions, to wander without end." It seems that the Sphinx has led us into the labyrinth and left us there. How to find our way out? In *Nature*, Emerson must repeatedly turn back to recover the material universe his idealism annihilates—as he poignantly adds, he loves nature as a child loves her, and "I do not wish to fling stones at my beautiful mother, nor soil my gentle nest." So as necessary to Emerson as emptiness is plenum, the material richness, the dazzling, splendid multiplicity of nature's physical objects.[79]

As David Van Leer details, Emerson's terms and argument are Kantian: "The real 'end' of nature, then, is to fulfill Kant's objectivity requirement—to allow for the possibility of objects separate from the experience of them," for objects alone make experience possible. The consciousness of self is bound up with the existence of things. For Emerson, this very division opens up the destined place and purpose of man, which is, once again, to "marry" or reintegrate the two, pure mind and material matrix, thereby putting God back into nature: "Therefore . . . the Supreme Being, does not build up nature around us, but puts it forth through us, as the life of the tree puts forth new branches and leaves through the pores of the old. As a plant upon the earth, so a man rests upon the bosom of God; he is nourished by unfailing fountains, and draws, at his need, inexhaustible power." In a trope that returns to Cudworth's original argument, if nature is dead without man, only through man is nature animated—and this is the "Theory of Animated Nature" proposed by Emerson.[80]

After proposing that science's one aim is "to find a theory of nature," Emerson adds: "Whenever a true theory appears, it will be its own evidence. Its test is, that it will explain all phenomena." Emerson's own theory that nature is animated by man depends on the self-evidencing interplay of mind and matrix, whereby the external world is necessary for the mind's realization, and mind or concepts are equally necessary to assemble a world of dead atoms into living meaning. The ascending chapters or "cantos" in *Nature* describe the necessary steps or stages in this marriage and prophesy its ultimate fruit, man, with his divinity fully recovered. The stages, then, describe a process of growth, keyed to the underlying organic metaphor whereby God puts forth nature through us and we can draw on unfailing fountains of power. The metaphor of growth culminates in the triumphal assertion that mankind has, at last, grown up: with the progress of civilization we are no longer children, wise with the unconscious wisdom of animals; nor the troubled and contending adolescent, struggling through darkness toward the light of reason; but adults, approaching the crest of our powers, the globe at our feet, the stars at our back, the powers of nature in our hands and futurity before us.[81]

To get us to this point, Emerson leads us through a long essay in two parts. The first part takes up the "uses" of nature, distilling the many issues raised in his early science lectures into four chapters: "Commodity," "Beauty," "Language," and "Discipline." Commodity, the lowest use, here is treated perfunctorily with a reference to the natural theological argument that all the parts of nature "incessantly work into each other's hands for the profit of man." In the chapter "Beauty," Emerson wrestles with the three-way conflict between the sensual beauty of nature, which, with all its delights, all too easily becomes mere show; a deeper beauty, which is first the mixture of human will, or "virtue," with nature, by which "he takes up the world into himself"; and human intellect, or truth, by which we seek out "the absolute order of things as they stand in the mind of God, and without the colors of affection." The three aspects correspond to Emerson's trinity of beauty, goodness, and truth, "different faces of the same All."[82]

The important chapter "Language" develops the idea that since thought was the origin of nature, "Nature is the vehicle of thought," connecting us back to the Creator through the three steps of word, fact, and spirit: word is sign of fact; natural fact is symbol of spiritual fact. The mirroring of things and thoughts guarantees the analogous relationship by which each is the means to understand the other: "All the facts in natural history taken by themselves, have no value, but are barren like a single sex. But marry it to human history, and it is full of life." In the still stronger form, language (which, as he will shortly show, is limited at best) drops out of the equation altogether, in favor of a still more "radical correspondence" between fact and spirit (and in which the saying about the axioms of physics, taken so long ago from de Staël, will finally come home): "The world is emblematic. Parts of speech are metaphors because the whole of nature is a metaphor of the human mind. The laws of moral nature answer to those of matter as face to face in a glass. 'The visible world and the relation of its parts, is the dial plate of the invisible.' The axioms of physics translate the laws of ethics." Following the path from the fact as the "terminus" of spirit back to the original spiritual sense will disclose the world to us as "an open book," such that—quoting Coleridge—" 'every object rightly seen, unlocks a new faculty of the soul.' That which was unconscious truth, becomes, when interpreted and defined in an object, a part of the domain of knowledge,—a new weapon in the magazine of power."[83]

The evocation of Bacon's axiom that knowledge is power leads to the chapter "Discipline," the conclusion of this section, where it becomes clear that the highest use of nature is to "educate" us. Here Emerson introduces Coleridge's crucial distinction between the understanding, which "adds, divides, combines, measures"; and Reason, which "transfers all these lessons into its own world of thought, by perceiving the analogy that marries Matter and Mind." Whereas understanding merely *adds,* Reason *marries.* The discipline of nature teaches two lessons. First, to truth it teaches the lesson of power: man can reduce all things under his will, for "nature is thoroughly mediate. It is made to serve. It receives the dominion of man as meekly as the ass on which the Saviour rode." Human reason gradually converts all the kingdoms of nature to the single kingdom of his will, "until the world becomes, at last, only a realized will,—the double of the man." Second, to goodness it teaches the lesson of morality: "All things are moral," for "the moral law lies at the centre of nature and radiates to the circumference. It is the pith and marrow of every substance, every relation, and every process." Through this hub-to-spoke vision of nature, every part, every being in nature, no matter how small or how far out on the uttermost rim of the world, embodies and mirrors the whole of nature: "Each particle is a microcosm, and faithfully renders the likeness of the world." Here is where language, finally, fails. As "finite organs" words cannot contain the whole, whereas action, "the perfection and publication of thought," becomes a superior kind of language, one that is higher or divine.[84]

Having dissolved nature into words, and words into action, Emerson must in the concluding half of his essay take up the problem the first half has created: If nature's final end is to educate, or discipline, humanity through the agency of law, does physical nature actually exist? The chapter "Idealism" examines this "noble doubt" and shows why culture answers in the negative. No, nature's only existence is phenomenal, not real. After Kant's Copernican revolution, nature is, so to speak, a cultural construction. First, motion itself—even a coachride across town—shows nature to be a spectacle, "wholly detached" from the viewer. Second, the poet makes the spectacle of nature dance and spin: "He unfixes the land and the sea, makes them revolve around the axis of his primary thought, and disposes them anew." Or, again paraphrasing Bacon, "The sensual man conforms thought to things; the poet conforms things to his thoughts." For the poet, then, nature is "ductile and flexible," its objects the words of Reason, which his imagination can freely dispose anew: "The imagination may be defined to be, the use which the Reason makes of the material world."[85]

Here Emerson comes to the nexus, the point where the two great realms, poetry and science, converge: "The true philosopher and the true poet are one, and a beauty, which is truth, and a truth, which is beauty, is the aim of both." Both dissolve matter by thought. So, third: in science, that thought is law, and once it is seized, it tosses matter aside "like an outcast corpse." Fourth, all intellectual insight lifts man out of matter into the realm of "immortal necessary uncreated natures, that is, upon Ideas," where we are "nimble and lightsome" and suddenly seem to tread on air; and, fifth and finally, religion and ethics "both put nature under foot." It is at just this point that Emerson hesitates, fearing to fling stones at his "beautiful mother"; nevertheless, here he closes by reaffirming his conviction that culture does dissolve nature and disposes of it according to its own needs. Hence the enormous advantage of idealism, "that it presents the world in precisely that view which is most desirable to the mind," thereby making idealism the only possible view that Reason can take—literally, the only "rational" view of the world.[86]

Yet in "Spirit," Emerson hesitates again. There is a still more serious objection to idealism. To be true, his theory must be "progressive." That means there must then be something left outside the all-dissolving power of culture to act as a guide, rather as the shadow on a sundial that points "always to the sun behind us" must be cast by an object. Just as consciousness cannot exist without objects, objects must exist outside consciousness: "Truth is, in fact, that without which experience could not itself exist—less what we choose to believe than what permits the notions of choice and belief, even of personality." From his first assertion that moral order was possible only in a world of *relation,* a relation begun in the separation of Creator and created, Emerson understood that education, the self-generating experience, was possible only in the "dynamic interplay" of self and other, "ME" and "NOT ME." Nature must be respected, then, as "the apparition of God . . . the great organ through which the universal spirit speaks to the individual." Idealism must be qualified, lest it abandon us in that labyrinth of our perceptions and make nature foreign to us. "Let it [idealism] stand then," Emerson suggests, "in the present state of our knowledge, merely as a useful introductory hypothesis"—useful because it distinguishes soul and world—or, to return to the beginning, "ME" and "NOT ME."[87]

Yet the former will never dissolve the latter; it appears that our power is limited after all. As Emerson has suggested at the beginning, nature is finally *not* "subjected to the human will. Its serene order is inviolable by us. . . . It is a fixed point whereby we may measure our departure." One moment we are rebuilding nature in our own image; the next, nature is inviolable. Emerson tells us that we are nowhere foreign, then that we are everywhere "strangers in nature" from whom the very animals run. How can both be true? They seem contradictory, yet somehow, both must be true.[88]

The answer, only implied in Emerson's introduction, is supplied in his conclusion, when, as Van Leer observes, Emerson turns from epistemology to theology, from reason to faith, to show the power of transcendental idealism over empirical realism. It is nature's very inviolability that makes possible the exhilarating vision of "Prospects," *Nature*'s conclusion, which looks ahead to nature's entire reconstruction. "Empirical science," Emerson begins, "is apt to cloud the sight, and . . . to bereave the student of the manly contemplation of the whole." The error is corrected by building science "upon ideas," that is, after the fashion of the Paris Muséum, where the cabinet of natural history shows "the relation of the forms of flowers, shells, animals, architecture, to the mind"; and also by honoring the "humanity" of science, which will not overlook "that wonderful congruity which subsists between man and the world," of which man is head and heart. The "half-sight of science" is, then, corrected by opening the other eye of poetry in a binocular vision that restores to the world "original and eternal beauty."[89]

For, as Emerson continues, "The ruin or the blank, that we see when we look at nature, is in our own eye. The axis of vision is not coincident with the axis of things, and so they appear not transparent but opake." Reasserting the vitality of his chosen role, Emerson warns that "he cannot be a naturalist, until he satisfies all the demands of the spirit." But should he do that—celebrate that "marriage," be resolute to detach the object "from personal relations, and see it in the light of thought" and at the same time "kindle science with the fire of the holiest affections, then will God go forth anew into the creation." We shall "come to look at the world with new eyes": it will be the same old world yet, but Emerson's "New Organon" shows us how to be visionaries, to unfix it and revolve it around the axis of our desire, rendering it as transparent as a crystal globe. "Build, therefore, your own world," the voice of Emerson's Orphic poet commands, and Bacon's vision, the kingdom of man over a reclaimed Eden will be at last realized: "The kingdom of man over nature, which cometh not with observation,—a dominion such as now is beyond his dream of God,—he shall enter without more wonder than the blind man feels who is gradually restored to perfect sight."[90]

Emerson's is a heady, a deeply inspirational prophecy—wind under the wings of America. For most remarkable at last is its materiality: these are marching orders to scientists and poets to join ranks and remake nature according to what will become, increasingly, a nationalist vision. Ironically, *Nature* includes a manifesto for modern industrial progress. The material world we inherited from the twentieth century, for better and for worse, has gone far toward fulfilling Emerson's vision of making the world "only a realized will,—the double of the man." We are said to be emerging into a posthuman, a "virtual," world, where the physical mediations that hobbled Emerson will be replaced by their weightless electronic substitutes. One might think we were still driven by Emerson's early dream that "all this outward universe shall one day disappear, when its whole sense hath been comprehended and engraved forever in the eternal thoughts of the human mind." Decades later, in 1863, Emerson would observe that "our civilization and these ideas are reducing the earth to a brain. See how by telegraph and steam . . . the earth is anthropized, has an occiput, and a fist that will knock down an empire. What a chemistry in her magazine." In *Nature*, he had written that every object crystallized into truth becomes knowledge, "a new weapon in the magazine of power." Of all the weapons in that magazine, was any more powerful than the method of nature?[91]

THE ANATOMY OF TRUTH: EMERSON'S ECSTATIC SCIENCE

The riddle of the Sphinx was how to master the whole of a nature accessible to mastery only piece by piece. The poet's solution was lonely, ecstatic surrender to divine law, whereas the scientist's was a painstaking, empirical social practice. The hallmark of Emerson's scientific method was the marriage of both: ecstatic surrender *and* cool intellectual analysis. What is extraordinary here—although on reflection it should hardly be surprising—is that two such apparently incompatible practices could be the left and right hands of a single strategy. That strategy powered tremendous technological change even as it rationalized that change as part of natural progress, neither more nor less than the unfolding of God's great design. The double union of hot ecstasy and cool intellect successfully merges poetry, religion, and science to forge the dynamic engine of American ideology. Nowhere are the terms of that paradox posed so sharply or urged so keenly as in Emerson's 1841 oration, "The Method of Nature," delivered at Waterville College in Maine.[92]

By 1841 Emerson knew himself to be a rising star in the intellectual firmament. Thanks to the inheritance of Ellen's estate, he had secured enough income to fund his projected career as a writer and lecturer, to remarry in 1835 (to please his poet's ear he asked his bride, Lidia Jackson, to rename herself "Lidian"), and to move into a large house in Concord where they would spend the rest of their days. Waldo had planned to share the house with his brother Charles and his fiancée, Elizabeth Hoar, but Charles died of tuberculosis in May 1836, less than two years after his brother Edward's death from the same cause. In the five years since publishing *Nature*, Emerson had become the center of the Transcendental Club, delivered "The American Scholar" and "The Divinity School Address" at Harvard, preached his last sermon, helped launch a new literary journal (*The Dial*) with Margaret Fuller, invited Thoreau to live with his family, become a father (to Waldo Jr. and Ellen, with Edith on the way), and published his first book, *Essays*. Clearly, he had matured into the true American Scholar, and "The Method of Nature" opens with a congratulatory acknowledgment of his new status: "Where there is no vision, the people perish. The scholars are the priests of that thought which establishes the foundations of the earth." Yet this particular priest, for all his material prosperity, was entering a period of crisis, a turning point between the radical optimism of *Nature* and the darker vision of the later Emerson.[93]

A number of commentators have located that turn here, in "The Method of Nature." Before 1841, David Jacobson argues, Emerson still could imagine that human reason organizes the universe, but now, "We no longer are wed to nature; it no longer belongs to us." The once-redemptive "humanist synthesis" sublimates to "antihumanism" and a new "practical philosophy of obedience." What Jacobson calls "obedience," Jonathan Levin calls "abandonment": Levin points to this lecture's emphasis on "not nature but its method," the laws that organize nature's manyness into oneness. Swept up in this mystical stream, we can only abandon ourselves in "ecstasy," a surrender that repays the loss of our individuality in the new power of "relationality," resulting, ultimately, in a science learned in love, a science "practiced with an eye to irreducible human interests and emotions." David Robinson argues even more eloquently for Emerson's turn, in this period, toward human interests and ethical engagement. For Robinson, the ecstasy of "The Method of Nature" finally proved "unstable and self-defeating," and Emerson responded by turning outward, away from individual ecstasy and toward a growing recognition of relational ethics and the power of social forces—in effect, reformulating "self-culture" into ethical pragmatism. As both Jacobson and Robinson assert, this shift, however named, was neither a reversal nor a defeat, but a completion of tendencies evident throughout Emerson's earlier work. As early as the 1820s, Emerson had learned in life's battle to convert truth from opponent to ally. After the ebullience of *Nature,* "The Method of Nature" represents a return and intensification of Emerson's older and darker optimism, submission to law as the only path to command.[94]

Five years before, Emerson had dispatched the question "To what end is nature?" with the confident answer *us*, or at least those of us who can keep nature "under foot," a stable magazine of power. Now, in 1841, nature has no end, or at least none that can be declared. In the "rushing stream" of "perpetual inchoation," "We can point nowhere to anything final," only to nature's constant metamorphosis. "To questions of this sort, nature replies, 'I grow, I grow. . . . I have not yet arrived at any end.' " No method of logical analysis will suffice in a universe in which "this refers to that, and that to the next, and the next to the third, and everything refers." While endless nature is a work of ecstasy, to be represented by a circular movement, we, her alleged interpreters, are but limited beings who pursue "straight line[s] of definite length," blind to the universal unless "housed in an individual." The age has gone wrong: "We are a puny and a fickle folk," diseased with avarice and hesitation; "Nothing solid is secure; everything tilts and rocks." Even Bacon's crystal globe has failed. "The crystal sphere of thought is as concentrical as the geological structure of the globe. As all our soils and rocks lie in strata, concentric strata, so do all men's thinkings run laterally, never vertically." The globe resists us, and not one man can pierce downward, to "the core of things." For method, then, instead of mastering nature we must abandon ourselves to the flow. The genius does not will his success but is "hurled into being as the bridge over that yawning need, the mediator betwixt two else unmarriageable facts." To succeed he must become transparent, the "channel through which heaven flows to earth," surrendering to "an ecstatical state." The method of nature, then, is ecstasy, self-surrender; the solution to the riddle posed by individuality, embodiment, is disembodiment, "ek-stasis" or out-standing, no longer *in* our bodies but *beside* them.[95]

This method comes with a cost. One must fend off the provoking seductions of the actual world, while cleaving to the severe austerity of the law that gave those objects their meaning. Beauty and truth become locked in a kind of hostile dance, a tango which beauty sustains even as it seeks to betray truth, which yet must always prevail: "Truth and Beauty," Emerson wrote—Law and Body, Spirit and Matter—"always face each other and each tends to become the other." This dynamic lures us in: Beauty pleads with us to follow her, for she leads us to Truth, but in the face of Truth she vanishes, a body burned away by the fire of meaning. Once the entire world has burned away, only Truth will remain, infinite, absolute, weightless, objective.[96]

In other words, while the beauty of physical nature evidences the law that would otherwise be invisible and unknowable, obedience to the law means resisting her seductions. Here Emerson's model has to be science, for it is science that knows how to submit without loss of control. As he proclaims, "Every star in heaven is discontented and insatiable. Gravitation and chemistry cannot content them. Ever they woo and court the eye of every beholder. Every man who comes into the world they seek to fascinate and possess, to pass into his mind, for they desire to republish themselves in a more delicate world than that they occupy." Who has attended to the siren song of the stars? Not poets but men of science: "Newton, Herschel and Laplace"—here not just men of science after all, but poets, whose writings fill the finer world of rational souls with their fame. But take care! Objects of nature are "beautiful basilisks" who "set their brute, glorious eyes" on us, willing us, seducing us to take them up into mind; and so man must be "on his guard." Thus the promise of science is attended with peril. The genius of science must capture the object, take it into mind, without being captured *by* it. The challenge is unremitting, for the stars, though always present, are ever "inaccessible," as are all natural objects "when the mind is open to their influence."[97]

How, then, can these beautiful, tantalizing objects be safely captured and possessed? Man must stand apart from nature, look on her with "a supernatural eye. By piety alone, by conversing with the cause of nature, is he safe and commands it." This demands a practice of surrender, not so much self-command as a kind of disciplined release: "And because all knowledge is assimilation to the object of knowledge, as the power or genius of nature is *ecstatic,* so must its science or the description of it be." Such out-of-body practices are the ascetic achievements of a select few, the "scholars" who are "the priests of that thought which establishes the foundation of the earth." For, "When all is said and done, the rapt saint is found the only logician." In short, only the disciplined ecstasy of the rapt saint-scholar-priest who sees through things to thought and commands the Logos of creation can keep the stars from commanding us. Only the disembodiment of truth will protect against the body of nature.[98]

The key is careful subordination of nature to intellect even in the process of joining them, in the union constantly metaphorized as their "marriage." Newton, the first in Emerson's trinity of scientist-poets, was universally credited as the first to "marry" heaven and Earth by showing that the reign of natural law extended through and embraced or encircled them both. Such a marriage is set about with the rigorous sanctions of law: it must be a chaste and (literally) Platonic wedding, out-of-body, "ecstatic." The objects of nature that "woo and court" man "seek to penetrate and overpower" his nature, seek to "cause their nature to pass through his wondering eyes into him, and so all things are mixed." It is to guard against such "mixtures" that Emerson erects a protective barrier, a prophylactic against the unchaste union that would demand a surrender to the body, dissolving our human nature and precipitating our selves as "bright sediment" in the world, no longer the cultivators of nature but part of its "enchanted dust." Such surrender to the body evokes the terror of mortality, of "Necessity," of "Fate": behind it lies the sober knowledge that only in the grave is our organic union with nature complete.[99]

The alternative is another and far more cheerful kind of fatality: "The poet must be a rhapsodist: his inspiration a sort of bright casualty: his will in it only the surrender of will to the Universal Power." This directive derives its strength from Bacon's inaugural insight that "the chain of causes cannot by any force be loosed or broken, nor can nature be commanded except by being obeyed." Emerson uses Bacon's insight to draw science and literature into one united enterprise through the notion of "action" in the

Emerson's method

Can do Science both ways

Man altering Nature

early biology

world, action that by submitting to nature acquires the power to reshape the world not just conceptually but actually. Emerson was theorizing man as a force of nature because he was one of the first intellectuals of his age to comprehend the degree to which man was altering, even remaking, nature itself. The world that was emerging in the early 1800s was a novel natural-technological complex in which the traditional boundaries between man and nature were being dissolved on a global scale; despite everything, "all things" *were* being "mixed."[100]

To separate these volatile mixtures, Emerson spent much of his career formulating and refining a double metaphor that provided him with the twinned resources of fate and freedom, necessity and power—analogues to Bacon's pairing of command and obedience. Both sides of the pairing of body and law, or what I will call the "organic" and "gnomic" halves of the metaphor of organicism, were drawn from science: "organic" from the developing sciences of the Earth (geology, comparative anatomy, physiology, biology, embryology), all of which were struggling to define themselves against the merely observational and descriptive study of natural history; and "gnomic" from the well-developed tradition of natural philosophy (physics, astronomy, chemistry), which was authoritatively based in self-evident laws of reason. As discussed in chapter 2, the goal of natural science was widely taken to be the conversion of natural history into a true science, such that Cuvier, for instance, had called for "the Newton of natural history," a prescient seer who would find in the chorus of variable natural forms the single laws of their formation and regulation, and so, as Newton did, marry Earth and heaven under one rule.

Against the mixing of all things, Emerson initiated the double metaphor of organicism by invoking the act of separation. As he stated in "Intellect," "The considerations of time and place, of you and me, of profit and hurt, tyrannize over most men's minds. Intellect separates the fact considered from you, from all local and personal reference, and discerns it as if it existed for its own sake. . . . Intellect is void of affection, and sees an object as it stands in the light of science, cool and disengaged." The act of separation itself detached actor from object. Cuvier had expressed perfectly the way the process of offering up an object to the light of science simultaneously unified and divided: "Every organized being forms a whole, a unique, and perfect system, the parts of which mutually correspond, and concur in the same definitive action by a reciprocal reaction. None of these parts can change without the whole changing; and consequently each of them, separately considered, points out and marks all the others." In this organic system, all parts required each single part, and each single part "requires them reciprocally; and beginning with any one, he who possessed a knowledge of the laws of organic economy, would detect the whole animal." The act of analysis both verified the organic status of the examined body *and* separated Cuvier's intellect from the bodies he handled—that is, separated the physical Cuvier, who was himself embodied and hence subject to the very laws he pronounced, from the "higher" Cuvier, who pronounced the laws of embodiment. The ambivalence is fundamental to the double metaphor. The embodied self "faces two ways," on the one hand toward closure, boundaries, and determination by law, and on the other toward openness, experience, growth, and transgression of the law. The social model sketched out here is thus a bundle of contradictions, and as the model is precipitated into and explored through various metaphors, the contradictions are reflected, animated, manipulated—but never resolved.[101]

The reason for revisiting the metaphor of organicism in Emerson, then, is not to ask whether his own work is "organically unified," but to ask how and why he deployed this particular metaphor, so deeply implicated in the biological sciences and more broadly in the fundamental conception of all science, to articulate the relationship of self, society, nature, and God as a reciprocity between part and whole, a means simultaneously to liberate and to contain the self. As indicated earlier, metaphor can act as the "medium of exchange" both within science and between science and other discourses, creating an "ecological" network. Emerson liked to say that "the whole of nature is a metaphor of the human mind," opening all of science as a public greenhouse within which to store and nourish moral metaphors—a kind of tropical jungle under glass, an updated version of Bacon's crystal globe.[102]

But to invoke another of Emerson's favorite sayings, one with wide currency in the nineteenth century, truth is one, error many. Organicism was the dominant metaphor for truth as one and coherent, Coleridge's reassuring "unity in multeity," the integrating wholeness that excluded error and signified a lawful and lasting "marriage" of mind and nature. The metaphor for metaphor itself is often marriage, as Dale Pesmen points out, including the prohibition against mixed metaphor as a kind of miscegenation or monstrous marriage. Such a prohibition "stems from a wider ideology of coherence," which "governs our judgments of the 'truth,' 'validity,' and 'realism' of pictures of the world and distinguishes things we *can* think about from things we find objectionable and/or impossible to think." Mixed metaphors call attention to their own metaphoricity, shattering the illusion that language is a transparent channel of truth. Emerson, too, wanted a pious marriage as a way to achieve coherence and to avoid "mixture," and he recognized science as the authoritative model for such a marriage, for science constructs wholeness in the thinking subject, in the material object, and in the method of inquiry. But an inquiry into the anatomy of organicism shows that its coherence is of a very special kind: turned on itself, single in the way of a Möbius loop, each half covering and yielding to the other in a totality that defies inspection as a whole—the whole of modernism in the palm of the hand. Organicism presents a technology for integration of a very mixed world, while itself being mixed to the core.[103]

Modernity is inseparable from the metaphor of organicism: "Modernity is inaugurated when the mind becomes identified as essentially constructive and when, as in Vico and Kant, our knowledge is restricted to the things we make ourselves." Put simply—no doubt too simply—in modernity, the organizing idea (or the idea that renders nature "whole," organized, or "organically" connected) is identified as our own rather than God's, although Emerson's modernism preserved the symbiosis of science and religion by identifying the mind of man with the mind of God. As Emerson said, "We can never be quite strangers or inferiors in nature," for it is "flesh of our flesh"; but neither can we be strangers to God, whose word is our word. We know we have spoken that word when it organizes chaos into meaning. The poet-priests of science—Newton, Herschel, and Laplace—had learned to speak the language of God, which lesser men, through their smaller but still precious spiritual acts, have translated into waterwheels, railways, ships, trade, and mechanical craft, the forms that organize material nature according to our same governing idea. The greater the organic or organizing power of an idea, the truer it is. Very powerful ideas organize enormous networks of objects, people, and relations; it can even be said that they "construct the real."[104]

Organicism itself is a very powerful idea, as Coleridge realized when he made it the standard Romantic figure for the dynamic organizing power of Reason (over mere aggregative understanding) and Imagination (over the associative fancy). The key word here is *dynamic*. Organic power cannot rest, for stasis means death. It realizes itself only in the action of organizing, or breaking down and assimilating to build itself from within outward, as a plant is said to do. For Coleridge, dynamic organicism was a powerful repudiation of eighteenth-century materialist science—although an object being colonized by an organic force might protest that being assimilated by a dynamic whole is not noticeably superior to being positioned as a cog in a materialist machine. In either case, the will of the individual part is subordinated to the will of the designer of the whole. Organicism typically rationalizes its form of coercion as the true form of freedom, freedom through obedience (in the old Puritan formulation) to the internalized will of God, or, in the Baconian version, command of nature by obedience to her. Organicism quite literally "rationalizes" mechanism—re-creates mechanism by a sweeter name, a kinder, gentler machine, a machine that effects change not by outer force but by inner power.

Historically, Romantic organicism synthesized Enlightenment mechanism with the dynamic new natural science, defining nature as dead matter in order to animate it with supernatural spirit. Coleridge borrowed the metaphor he needed from the new biological science. Whereas Plato had said that works of art were "organized" wholes rather like creatures, Kant had suggested that creatures were "organized" wholes rather like works of art. Whereas Kant had gone on to reject this analogy (thereby making possible the science of biology), Coleridge folded Kant's analogy back into Plato, to assert that organisms *were* works of art, God's art, and organicism the natural, therefore deeply binding, principle of social organization. Kant's organisms were self-organizing and went their own incalculable way, but Coleridge's organisms existed only by order of a higher intelligence. Plato's poetic metaphor thus returned by way of the new biology, bearing now all the alienated majesty of science; that is, it was no longer merely metaphor. It was truth.[105]

Like Coleridge, Emerson owed to science the organizing idea that gave his social theory its structural support: the law of nature and the laws of the mind are finally one. We can trust the order of Creation, because its order is also ours. In this self-similar, self-mirroring universe, "Whenever a true theory appears, it will be its own evidence. Its test is, that it will explain all phenomena." Phenomena flicker into life in the light of the theory that makes them visible, that makes them organs of its truth. Such an organizing idea explains all phenomena by connecting them organically into a whole so open that it will literally show itself—self-evident because our organizing idea has conjured exactly its own best evidence. This, then, will be the "method of nature." But we need a sign, something outside us, something physical, to guide us. Emerson had to construct an exemplary object that by embodying its own constructive law was utterly self-evident, perfectly "transparent" to reason. Such an object would secure the ultimate convergence of nature and mind, truth and beauty, science and art; it would mark the end point at which they become interchangeable and in effect the end point of ultimate explanation: the smallest would contain the largest. We would see a world in a grain of sand, and eternity in an hour.[106]

Organicism thus opens into a second, complementary figure, which I call "gnomicism." The word compresses three facets of this figure: first, "gnomic" sayings are, like Blake's utterance, compact to the point of self-evidence, refusing dissent. They proffer the universe in a nutshell, wisdom in a wisecrack, hard-won souvenir of some ecstatic flash of insight. Second, gnomic figures are "gnomonic" in the technical sense. The gnomon was defined by Hero of Alexandria (in the first century A.D.) as "any figure that, when added to an original figure, leaves the resultant figure similar to the original," like a carpenter's square or a seashell. It accomplishes this neat trick through growth by proportional increments, the basic principle of all organic growth. Third, they are "nomian" in that they give the law, or "nomos," to themselves; hence their self-similarity. They literally cannot be otherwise and still be themselves.[107]

Gnomic figures appear to offer a fixed point in that world in which "everything tilts and rocks." In a world of social constructions and bootstrap universes, they alone cannot be built: no mere diligence "can rebuild the universe in a model, by the best accumulation or disposition of details,

Unfasten the Power of the word — Key.. A

yet does the world reappear in miniature in every event, so that all the laws of nature may be read in the smallest fact." The gnomic impulse is characterized by the urge to signify infinitely, to collapse the greatest into the smallest, to see, as Blake said, "a world in a grain of sand"—or in the branchings of a tree or an ice crystal, the ramifications of a leaf, the whorls of a seashell or a spiral nebula, the globe of a waterdrop or of the whole Earth, the rings of atomic or planetary orbits, the spiral of a climbing plant, a twist of DNA. As Emerson said in *Nature,* "The fable of Proteus has a cordial truth. Every particular in nature, a leaf, a drop, a crystal, a moment of time is related to the whole, and partakes of the perfection of the whole. Each particle is a microcosm, and faithfully renders the likeness of the world."[108]

Such Romantic figurations haunt us still, and, as ever, they arrive in the popular imagination from the cutting edge of science. In Einstein's magical equation $E = mc^2$, we see, as Roland Barthes pointed out, "all the Gnostic themes: the unity of nature, the ideal possibility of a fundamental reduction of the world, the unfastening power of the word," the key that unlocks total knowledge. Emerson himself said that the intellect can pervade and dissolve the "solid seeming block of matter" with a thought: "In physics, when this is attained, the memory disburthens itself of its cumbrous catalogues of particulars, and carries centuries of observation in a single formula." Later in the twentieth century, Einstein's mysterious formula was joined by a mystical sign, James Watson's "double helix," Richard Dawkins's "immortal coil." In a typical popular treatment, the architect György Doczi wonders: "Is it pure coincidence that on the molecular level the joint three-dimensional spiral pattern of the double helix—matching the double snakes of Hermes' magic wand—was found a few years ago to be the true shape of the DNA molecule, which contains within its miniature coded pattern the master plan of the entire future development of living organisms?" Recent additions to the pantheon include the ubiquitous image of the Earth as a grain of blue sand in a black void, the Mandelbrot fractal's recursive plunge into infinity, and the human genome, which threads all humanity onto a staff of four notes, G, A, T, and C.[109]

The goal of gnomic science is to grasp the infinitely large by revelation of its principle of production. In Emerson's words, "Nature shows all things formed and bound. The intellect pierces the form, overleaps the wall, detects intrinsic likeness between remote things, and reduces all things into a few principles." Emerson was encouraged in this belief by John Herschel, who had ended his *Preliminary Discourse* by observing a kind of double movement in science: first, "every new discovery in science brings into view whole classes of facts which would never otherwise have fallen under our notice at all," leading to a "constantly extending field of speculation"; simultaneously, the steady advance of science discovers generalizations of an ever higher and more inclusive order, and "every advance towards generality has at the same time been a step towards simplification." Nature appears complicated only when we are lost in the maze of particulars, but when we rise to the "commanding view . . . we never fail to recognize that sublime simplicity on which the mind rests satisfied that it has attained the truth." Emerson imagined this movement carried to its end point, where the few principles that contained much would finally resolve back into the one that contained all: "a rule of one art, or a law of one organization, holds true throughout nature"; "The whole code of her laws may be written on the thumbnail, or the signet of a ring."[110]

Emerson's gnomic science leaned very hard on that word *law.* He took his concept of law from the Aristotelian tradition that proclaimed laws of nature to be "causal and constitutive" rather than empirical generalizations, aligning Emerson with the minority view of scientists of his day but in the very good company of Coleridge and Herschel. In his discussion of the word *law,* though, even Herschel had cautioned that its use meant only that *we understand,* rather than that the *material universe obeys,* certain rules: "The Divine Author of the universe cannot be supposed to have laid down particular laws, enumerating all individual contingencies, which his materials have understood and obey,—this would be to attribute to him the imperfections of human legislation,—but rather, by creating them, en-dued with certain fixed qualities and powers, he has impressed them in their origin with the *spirit,* not the *letter* of his law"—making all further consequences "inevitable"—although the conservative Herschel hastened to add that the Creator still directly maintained and energized this vast system with his will. However, by equating letter and spirit, Emerson slid from law as human understanding back to law as divine legislation, the mind or spirit organizing obedient matter: "That which once existed in in-tellect as pure law, has now taken body as nature." In the same way, the "gnomic" principle slides into the "nomian" principle as the word of God or the "Universal Spirit," the letter or Logos that determines the unswerv-ing logic of all creation. "All our progress is an unfolding, like the veg-etable bud," said Emerson. We ever become more of what we already are.[111]

"I grow, I grow," says nature. The vegetable bud unfolds according to the principle of organic growth, by which the new builds on and incorpo-rates the old. In the same way, the mollusc grows by adding a ring of new material to the edge of the old, as does a tree, a ram's horn, or an elephant's tusk. Aristotle had noticed that certain things "suffer no alteration save in magnitude when they grow," leading Hero of Alexandria to his formulation about additive growth, which leaves the resulting figure unchanged. Such a rule can produce any number of figures, each one a gnomon, all of them self-similar at any scale (figure 1). Mathematically speaking, such self-simi-larity is the result of what D'Arcy Thompson called the "simplest of laws of growth," proportional addition, by which every addition becomes the gnomon of the original form. In one of the more elegant examples, if growth takes the ratio of $1:\sqrt{2}$, the figure becomes its own gnomon (figure 2). But in all gnomonic forms, all stages of generation will be self-similar. The classic instance is the chambered nautilus (figure 3), as celebrated by Emerson's friend Dr. Oliver Wendell Holmes in 1858:

> Year after year beheld the silent toil
>   That spread his lustrous coil;
>   Still, as the spiral grew,
> He left the past year's dwelling for the new,
> . . . .
>
> Build thee more stately mansions, O my soul,
>   As the swift seasons roll!
>   Leave thy low-vaulted past!
> Let each new temple, nobler than the last,
> Shut thee from heaven with a dome more vast,
>   Till thou at length art free,
> Leaving thine outgrown shell by life's unresting sea!

In "The Chambered Nautilus," Holmes exploited the analogy between nature and mind to sound a moral lesson in self-culture, whereby the soul grows by steady increments through the ladder of advancing but isomorphic forms, discarding each for the next, and finally all for the heaven of spirit. As with Emerson's poet-scientists, the goal all along had been to deploy the discipline of form toward the end of liberation from form's necessary vulnerability. The lesson looks to all three phases of Emerson's fundamental trinity of beauty, truth, and duty, for not only is the nautilus beautifully vaulted in pearl and suggestively chambered; it is also the most perfect example in nature of a logarithmic spiral, whose generating proportion is none other than the Golden Rectangle, on which the Greeks constructed the Parthenon and the ideal human form. Here is exactly what Emerson, in *Nature,* said he most wanted of science: that it give not mere details but declare "the relation between things and thoughts . . . the *metaphysics* of conchology, of botany, of the arts, to show the relation of the forms of flowers, shells, animals, architecture, to the mind, and build science upon ideas."[112]

The gnomon appears in another Transcendental metaphor that superficially looks very different: the sundial. The triangular plate whose shadow points to the time of day is perhaps the best-known gnomon of all, yet it is not the object itself, but the shadow it casts, that indicates the direction of the sun (figure 4). The original Greek carried the meaning of "interpreter" or "discerner," from *gignoskein,* "to know": the gnomon is thus one who knows—who uses shadows to discern the truth. As Emerson avowed in 1839, "I call my thoughts The Present Age, because I use no will in the mat-

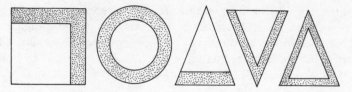

Figure 1. Gnomonic figures: self similarity at any scale. Drawing by author.

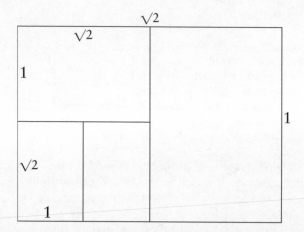

Figure 2. The figure its own gnomon: growth by the ratio of $1:\sqrt{2}$. Drawing by author.

ter, but honestly record such impressions as things make. So transform I myself into a Dial, and my shadow will tell where the sun is." If the gnomon of a sundial is turned parallel to the axis of the Earth, the direction of its shadow at any given hour will always tell the true time, regardless of the changes of the seasons. "Gnomonics," then, is the art of constructing sundials, aligning the variable surface of the Earth with the changeless heavens. Long after clocks and watches had become commonplace, sundials were still routinely used not only decoratively, but practically, as the unerring and celestial standard by which man's earthly and inconstant time machines were set and corrected—just as the New England housewife of old regulated her daily chores by the noon mark on the kitchen floor. When Emerson's friend Bronson Alcott, the Orphic Sayer himself, named the Transcendentalist's new journal *The Dial*, he evoked both the sunlight of the present and the shadow of time's passage, marked by the faithful whose axis of vision was, as Emerson had written, coincident with the axis of things.[113]

The infinite self-similarity of such figures allows Emerson, in "Circles,"

Figure 3. Chambered Nautilus (*Nautilus pompilius*), cutaway view. Drawing by author.

to make a characteristically gnomic observation: "Our life is an apprentice-ship to the truth, that around every circle another can be drawn; that there is no end in nature, but every end is a beginning; that there is always an-other dawn risen on mid-noon, and under every deep a lower deep opens." Hence "the natural world may be conceived of as a system of concentric cir-cles," perpetually sliding or telescoping into one another, and objects are not really objects "but means and methods only,—are words of God, and as fugitive as other words." Objects, in this Platonic vision, are shadows, but right vision sees them point to the truth: in *Nature*, Emerson quoted Swedenborg's saying, "The visible world and the relation of its parts, is the dial plate of the invisible." Hence at the center of the telescoping circles of divine language is man, Emerson's Central Man, carrier of the divine. This trope, too, follows a technical metaphor, for a gnomonic projection, natu-rally, "is a projection of a sphere in which the centre of sight is the centre of the sphere." Every rainbow, then, and every horizon, is a gnomonic projec-tion. And so, inevitably, from center out, the logic unfurls like a vegetable bud: everything proceeds "from the eternal generation of the soul," which, like man, in whom that soul is manifested, "is a self-evolving circle," rush-ing perpetually outward, "wheel without wheel," every law only a particu-

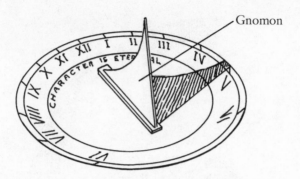

Figure 4. A typical horizontal sundial. Drawing by author.

lar fact to some more general law, ever to be subsumed in some still wider circle.[114]

Thus one has perpetual growth, limited only by the energy of the source; perpetual growth, perpetually without change; every object the seed of itself, everything multiform but everywhere the same. Furthermore, the growing object inscribes in itself its own past, like the rings of a tree, embodying memory as timeless form, precipitating time as an eternally present text that can be read just like the strata of the Earth. All of nature, Emerson says, is just such a living record of the past, a chronicle of lawful Creation; or, in "The Method of Nature," that nature is "the memory of the mind"; or, more particularly, in "The Conservative": "Throughout nature the past combines in every creature with the present. Each of the convolutions of the seashell, each of its nodes and spines marks one year of the fish's life, what was the mouth of the shell for one season, with the addition of new matter by the growth of the animal, becoming an ornamental node." The leaves that cool us are the growth of a summer, but the "solid columnar stem . . . is the gift and legacy of dead and buried years." Thus we ornament ourselves with the fashion of the moment, even as we grow beyond it, using the energy of the present to build the tradition that will unite future and past; the leaves of the hour build and layer our duration for the ages.[115]

Such operations as these mark a boundary zone between nature and culture, or science and art, drawn together by the force of belief in cosmic order. This is not just science, its disciples proclaim; it is art. Or, this is not merely art, it is "Sacred Geometry," "one revelation of the spirit of Nature." The more closely either science or art approaches this common boundary, the more completely are the arbitrary choices and flawed instruments of representation transcended to reveal the essence common to both. Thus to behold such gnomic natural objects is to witness a nature so pure that, as Emerson tirelessly reiterated, it literally embodies in matter the divine idea or formula of its genesis, revealing nothing less than the harmonic proportions that create and govern the universe. Or, it is to witness an art so pure that it coincides with the primal generative forms of nature, revealing the mind of the Creator as identical with the mind and heart of man. That is, these objects are held to be simultaneously pure art and pure nature. They constitute a unique class of entities in which neither aspect is sullied by mixture with the other—not art by the accidental flaws of particular natural objects, nor nature by the artifactuality of the presentation. They display the polar opposition of nature and art united as one, a single inextricable wholeness, paradoxically at once both pure and hybrid. Hence their mystical status, their revelatory force: the seal of truth is beauty, beauty is the sign of truth, and together they sweep all contingency aside. James Watson famously said of his DNA molecule that it was too pretty not to be true. The siren song of gnomic science compels precisely because it demands yet defies analysis, all the while sweetly condemning the skeptical analyst's unbeautiful soul.[116]

Although these two figures, organic and gnomic—tree and nautilus—overlap and complement each other, they are not quite synonymous. How, then, are they different? What does the "organic" metaphor leave out that the "gnomic" metaphor supplies? The organic metaphor relates various components as constituents of a whole, constituted by some outside force, will, or intelligence. Its analogue with living processes imposes a chronological structure on its objects, which are seen to unfold through timeless cyclic stages in the natural process of birth, growth, senescence, death, rebirth. But the gnomic metaphor collapses all parts into self-similar versions of the whole, and it suspends all movement in time as well: it creates its object as a motionless container for all potential motion. All its stages are the same stage. Organicism promises growth without change; gnomicism delivers on that promise. Organic objects express gnomonic principles of growth by proportional addition; gnomic objects do not need to be biologically organized or "organic" at all.[117]

Gnomic objects do, however, need to be microcosmic, even as the goal of the gnomic saying—the aphorism, epigram, proverb, or formula—is to express much in little, *multum in parvo*. Susan Stewart observes that "there are no miniatures in nature," for the miniature is a product of cultural operations. Sayings, and by implication all other gnomic objects, gain this quality by being abstracted from their context and made to "transcend lived experience and speak to all times and places. The *multum in parvo* is clearly rooted in the ideological; its closure is the closure of all ideological discourse." It becomes "monumental, transcending any limited context of origin and at the same time neatly containing a universe." Gnomic objects, too, are products of the act of attention, which removes the object from its context, abstracts its law of generation, and from that law generates a world that is thereby rendered both closed and total, infinitely controlled and contained.[118]

The organic or organized whole, by contrast, transforms the body itself into a miniature. It is *we* who are organized and contained: "Just as we have emphasized the relation of the miniature to the invention of the personal, so must we finally emphasize the relation of the gigantic to the invention of the collective. For the authentic body of the giant marks the merger of the self-as-part with an ideological whole." Stewart's meditation on the miniature and the gigantic suggests that the body itself is the hinge. Indeed, Emerson's endless series of telescoping circles is arrested here, where our intimate experience of embodiment marks a limit. As for Cuvier the scientist, the human body is in effect ambivalent, facing two ways: in one direction, it is complete in itself, the sum of the universe, a gnomic microcosm; in the other, it is radically incomplete, a fragile "part" or organ of the gigantic macrocosm. To think gnomically is to contain, but to think organically is to be contained, to be assimilated or swallowed as a constituent of the collective whole. To be gnomic is to give the law; to be organic is to be given by the law. In the late essay "Fate," Emerson proposes to face both ways: we who are crushed by fate can turn and master fate by converting it into power. "Now whether, seeing these two things, fate and power, we are permitted to believe in unity? The bulk of mankind believe in two gods." Emerson wishes to show that they are, finally, one, the "Blessed Unity which holds nature and souls in perfect solution, and compels every atom to serve an universal end," and by which we are the more free as we take unto ourselves the power of necessity.[119]

But the body cannot face both ways at once. To visualize this is to recognize that this is a mixed or monstrous metaphor, figuring a freak or a Janus-god. If the body—integrated, coherent, therefore whole—is the metaphor for truth, then what is at stake here is the coherence of our cultural model of truth as single, integrated, and integrating. The gnomic metaphor solves this difficulty by suggesting that lesser truths will nestle inside greater truths in a hierarchy of coordinate classes, as at the Paris Muséum. But to inhabit a living body that can face only one way at a time is to know (or fear) that truth is also a matter of perspective, and that shifting perspective can put enormous strain on the coherence of a single truth.

For example, Emerson could look at the death of a loved one as the gnomic growth from the "low-vaulted past" to the "new temple, nobler than the last" of Holmes's poem; such, as he counseled in "Compensation," is "the natural history of calamity": "The changes which break up at short intervals the prosperity of men are advertisements of a nature whose law is growth. Every soul is by this intrinsic necessity quitting its whole system of things, its friends, and home, and laws, and faith, as the shell-fish crawls out of its beautiful but stony case, because it no longer admits of its growth, and slowly forms a new house." Hence the compensations of calamity: "The death of a dear friend, wife, brother, lover" seems a privation at first but later "assumes the aspect of a guide or genius," terminating one's old life and opening a new one "more friendly to the growth of character." In *Nature,* Emerson had ended the bracing chapter "Discipline" with the lesson that "when much intercourse with a friend has supplied us with a standard of excellence . . . when he has, moreover, become an object of thought, and . . . is converted in the mind into solid and sweet wisdom,—it is a sign to us that his office is closing, and he is commonly withdrawn from our sight in a short time." In this way Emerson reconciled himself to the deaths of his first wife, Ellen Tucker, and of his brother Charles. Yet what of their perspective? The bitter truth is that circumstances do not allow everyone to seize the law and become his own oracle.[120]

To point this out, however, is not to give the lie to the metaphor but to suggest why it has been so effective. It supplies resources not only to the mourning Emerson, whom it instructs to seize the law and build a new life, but also to the dying Ellen and Charles, who, having surrendered to the law, are then "free" of their "outgrown" shells. In effect the metaphor allows each perspectival truth to cover for the other, reconciling coherence and mortal division: all nature dies, but nothing, in the end, really dies. It is a "paradox": no single point of view can withdraw far enough away to see both sides at once. The solution lies in surrender to the method of nature, for as disembodied minds, we can move freely up and down the chain of being. In "Intellect," Emerson wrote:

> Water dissolves wood, and iron, and salt; air dissolves water; electric fire dissolves air, but the intellect dissolves fire, gravity, laws, method, and the subtlest unnamed relations of nature, in its resistless menstruum. . . . How can we speak of the action of the mind under any divisions, as of its knowledge, of its ethics, of its works, and so forth, since it melts will into perception, knowledge into act? Each becomes the other. Itself alone is. Its vision is not like the vision of the eye, but is union with the things known.[121]

The human body thus becomes a threshold, both contained and containing, the one figure that is simultaneously gnomic and organic according to whether one is subject, or object to the organizing idea; the designer of the machine, or a cog in the works. As the foregoing suggests, the pair are profoundly gendered. Organic embodiment is the hybrid offspring of the marriage of passive vegetative female nature with active organizing male mind, whereas gnomic modes absorb and abandon embodiment altogether, dissolving nature in a transcendence that erases difference, hence erases the feminine. The organic object may be, perhaps must be, female, in the way of ships, mother countries, and "mother" nature herself. By contrast, the gnomic object overlooks gender in the same way that the word *man* evacuates gender to assert universality. However, recall that the gnomic-organic pairing offers not an absolute binary categorization but a performative process. Thus each is not divided from the other but acts as a resource to the other.

Emerson both accepted and resisted organicism's coercive force on the individual. As we have seen, "The American Scholar" opens with a grotesque image of a society blasted into fragments: "The state of society is one in which the members have suffered amputation from the trunk, and strut about so many walking monsters,—a good finger, a neck, a stomach, an elbow, but never a man." In contrast to Cuvier's legendary ability to read the whole animal from a fossil fragment, in this nightmare universe each severed part fails to bespeak the whole. Interestingly, Emerson locates the solution in the corporate collective: all partial men gathered together will make One Man. But cannot *one* man make one man? Regrettably, no, for human limitations prevent it. So to each man goes his delegated office, such that "the scholar is the delegated intellect," Man Thinking—the brain of the social body. In the organic whole, individuals are no longer microcosms, self-contained gnomic atoms that point to the self-similar macrocosm, but instead acquire meaning as functioning parts (or "organs") of the whole, a whole greater than all of them put together. The social body thus "incorporates" many individuals, bonding them into the transcendent body politic, which designates, coordinates, and disciplines the multitude of lone dependent bodies. As the plant grows by assimilation, the growth of this corporate society is accomplished by the active assimilation of new individuals, both human and nonhuman, in an ever-widening circle of action: "I run eagerly into this resounding tumult. I grasp the hands of those next me, and take my place in the ring to suffer and to work, taught by an instinct that so shall the dumb abyss be vocal with speech. I pierce its order; I dissipate its fear; I dispose of it within the circuit of my expanding life."[122]

Emerson's powerful and reassuring vision both derived from and fed back into the ongoing myth of America as less a place than a dynamic process, Crèvecoeur's "melting pot," merging the many into a new and progressive unity. Christopher Newfield mounts a major critique of Emerson on a parallel argument, naming this paradox the "Emerson Effect": "individual autonomy and public authority vanish together before unappealable laws, but this leads to the enhancement of freedom." Howard Horwitz also analyzes this liberating deference to transcendent authority: "The dream of the trust, shared even by its radical opponents, is to become a powerful person by not being an agent, or rather by being merely the agent or instrument of transcendent forces." Both Newfield and Horwitz understand the way in which this curious formation both protects and eliminates the individual, though neither refers to the grounding of this cultural formation in the successes of modern science. The growth of this myth of growth correlated with the material expansion of the United States, whose widening networks assimilated ever-greater numbers of individuals, human and nonhuman, into the organic whole of the American state, through the circulation of Emersonian Spirit in the form of knowledge and capital on the tracks of the transportation network. No wonder Thoreau was so fascinated, and so repelled, by telegraph wires and railroad tracks.[123]

The double metaphor of organicism is thus modernism's icon, holding its deep paradoxes in restless suspension. We submit as organs of the transcendent whole; we command as agents of transcendence. By folding the two into mutually exclusive sides of a single process, we can be said to discover our truths and to make them, too, to construct nature and to submit to nature's transcendence. Each covers for the other. Nevertheless, for all the effort Emerson made to turn to nature and natural law—science—for the pure and stable truth and single law that lay beyond the shifting sands of American society, what he came away with was nothing more nor less than inscriptions written by men in laboratories, bodies designated by men in museums, and machines made by men at workbenches, all the most social of sites. Humanity realizes itself through nature; nature manifests itself as human resource. Despite the prohibition on mixtures, we separate and purify humanity and nature in order to marry them: there is purity nowhere.[124]

As Emerson said, one studies nature to study mind. Intellect "existed already in the mind in solution: now, it has been precipitated, and the bright sediment is the world." What we cannot study directly, then, we can study mediately, using the "bright sediment" that is nature to study the mind, even as "we explore the face of the sun in a pool, when our eyes cannot brook his direct splendors." Here is the last piece, the doubling of organic immanence and gnomic transcendence, the act of purification that after all makes this marvelous hybrid possible: in such instances of revelation, when in the pure harmony of the gnomic object the mind within recognizes the Mind without, we are no longer ourselves, our daily, divided, worldly, social and fragmented selves, but "beside" ourselves—with joy—in a transport of ecstasy.[125]

In this state of ek-stasis, "out-standing" or standing aside from ourselves, we are split from ourselves to join in an ecstatic union with God, seeing ourselves now as the singular soul or "ME" might were it to lift from and look down upon the (collective social) body, the "NOT ME," realizing a double consciousness that comprehends that the sun in the pool is not the sun in the sky, the stars in the heavens not the stars in the soul. "For the truth was in us, before it was reflected to us from natural objects; and the profound genius will cast the likeness of all creatures into every product of his wit." That gnomic genius will not read, but be, the gnomon that, in facing the sun, casts the world in the shape of his shadow; he will know not to mistake reflection for reality, shadow for substance, image for reality. In rare moments of ecstatic vision, he will even displace his earthly body to see the world as God sees: "Our globe seen by God is a transparent law, not a mass of facts. The law dissolves the fact and holds it fluid." Such ecstatic displacement is necessary, for "the field cannot be well seen from within the field." From the field, one risks mistaking the part for the whole, the local for the global—of being seduced by the stars. So one must aspire to rise

Great Scientist is a poet

above the field and to become the spectator of oneself in the exhilarating disconnection of transcendence.[126]

Thus it makes sense that Emerson's own singular initiating revelation of the wholeness of nature happened not in "pure" nature, which would in this scheme be the "empty" American wilderness, but abroad in that most busy and mixed of places, the Paris Muséum, where all the plants and animals were laid out and partitioned according to their assigned place in the order of nature. In the Muséum Emerson witnessed the generative equation of life in all its gnomonic extensions, while the complex of museum, garden, and zoological park both embodied, and stood beside, the world. Both garden and laboratory are ecstatic spaces in which the scientist leaves his self behind to penetrate to the selfless realm of law, of mind. In such a space, controlled and separate, one can seize the generative principle and so take all the infinite world into the self, lift the very stars themselves from matter to mind.

Although Emerson did not become a scientist in the conventional sense, he went on to enfold science and literature in a wider vision, constituting the modern for America and thereby playing his part in creating the unbearable vacuum left in the wake of what he would call the "stupendous antagonism" of nature and man, and in filling that vacuum with the inrushing of the genius whose pure intermediation could translate each to the other. In the very next moment, the same genius who marries nature with man rushes to remove man from nature:

> As a ship aground is battered by the waves, so man, imprisoned in mortal life, lies open to the mercy of coming events. But a truth, separated by the intellect, is no longer a subject of destiny. We behold it as a god upraised above care and fear. And so any fact in our life . . . disentangled from the web of our unconsciousness, becomes an object impersonal and immortal. It is the past restored, but embalmed. A better art than that of Egypt has taken fear and corruption out of it. It is eviscerated of care. It is offered for science.

Thus mysticism and science meet. The hot ecstasy of revelation severs earthly connections, then slides into the serene glow of objectivity: "Intellect is void of affection, and sees an object as it stands in the light of science, cool and disengaged. The intellect goes out of the individual, floats over its own personality, and regards it as a fact, and not as *I* and *mine*." Freed of earthly connections, the fact is ready for the anatomizing eye of science, which, piercing to its law, will dissolve its substance, translating it from Earth to heaven, matter to mind, terrene quotidian to extraterrestrial power.[127]

Thus is gnomic science also nomian science, the science of law, or *nomos*, always able by inscribing a wider circle to rein the errant fact and the antinomian impulse into the circle of reason. As Emerson's ideas developed, he came to see this as a crucial social function of science. His last science lecture, "The Humanity of Science" (delivered in December 1836), takes up themes familiar from the earlier lectures and uses them to introduce his ambitious new lecture series, *The Philosophy of History*. Among the familiar themes, one gains new emphasis. Poet and scientist should join so that each might correct the other's excesses, which in this case include both the false conclusions of poetry and, much worse, the "unhallowed and baneful" science resulting when the man of science is but the slave of nature, "as happened signally in philosophic France." From this Emerson derives the important new idea that men of science should write for a popular audience. Although the benefit to the laboring man might be negligible, the benefit to science will be significant: "It will be the effect of the popularization of science to keep the eye of scientific men on that human side of nature wherein lie grandest truths." Why should only the poet and the priest—and not Newton and Laplace—"bring the oracle low down to men in the marketplace"? National education and the press will bring the judgment of great numbers of men to "the experiment of the philosopher" and serve to check "the whims and spirit of system of the individual."[128]

More telling still, though, is the benefit to society: "The highest moral of science is the transference of that trust which is felt in nature's admired arrangements, light, heat, gravity,—to the social and moral order. The first effect of science is to stablish the mind, to disclose beneficent arrangements, to remove groundless terrors." For "the survey of nature irresistibly suggests that the world is not a tinderbox left at the mercy of incendiaries. No outlaw, no anomaly, no violation, no impulse of absolute freedom is permitted to exist; that the circles of Law round in every exception and resistance, provide for every exigency, balance every excess." Man, that most errant of creatures, is above all else rationalized by the rule of harmonic growth, which assures that his waterwheels and railways, canals and factories, are not violations of the old order (that whiff of French gunpowder), but the emergence from the old of the new, the modern order. By 1841 Emerson could conclude, in "Art," that "the boat at St. Petersburgh, which plies along the Lena by magnetism, needs little to make it sublime. When science is learned in love, and its powers are wielded by love, they will appear the supplements and continuations of the material creation." Modern science takes its place not as the conquering opponent of nature but as nature's destined extension, its supplement and continuation. This does more than "stablish" the mind, transferring trust in the truth of nature to the social and moral order, founding what I call the "culture of truth" in America; this marks the historical moment when man defines himself as the controller of the environment, reversing a historical relationship that had held for millennia. Man no longer merely held the gnomic law in mind, an achievement of art and intellect. Just a decade before, nature had been an ageless and eternally balanced economy set in motion by the laws of nature and nature's God—God's perfect and unchangeable design. Abruptly, man became the agent of the law, empowered to dissolve the Earth into the stream of his desire.[129]

In short, man is shown by Emerson's poetic science to be nothing less than the necessary gnomon to nature, the completing figure through which the generative law of creation extends itself upward and outward, in the ecstatic spiral, ever to the next level of ascension. In man the old series of nature concludes. The new, modern order takes up creation where the old, premodern order of nature gave out, incorporating nature by marking it off, sealing it shut, taking it up into the mind, and finally precipitating it as spirit, as art, as the Law finally realized. In Emerson's time, polarity became the engine of power, generating the irresistible upward push of the unstoppable life force that America would crown and transform, America which was "modern" because Americans were not just one people among many, but that singular people who completed the last circle of nature and opened the next circle beyond.

CHAPTER FOUR

••

Global Polarity and the Single Life

ONE PLUS ONE EQUALS ONE

Emerson's science divided the world into body and law, matter and spirit, nature and humanity, a primal polarity that energized the universe. Ironically, unity demanded duality: as he speculated in 1834, "All grows up from plus and minus." When Emerson needed to search for a single, primary generative principle from which to derive the universe, whether physical or moral, he did not have far to seek. For a post-Kantian intellectual, the question came with its own self-evident answer: polarity. Back in June 1827 Emerson had witnessed what happened when a magnet was laid across a heap of steel filings, and the image of reciprocal alignment of axis to axis in a field of force stayed with him for life. It seemed a key that unlocked the great mystery of the universe, *how* natural and moral truth were aligned. The great soul—the *magnetic* soul— "converts calamity to knowledge; knowledge to power; hope to happiness."[1]

Emerson's reading confirmed that polarity was the keystone that divided, connected, and stabilized the arc of the universe. The German philosophers, for instance, knew that the necessary trinity of God was the Infinite, the Finite, and Creation, the passage between them. "Unity says Schelling is barren. Duality is necessary to the existence of the World. Shall I say then that the galvanic action of metals foreshows from afar the God head, the zinc the metal & the acid; or the marriage of plants the pollen, the ovary, & the junction?" Two years after writing these words, in 1837, Emerson was willing to go still further: "Polarity is a law of all being. Superinduce the magnetism at one end of a needle, the opposite magnetism takes place at the other end. If the south attracts, the north repels. To empty here, you must condense there. Light, shade; heat, cold; centrifugal, centripetal; action, reaction. If the mind idealizes at one end perfect goodness into God coexistently it abhors at the other end, a Devil."[2]

It all seemed almost too simple, to roll so much into such a tight little ball. "The Wonder [is] perpetually lessened," Emerson protests, when such marvels are shown to be the result of "simple combination." The shell, that marvel of gnomic design, is demystified: it is "not one effort but each knot & spine has been in turn the lip of the structure.... The part that was builded instructed the eye of the next generation how to build the rest." The wonder of the magnet, too, is diminished when the way it wheels to the north and clings "to iron like one alive" is shown to be "only one instance of a general law that affects all bodies & all phenomena: light, heat, electricity, animal life." However, although knowledge may have destroyed *wonder,* now the *power* is his, to "throw myself into the object so that its history shall naturally evolve itself before me."[3]

Such expressions of disappointment evaporated with Emerson's growing awe at the totality of explanation offered by polarity. As a method that organized literally everything there was, polarity became irresistible. Yet the more irresistible it became, the more disturbing, too, for the impulse of polarity was to draw a line down the center of the universe, sundering it like an apple split in two. "A believer in Unity, a seer of Unity, I yet behold two," Emerson lamented. "Cannot I conceive the Universe without a contradiction?" The answer was yes—and no. Ultimately the impossible agony of a universe split through its equator into Fate and Power, Nature and Mind—and all the other "bipolar unities" that by then had crystallized around his figure of the magnet—precipitated into the superheated language of "Fate": "If Fate follows and limits power, power attends and antagonizes Fate. . . . Man is not order of nature, sack and sack, belly and members, link in a chain, nor any ignominious baggage, but a stupendous antagonism, a dragging together of the poles of the Universe." The poles of a magnet can never be joined. They repel each other with infinite force. Emerson's Man makes the torment of Sisyphus look like a half-time job, for at least Sisyphus could rest while the stone rolled back downhill. But there was no rest and no escape once Emerson rejected Chance for single and all-generative law. Polarity would be the "systole and diastole of the heart," the same heart that was, and could be, the only true prophet.[4]

POLAR COORDINATES: KANT TO GUYOT

Polarity was not merely one of Emerson's affectations. It remains one of the most powerful metaphors of modern science, intimately tied to the concept of science itself. Around the turn of the nineteenth century, the metaphor of polarity became a key site of exchange between the physical and biological sciences, Romantic literature, religion, and social theory. Efforts to ground knowledge and social practice in some ultimate reality were self-consciously based on polarity as nature's prime generative principle. The genius of polarity is that it defines opposites as mutually exclusive but also as mutually invested in each other—like the magnet, which cannot be imagined as all negative, with the positive somehow removed. Polar opposites form a co-producing system that can be divided but never detached. Furthermore, the visible figure of the magnet is only half the picture. The completing half is composed of the energy that circulates in the figure's environment, connecting pole to pole and orienting everything surrounding the magnet along its lines of force. Thus the polar figure organizes an entire environment around itself *and* sets the components of that environment in motion. By "organizing" matter, polarity turns inorganic matter into organized life; it organizes that life into "individuals," then organizes those individuals into whole societies. All levels are structured by the same uniting antagonism, the same bipolar unity. If metaphor is technology, then polarity is the metaphoric technology of modernism itself.

Immanuel Kant's *Metaphysical Foundations of Natural Science* sought to replace the old Newtonian mechanico-corpuscular science with a new, dynamic mode of thought. The old Newtonian mode had explained matter as atoms of impenetrable density in absolutely empty space. This treated "matters" as machines, "i.e., as mere tools of external moving forces." According to Kant, this mode had dominated science from Heraclitus to Descartes and on to his own day only because it had proved so very convenient to mathematics. It was time for a more realistic picture of the universe: and so Kant's "dynamical" way replaced atoms and the void with the dual forces of attraction and repulsion. By repulsion, matter disperses itself through space; by attraction, it gathers itself into bodies. Matter must be a balance of these two inseparable forces, for without attraction, matter would disperse to infinity, and without repulsion, matter would contract to a mathematical point. By these two forces, then, all space is filled "and yet filled in varying measure"; no space can be absolutely empty. Kant offered his own dynamic alternative as "far more suited and more favorable to experimental philosophy," for if matter was understood as a balance of varying degrees of opposed forces, measurements of those forces could lead "directly to the discovery of the moving forces proper to matters," and ultimately to the experimental determination of their governing "laws."[5]

Kant believed his dynamical natural philosophy was superior because it promised to be fruitful for the scientific researcher. However, he also communicated a certain contempt for the old view of matter as a "mere" tool or machine, and excitement for the view which derived the variety of matter "from the proper moving forces of attraction and repulsion"—that is, which derived the motive power of matter as arising from within, not imposed from without. Why was motive power—motion—so foundational? Because to Kant "the fundamental determination" of an object "must be motion, for thereby only can [the external] senses be affected. The understanding leads all other predicates which pertain to the nature of matter back to motive; thus natural science is throughout either a pure or an applied doctrine of motion." This a priori cognition—that only through motion can an object be determined, can we know it *as* an object—makes dynamism the condition of knowing the universe. Only as matter moves can we know it. Finding the laws of movement is therefore the goal of physics, the purest of the "rational" sciences.[6]

The idea that polarity was the fundamental first principle of nature became the basis of Schelling's *Naturphilosophie*. From Kant, Schelling derived the idea that "matter is an equilibrium of active forces that stand in polar opposition to one another"—or, more precisely, *dead* matter was in "equilibrium." *Life* was essentially out of balance, a "struggle of divided forces . . . and for this reason alone we regard it as a visible analogue of the mind." The first business of the natural philosopher was *"to go in search of polarity and dualism throughout all nature,"* and that principle governed Schelling's *Naturphilosophie:* all natural phenomena—combustion, light, air, electricity, magnetism, matter, and chemical process—could be explained through the interaction of polar forces. The most obvious instances led to the greatest success of the *Naturphilosophen* and one of the nineteenth century's greatest scientific breakthroughs, when Oersted in 1820 succeeded in demonstrating that electricity and magnetism are fundamentally the same, united through the law of electromagnetism.[7]

However, the *Naturphilosophen* who followed Kant made one major change in his philosophy, to correct the rigid division Kant had made between the polar opposites of mind and matter, self and object. Schelling berated Kant for utterly severing body from mind, for it was obvious to him that if Kant were right, matter and idea could have no contact at all, no reciprocal influence. How, then, could matter ever enter the mind? As J.B. Stallo, the interpreter of *Naturphilosophie* to America, argued, if Kant were right, then we could not even *see* nature: "The mind would see a void in nature, if it did not recognize itself there; it would gaze upon the utterly Dark, upon an absolute night, upon an abstract vacuum. No reciprocation of any kind between the radically Heterogeneous." The answer was that matter *was* mind. The universe began when the primal absolute unity *self*-divided into subject and object, twins separated at birth that are ever after striving to rejoin.[8]

In short, to the *Naturphilosophen,* a valid theory of knowledge could be established only if "the self or self-consciousness" were taken to be the "ground of all reality." In effect, self-consciousness bridged the polar antagonism of mind and matter by making the "Other" part of itself. In Coleridge's *Biographia Literaria*—a book that was at Emerson's elbow from 1826 onward—there really *is* no Other if all is mind to begin with. Yet there is still that originating self-division of primal unity into subject and object, which must be rejoined—and that is the function of *knowledge.* Coleridge claimed that "all knowledge rests on the coincidence of an object with a subject." While objective nature and subjective self are truly *polar* conceptions, existing "in necessary antithesis," the act of knowledge unites the two so that they become "coinstantaneous and one," with no possible priority of one over the other. Moreover, if knowledge itself has "two poles reciprocally required and presupposed, all sciences must proceed from the one or the other, and must tend toward the opposite as far as the equatorial point in which both are reconciled and become identical." Knowledge, then, is bipolar, too. Natural science proceeds from the objective pole and moves toward the subjective; transcendental philosophy proceeds from the subjective and moves toward the objective. At the equatorial center that is their mutual goal, nature will be so spiritualized that it will disappear: "The highest perfection of natural philosophy would consist in the perfect spiritualization of all the laws of nature into laws of intuition and intellect. The phaenomena (*the material*) must wholly disappear, and the laws alone (*the formal*) must remain." At this point the "theory of natural philosophy" will be complete, when all nature is rendered as sheer "intelligence" or "self-consciousness"—or, as Emerson put it, once all nature is taken into the mind, "all external grace must vanish."[9]

Paradoxically, though, according to Coleridge, this equatorial zone at which subject and object merge can be reached only by a continued and careful expulsion by each pole of its opposite. The natural philosopher must strive to be "objective," avoiding "above all things the intermixture of the subjective in his knowledge, as for instance, arbitrary suppositions . . . occult qualities, spiritual agents, and the substitution of final for efficient causes." The border police are just as active on the opposite, subjective side: "On the other hand, the transcendental or intelligential philosopher is equally anxious to preclude all interpolation of the objective into the subjective principles of his science, as for instance the assumption of impresses or configurations in the brain. . . . This purification of the mind is effected by an absolute and scientific skepticism to which the mind voluntarily determines itself for the specific purpose of future certainty." Only thus is "certainty" or truth to be reached, through ritual purification by each of all traces of the other: mind must not be confused with brain. Coleridge concludes his discussion by demonstrating the Platonism that undergirds this polar division between objective and subjective. The man who believes he sees a table can never prove that table's objective existence, or that another sees the same actual table as he. What he can be "*certain*" of, however, is that from what it is that he *does* see, "the phantom of a table," he may "argumentatively deduce the reality of a table, which he does not see." This "real" table exists a priori in all our minds, and by it we severally recognize the "phantom" and shifting forms of sense upon which we may safely lay our books and our dinner plates. The law of gravity is "real" in precisely this way (whereas the dinner plate it smashes is merely an accident).[10]

It will hardly make the headlines that the Romantics distinguished objective and subjective, with far-reaching consequences. One of the most relevant consequences here is the supposed split of "subjective" Romantic poets from the dry and "objective" science that so repelled them, a myth that has blinded several generations to Emerson's importance as an interpreter of science. Criticism is full of commonplaces like the following: "Wordsworth and Coleridge gladly left reason, science, and objectivity to the dehumanized empiricists and exalted imagination as a more human means of achieving a higher truth." This kind of statement obscures the crucial role of Coleridge and other Romantics in instituting and maintaining the dichotomy between "objective" and "subjective"—terms that Coleridge himself coined; his thumb is on the scales of every dispute weighing "subjective" and "objective" knowledges against each other. Coleridge needed "objectivity" because to him, the subject could know itself only by realizing itself as the *object* of its own knowing, as if light generated the objects by which it was seen: the subject "becomes a subject by the act of constructing itself objectively to itself." So the subject can exist at all only in "antithesis to an object," and knows itself by self-duplication or self-representation in an object, which it then embraces as part of or identical with itself. To know ourselves, we make the world our mirror, so the world shows no other image than our own. Or, as Emerson said, "The truth was in us, before it was reflected to us from natural objects; and the profound genius will cast the likeness of all creatures into every product of his wit."[11]

Similarly, Coleridge defined the principles of art by "objectifying" science in order that the "subjective" spirit can "know itself." He stated that his work in the *Biographia Literaria* referred "solely to one of the two Polar Sciences, namely, to that which commences with and rigidly confines itself within the subjective, leaving the objective . . . to natural philosophy, which is its opposite pole." Self-consciousness is the result of the "coincidence" of natural and transcendental science, and the goal, from whichever side one starts, is the same: "the absolute identity of subject and object." Science calls this "nature," but really, "in its highest power [it] is nothing else but self-conscious will or intelligence." Nature is neither more nor less than intelligence realized, and intelligence is "a self-development" that we may

> abstract . . . under the idea of an indestructible power with two opposite and counteracting forces, which, by a metaphor borrowed from astronomy, we may call the centrifugal and centripetal forces. The intelligence in the one tends to *objectize* itself, and in the other to *know* itself in the object. It will be hereafter my business to construct by a series of intuitions the progressive schemes, that must follow from such a power with such forces, till I arrive at the fulness of the *human* intelligence.

As should be obvious by now, Coleridge felt entirely comfortable borrowing metaphors from astronomy to discuss the formation of intelligence because he believed that ultimately he wasn't borrowing at all: science was the way by which intelligence realized itself. Astronomy—the very stars themselves—was just another manifestation of our thought. We must first see the stars in our mind before we can see them in the sky; and the stars we see in the sky are really our own mind made visible.[12]

Biographia Literaria was after all a "literary" biography; hence it developed only one of the polar sciences. Later in life Coleridge returned to the topic to develop the other pole: in "Hints towards a More Comprehensive Theory of Life," he showed how polarity in nature conspires to create "individuation" in organic life forms, a process that likewise reaches its apex in "the fulness of human intelligence." In accordance with Kant's definition of a true and lawful natural science, Coleridge proposed "a physiological, that is a real, definition" of life, rather than a mere "history." A physiological definition must consist "in the *law* of the thing, or in such an *idea* of it, as, being admitted, all the properties and functions are admitted by implication."[13]

He found his unifying figure in the magnet: polarity was the law of life. Even as Kant used the varying balance of polar forces to create matter as full and continuous space, Coleridge used the varying proportion of polar forces to fill up "the arbitrary chasm between physics and physiology." As he said, "The arborescent forms on a frosty morning, to be seen on the window and pavement, must have *some* relation to the more perfect forms developed in the vegetable world." Thus, life is manifested first in gravity as the power of unity, then in crystals as the individuation of unity, and so on up the ladder to man. According to Coleridge, life is "the *tendency to individuation*"; "and the individuality is most intense where the greatest dependence of the parts on the whole is combined with the greatest dependence of the whole on its parts." This increasing interdependence of whole and parts is wrought by the antagonism of polar opposites: "This tendency to individuate cannot be conceived without the opposite tendency to connect, even as the centrifugal power supposes the centripetal, or as the two opposite poles constitute each other, and are the constituent acts of one and the same power in the magnet. We might say that the life of the magnet subsists in their union, but that it lives (acts or manifests itself) in their strife." Thus the magnet—a unified power composed of contending forces—is the organizing figure for the concept of life as a struggle of opposites, a figure that moreover fills the "chasm between physics and physiology" by representing both in one.[14]

Coleridge has now laid the foundation for his cyclical or spiral pattern by which life ascends "as the steps in a ladder," pausing at each upward rung to expand in radiating circles of organic form. "All things strive to ascend & ascend in their striving," quoted Emerson approvingly from Coleridge's *Aids to Reflection;* or, as he put it in "Circles," "Step by step we scale this mysterious ladder: the steps are actions; the new prospect is power." Coleridge's treatment becomes increasingly complex as he expands his antagonistic polarities to a total accounting of space, time, and motion, mapping his theory across Cartesian coordinates, which become both the four cardinal points of the compass and the "four elemental forms of power."[15]

Having established that nature is constituted in the "endless strife between indifference and difference," Coleridge extends his grand theory to a survey of organic forms, establishing, for instance, that the vegetable and animal worlds "are the thesis and antithesis, or the opposite poles of organic life," then arranging the various animal forms in sequence—from corals to molluscs to insects, to fish, to birds, to quadrupeds, and finally to Man, in whom "the whole force of organic power has attained an inward and centripetal direction. He has the whole world in counterpoint to him, but he contains an entire world within himself. Now, for the first time at the apex of the living pyramid, it is Man and Nature, but Man himself is a syllepsis, a compendium of Nature—the Microcosm! . . . In Man the centripetal and individualizing tendency of all Nature is itself concentred and individualized—he is a revelation of Nature!" Man and nature emerge as the opposite poles of the universe. Nature all along has been struggling, through polar antagonism, to realize the ultimate polar opposite to itself, the global "subject" who would not only contain but even transcend objectivity itself.[16]

At this point Coleridge offers the bridge from science to society: he who is given organically by the law is also—voilà!—gnomic giver of the law. Thus the war of polarity produces *individuals:* "he who stands the most on himself, and stands the firmest, is the truest, because the most individual, Man. In social and political life this acme is inter-dependence; in moral life it is independence; in intellectual life it is genius." The most individuated man will be he who depends most on the social whole, and upon whom the social whole most depends. This calls for the kind of heroic individual who is such by virtue of his deep social engagement—not the isolated genius but the leader of nations, the great man whose greatness is a product of the needs of his people. Emerson's standard figure for this aspect of genius would be Napoleon, one of his "Representative Men." Coleridge's (and Emerson's) political ideal was the organic state, "organized" by the law, which designated to each individual his appointed place and role. Yet the law that designates us is not rigid or static but a *polar dynamism,* and thus the energizing constraints—the Foucauldian forms of discipline—grow ever more intense:

> As the height, so the depth. The intensities must be at once opposite and equal. As the liberty, so must be the reverence for law. As the independence, so must be the service and the submission to the Supreme Will! As the ideal genius and the originality, in the same proportion must be the resignation to the real world, the sympathy and the inter-communion with Nature. In the conciliating mid-point, or equator, does the Man live, and only by its equal presence in both its poles can that life be manifested!

Polarity demands that the pinnacle of individualism subsist in, and by way of, the pinnacle of interdependence. It is a paradoxical achievement, yet it patterns the texture of ordinary modern life: industrialized, technologized, divided and dispersed in ever-finer concentrations of ever-more-incandescent energy that demands an ever-widening network of interconnections to survive. Coleridge, in this obscure work of Romantic science, identified the formula for hypermodernity, and Emerson translated it into the idioms of America.[17]

As Kant had predicted, the concept of antagonistic or balancing forces was enormously productive. Among those who adopted it as the basis for their own systems of thought were Goethe, Lorenz Oken, J.B. Stallo, and Robert Chambers, all of importance to Emerson. Chambers was the most popular of them all. In 1844 this Scottish journalist published (anonymously) a best-selling narrative of natural science, *Vestiges of the Natural History of Creation,* an evolutionary bombshell that earned widespread notoriety as well as Emerson's deep admiration. Here, the polar metaphor takes the shape of a vertically oriented polar axis organizing a circulation of energy through a comprehensive field or system: that is, a tree. Goethe had seen all the diverse forms of plant life folded back into the leaf; Chambers sees all *life* in the leaf. Arborescence is everywhere: ice crystallization on a windowpane resembles "vegetable forms"; silver and mercury crystallizing on a piece of silver suspended in a solution of nitric acid and water create a form *"precisely resembling a shrub"*; "some of the most ordinary appearances of the electric fluid" present "vegetable forms." Such phenomena must indicate that plant forms are shaped by electric energies. Chambers finds it worthy of

especial remark, that the atmosphere, particularly its lower strata, is generally charged positively, while the earth is always charged negatively. The correspondence here is curious. A plant thus appears as a thing formed on the basis of a natural electrical operation—*the brush* realized. . . . In the poplar, the brush is unusually vertical . . . the reverse in the beech: in the palm, a pencil has proceeded straight up for a certain distance, radiates there, and turns outwards and downwards; and so on. We can here see at least traces of secondary means by which the Almighty Deviser might establish all the vegetable forms with which the earth is overspread.

Chambers's figure unites electromagnetism, chemistry, and organic form with the Earth itself, orienting entire systems of natural inquiry around the polar axis and "planting" it in the eye of the imagination as a great spreading tree.[18]

This axial magnetic tree organizes a complete landscape around itself, even as the magnet organized *its* environment, and the living tree—the fundamental figure of organicism—participates in the global circulatory cycle of air and water. Schelling explained air as the polar complementarity of oxygen (given off by plants and breathed in by animals) and carbon dioxide (given off by animals and breathed in by plants), tying man and tree together by their shared creation of the atmosphere, which mutually sustains them. Emerson's essay "Water" animated water as that friend and servant who "works when we sleep, circulates in our veins; is present in every function of life, grows in the vegetable, is a cement, and an engineer, and an architect, in inanimate nature." Water became the physical sign of the spiritual "fluid" whose "flow" regulates and supports all life and weaves it through the geography of the planet, from rain clouds and mountain streams to rivers and lakes to oceans that evaporate back into rain clouds whose rain feeds the tree whose roots seize the Earth and whose branches sweep the sky: "The circulation of the water in the globe is no less beautiful a law than the circulation of blood in the body." Polarity not only organizes the body; it organizes the entire process of life, opening bodies as systems of flow and environment as body exteriorized. Through the body flows the Earth itself, and the Earth reveals itself as one great body, eddying from pole to pole.[19]

The example of Arnold Guyot shows the uses of polarity as an organizing principle not just in Emerson's wider intellectual universe, but in his immediate neighborhood. Like Coleridge, Guyot deployed natural law, by which the many are generated by and resolved back into the One, as a model for social law. He is known, however, not as a political philosopher but as a geographer. Guyot came to America at the urging of his fellow Swiss immigrant Louis Agassiz. In 1849 he delivered the prestigious Lowell Lectures in Boston, in French, before joining the faculty of Princeton in 1854. The twelve lectures were translated by James Elliot Cabot and published as *The Earth and Man: Lectures on Comparative Physical Geography, in Its Relation to the History of Mankind;* the 1851 edition found its way into Emerson's library. Guyot's lectures were widely and well received, and he went on to write a series of graded geography textbooks for American schools (the model for geography texts for many years afterward) and to help institute the National Weather Service.

As the title *Earth and Man* suggests, Guyot, too, was interested in locating the relation of the individual to the collective. Polarity provided him with the unifying law of nature, and through the polar dynamics of individuation, "earth" and "man" became the two that are yet one, opposites united in unceasing strife that generates by division a quickening cycle of exchange whose circulation binds the ever-individuating Many into an ever-complexifying One. "Thus nature and history, the earth and man, stand in the closest relations to each other, and form only one grand harmony." To Guyot, the Earth is not *like* a body, it *is* a body; and the proper office of geography is not "coldly to *anatomize* the globe" by describing the "arrangement of the various parts which constitute it," but to study "the reciprocal action of all these forces, the perpetual play of which constitutes what might be called the life of the globe . . . its physiology." Otherwise geography is deprived of its "vital principle," by which Guyot means the principles of "connection" and "mutual dependence" by which the globe may be said to be *alive*.[20]

As Guyot reminds his readers, "We must elevate ourselves to the moral world to understand the physical world, which has no meaning except by it and for it." From the elevated viewpoint of the "moral world," the forms, arrangement, and distribution of the continental land masses take on new meaning. Not only do they reveal a plan; they are made for the human societies that populate them "as the body is made for the soul." As with Coleridge, man becomes the polar or dynamic opposite of nature—the soul of nature's body; as with Cuvieran comparative anatomy, apparently superficial or surficial differences will be shown to display deep internal differences that are united by an even deeper plan. In Guyot's case, defining life as a *"mutual exchange of relations"* establishes the basic requirement for life: inequality. "An exchange supposes at least two elements, two bodies, two individuals, a *duality* and a difference, an inequality between them, in virtue of which the exchange is established." This "inequality" excites "a mutual exchange between the bodies, in which each gives to the other what the other does not possess. To multiply these differences, to increase their variety, is to render the actions and reactions more frequent, is to extend and to intensify life"—and to extend the "evolution" or "development" of life, from diversity to organic unity. "Homogeneousness" marks the lower or savage state, whereas "diversity of elements" multiplies exchanges; thus out of an almost infinite specialization of functions arise the higher civilizations, to heights beyond the dreams of the wretched Indians in the Rocky Mountains. God distributes each of us to our place and assigns those "inequalities" that will bind us to one another in a proliferating and intensifying economy of exchange: inequalities exist that they may be orchestrated into harmony.[21]

All partial differences are summed up in a series of *"great contrasts,* two by two": land versus water, Old World versus New World, three northern continents versus three southern continents. So: the primitive ocean is inferior to the superior land. The coast, where extremes are united and tempered, is better than the interior. The Old World is superior to the New, for the greater contrasts of the temperate Old World have called out the vigorous activity of nature and of man, while the greater unity of the tropical New World results in the common physiognomy of American organisms, including its men, who all share a "coppery tint and a family likeness." Guyot's "law of contrasts" yields increasingly colorful differences: the Old World is "a mighty oak, with stout and sturdy trunk, while America is the slender and flexible palm tree, so dear to this continent. The Old World . . . calls to mind the square and solid figure of man; America, the lithe shape and delicate form of woman." The array of contrasts adds up to a clear conclusion, that although the Old World is superior, Old and New as two parts of one organic whole must commence "a life of interchanges, which will enrich them both." America is "glutted with its vegetable wealth, unworked, solitary"; the Old World is exhausted and "overloaded with an exuberant population, full of spirit and life." "As the plant is made for the animal," the

New World lies ready to be consumed, or consummated: "The two worlds are looking face to face, and are, as it were, inclining towards each other. The Old World bends towards the New, and is ready to pour out its tribes. . . . America looks towards the Old World; all its slopes and its long plains slant to the Atlantic, towards Europe. It seems to wait with open and eager arms the beneficent influence of the man of the Old World." Old World man, reawakened and reanimated, crosses the ocean to tread America's shore "enraptured." America, the daughter, grows to independence, and the two worlds contest "as power with power," until America, the bride, assents to "a well-assorted union, a true marriage." "Yes, gentlemen, the futurity and the prosperity of mankind depend on the union of the two worlds. The bridals have been solemnized." Now her "virgin soil" will be worked to gestate and give birth to "The American" as "the young man, full of fire and energy" as against the elder European's knowledge and wisdom.[22]

From the romance of Old World and New, Guyot turns to the great contrast between north and south. This will not be a story of the courtship and marriage of equals. Undeniably the tropical southern continents have the greatest energy and variety of life, yet this advantage does not argue for southern superiority. Continents are intended to serve *human* not natural life, and tropical man, far from being the "most beautiful of his species," is a degraded brute. Guyot emphasizes this point by presenting a series of drawings showing the declension of man from northern perfection to southern deformity. Man reaches his ultimate debasement at the poles, in the "miserable Bushman" of Africa, the South Australian ("the last degree to which ugliness can go"), the Pecheray of Tierra del Fuego ("the most misshapen" and "wretched" of the New World), and the Eskimo of North America. "Does not this surprising coincidence," Guyot asks, "seem to designate those Caucasian regions as the cradle of man, the point of departure for the tribes of the earth?" Thus it was not the equatorial south but the temperate north that became the theater of history, and the north that will now bring peaceful conquest to the whole Earth, realizing "the dream of the existing world," a dream "possible only in perfect obedience to the divine law, in absolute goodness."[23]

Ultimately, the formula is the same one that Coleridge derived in "The Theory of Life," but Guyot, a new immigrant to America, casts it in service to American nationalism:

> The founders of the social order in America are indeed the true offspring of the reformation,—true Protestants. The Bible is their code. . . . They submit from the heart to their Divine Leader, and to his law; this is the principle of order. Now the union of these two terms is free obedience to the divine will. . . . These, you will agree, gentlemen, are the sublime doctrines whence flow the religious, political, and social forms that distinguish America at the present time, from all the other countries on the globe.

In Guyot's world, "the privileged races" have the duty to bring the blessings and comforts of civilization and the gospel to the "inferior" races, helping them comprehend and perform their destined role as "the hands, the workmen, the sons of toil." Providentially, each northern continent has its own dependent southern continent "especially commended to its guardianship and placed under its influence." In the end, knowledge itself—"living, harmonious, knowledge"—is the union of Faith and Science, into a perfected faith that is "VISION," and this vision is of nothing less than the entirety of the globe brought under the union assured by the initiating premise of polar division, as deployed by those who best command the technologies it makes possible.[24]

In short, during Emerson's day the figure of polarity proliferated wildly. Built into it at the base—with Kant and Schelling—was the notion that rational cognition itself originates with polar distinctions, generating the "metaphysics" upon which is founded "physics" and, through physics, all science. Kant acknowledged that every doctrine in which "a whole of cognition" is ordered by principles is called "science. And," he continued, "since principles can be either of the empirical or of the rational connection of cognitions in a whole," natural science can be divided into "historical [empirical] and rational natural science." However, by defining nature as that which must derive its existence from its internal principle, Kant rejected the "historical doctrine of nature" as not scientific. Kant therefore denied the status of science to the *empirical* "connection of cognitions in a whole" because natural science—the science of the objective—must be *rational;* that is, the laws which underlie it "are cognized a priori and are not mere laws of experience." Forms of nature study that do not derive their existence from an internal principle, and hence can be pursued only empirically, are demoted to mere natural history and taxonomy.[25]

In the eyes of the Romantics, this move created, in one blow, the division between "rational holism" and "empirical holism" as modes of thought, and exiled "empirical holism" from true science, marginalizing it as natural history. In response, Alexander von Humboldt worked hard to reassert Kant's rejection of intellectual intuition and reclaim empirical natural science as a "science"—one reason why he remained in some ways curiously marginal to "mainstream" science. It is entirely in character that it should be Humboldt who registered one of the few votes in the literature Emerson read against the proliferation of polar explanations for all natural phenomena. It all reminded Humboldt of Aristotle, who also endeavored "to reduce all the phenomena of the universe to one principle of explanation": "All things were reduced to the ever-recurring contrasts of heat and cold, moisture and dryness, primary density and rarefaction—even to an evolution of alterations in the organic world by a species of inner division (antiperistasis), which reminds us of the modern hypothesis of opposite polarities and the contrasts presented by + and –. The so-called solutions of the problems only reproduce the same facts in a disguised form."[26]

What worries Humboldt is the self-mirroring quality of this form of cognition. Does it enable us to see only what we already know? Does "rational" science form, then, a closed loop? In Romantic theories of polarity, an originary unity gives rise to duality either by dialectical creation of its own opposite (as in Fichte and Hegel) or by self-division into subject and object (as in Schelling, Coleridge, and Emerson). Either way, the resulting pair must then be resolved back into a "higher" unity. For instance, in Emerson's polar universe, an originary God self-divided into nature and man, to be reunited as man takes nature back into mind; this coincidence of subject and object was possible at all because they were both, finally, manifestations of God. All difference was subsumed into originary sameness; finally, there were no real differences. Hence Emerson's universe was literally self-evident, contained by its own premises: "Whenever a true theory appears, it will be its own evidence." Or, as Coleridge said, we seek for "some absolute truth capable of communicating to other positions a certainty, which it has not itself borrowed; a truth self-grounded, unconditional and known by its own light."[27]

In his own geography, *Cosmos*, Humboldt tried to interrupt this loop at the moment of its formation by rejecting the very idea of a "rational" science:

> It is not the purpose of this essay on the physical history of the world to reduce all sensible phenomena to a small number of abstract principles, based on reason only. The physical history of the universe, whose exposition I attempt to develop, does not pretend to rise to the perilous abstractions of a purely rational science of nature. . . . The unity which I seek to attain in the development of the great phenomena of the universe is analogous to that which historical composition is capable of acquiring.

What Humboldt offered instead was a tapestry of "facts" achieved by a process he called "rational empiricism," or "the results of the facts registered by science, and tested by the operations of the intellect." In Humboldt's treatment, the world's data will not reduce to, but will whelm over, the smooth oscillations of the polar metaphor.[28]

But Humboldt was up against not merely polarity as a powerful structural metaphor, but polarity as *the* organizing technology, the key "idea" that "organized" nature into "rational" science. Ironically, Arnold Guyot declared himself to be Humboldt's disciple. He trumpeted his association with the great Humboldt at every turn but rejected Humboldt's merely historical continuity for one structured rationally across the endless dualisms of polarity. Humboldt's form of "geo-graphy" or Earth-writing was an attempt to let the particulars of the Earth write through him, whereas Guyot's geography writes the Earth, such that it mirrors, not just "man," but prosperous, white, American males. For polarity effectively colonizes the world, by "decomposing" its apparently unified forms, breaking them down, transforming and assimilating them according to the new polar logic. It was by this logic that Coleridge had further refined the image of the globe, evoked by terms like *poles* and *equator,* by distributing the globe across electromagnetic coordinates and mapping the "four elemental forms of power" onto the cardinal points of the compass. What Coleridge and Guyot created was a nature fully colonized by mind, an image of global totality. The only way to escape this polar system would be to leave the planet altogether.

There are two kinds of people, the old joke goes, those who divide the world in two and those who don't. The polar triad guarantees unity by building unity into its dual premise: first there is unity, but unity demands duality, and duality demands synthesis. As Coleridge said, "In life, and in the view of a vital philosophy, the two component counter-powers actually interpenetrate each other, and generate a higher third, including both the former." This way of establishing unity was, as Kant declared, a significant departure from the mechanical universe of atoms suspended in a void, for it made unity a function of a "dynamic" process. Motion and change became the conditions of existence, and polarity became the engine of change, for it initiated the unceasing circulation: excess expands into lack, lack draws to itself excess, seeking an equilibrium so precarious that it destabilizes in the very instant it is realized. Force was relocated from exterior to interior. Newton's impenetrable atoms were dissolved, and the Enlightenment's impenetrable bodies were opened up to the gaze, to all gazes. Life would be understood not by arranging its myriad forms on taxonomic grids, but by gridding the life force itself. Surface yielded to interior; anatomy yielded to physiology; the plenitude of living things to the single Law of Life.[29]

In sum, whereas theories of polarity subordinated or ignored what Humboldt called the world's "accidental individualities" and "essential variation of the actual," they enabled vitality, animation, motion, circulation, connectivity, exchange—all the possibilities implied by the fundamental economy of plus and minus, excess and lack, power and need. Kant's new dynamism set the world to music—a dance of measure, rhythm, and harmony—in ways that poets like Coleridge and Emerson would explore and that the physical geographer Arnold Guyot would extend to the cause of global harmony under the aegis of American dominance. Polarity made it possible to see the world through the divisions suggested by phenomena such as the magnet and electricity, to "organize" those divisions as inequalities, and to set the world in motion accordingly, as excess rushed to lack and lack attracted excess, opening the body and the world to each other. Kant predicted that his idea would be "productive." This must stand as the understatement of the age—Kant had reduced the very principle of production to nothing less than a law of science. To be sure, no one "invented" modernism. But once Kant showed the world how to look at nature in terms of polar forces, the cognitive formula that is modernism—purification into the poles of nature and society, translation into the equatorial zone of "knowledge"—now had all the authority of science, including all the science that Kant's "productive" theory might make possible.[30]

The metaphor is self-contained, its consequences packaged in its premise: polar exclusion demands and enables equatorial synthesis, in a progressive cycle that has the power to break down older structures, transform them, and assimilate them to itself. It "looks" unity into the world. One source of instability arose within natural science, as working scientists found their empirical methods insufficiently "rational" to count as proper science. Such scientists found themselves needing either to defend their enterprise as good science (like Humboldt) or to redefine science altogether (like Thoreau).

Another source of instability arises when "objects" turn into subjects yet find themselves nevertheless excluded, as subjects, from the synthesizing and universalizing gaze. The centrifugal power of polar exclusion spins such voices out of science into the "subjective" realm of literature. Frederick Douglass, for example, recalled his bewilderment at the mysterious power the white man had to subjugate the black man. Even as Douglass was famously writing his way to freedom, Guyot was projecting the image of the African Negro as the manifestation of the scientific principle that harmony requires the lesser to be servant to the greater. Guyot had ended his book with a vision of "the great tree of humanity, which is to overshadow the whole earth. It germinates and sends up its strong trunk in the ancient land of Asia. Grafted with a nobler stalk, it shoots out new branches, it blossoms in Europe. In America only, it seems destined to bear all its fruits." A few years earlier, in 1833, William Apess, a Native American of mixed ancestry adopted into the Pequot tribe, tried to refract this self-mirroring gaze in "An Indian's Looking-Glass for the White Man": "I ask: Is it not the case that everybody that is not white is treated with contempt and counted as barbarians?" To counter this truth, Apess discounts the "principle" that attached deep moral distinctions to surface differences: "but stop, friends—I am not talking about the skin but about principles. I would ask if there cannot be as good feelings and principles under a red skin as there can be under a white." His conclusion is barbed: "By what you read, you may learn how deep your principles are. I should say they were skin-deep."[31]

The principle of polarity erected an arborescent structure that distributed relationships vertically while joining all "organically" at the base, and it is here, Apess implies, that the ax must be wielded: to bring true peace to the Union, one must do up the wounds of the poor Indians, and "stop not till this tree of distinction shall be leveled to the earth, and the mantle of prejudice torn from every American heart." Emerson would follow polarity out along the lines first suggested by Coleridge, but when the principles called for the rational subordination of the lesser races, Emerson's armchair theorizing collided with the political realities of a United States dividing across its own equatorial center into the polar opposition of North and South. Despite all efforts to hold them together, the poles of the universe *were* coming apart. In the evacuated center was the self-evidently impossible figure of the American African. Could principle after all be deeper than skin? Did polarity hold as well a solution to the problem of America?[32]

"THE MIND GOES ANTAGONIZING ON": POLARITY IN EMERSON

From his earliest years, polarity became Emerson's primary organizing concept and his major port of entry to the concepts of science. The pattern was in place soon after he began keeping his journal, when he meditated on the fact that "contrast is a law" of mind and world, as well as on the coordinate facts that man is "a golden link in the system of things to unite unlike orders," a being mixed of "the Angel and the brute" subject everywhere to "Nature's . . . equal and pervading Compensations." His original, foundational concepts of Compensation, by which cause and effect were hinged together as two necessary phases of the same event, and Correspondence, the analogical system by which mind and nature mirrored each other, were swiftly captured and adapted into the more evidently "scientific" language of polarity. If, as Barbara Packer says, correspondence allowed Emerson "to transform nature into a text," it is science that taught how that text was to be read. Polarity became Emerson's central trading zone or site of exchange between science, poetry, and society, the universe's physical, aesthetic, and moral properties. As he proclaimed in "The American Scholar," "let me see every trifle bristling with the polarity that ranges it instantly on an eternal law; and the shop, the plough, and the leger, referred to the like cause by which light undulates and poets sing," and there will be no trifle and no puzzle, just the one design that "unites and animates the farthest pinnacle and the lowest trench."[33]

Polarity places man in the center, not as the imperial pinnacle to whom all things bow and around whom all nature orbits, but as the "golden link"—or, later, that "golden impossibility"—hurled into being by the contending energies of warring opposites, a marriage of irreconcilable differences. Life in this universe is by definition less a harmonious balance than a precarious instability perpetually on the verge of collapse. To live in such a world is to be whirled up into a perilous dance that cannot be suspended lest one be crushed by the jaws of fate—like the sailors whose constant attention just manages to keep the ship off the rocks. Gnomic science seemed to promise safe harbor, protection from the catastrophic ravages of chance. "There is no leap—not a shock of violence throughout nature. Man therefore must be predicted in the first chemical relation exhibited by the first atom. If we had eyes to see it, this bit of quartz would certify us of the necessity that man must exist as inevitably as the cities he has actually built." Yet the very conditions that open the vision of that safe harbor—the generative single law of polarity—bar us from entering it, for to enter is to be trapped by determinism. That first atom must already contain our unswerving fate. So freedom holds us offshore, perpetually tacking in the wind: "The mind goes antagonizing on," sighed Emerson in "Experience." From the late 1830s on he would work out in his essays the permutations of the central paradox set by polarity.[34]

One of the themes Emerson explored in his early essays is the circular nesting of cause and effect. On April 20, 1838, Emerson awoke from a night of "ill dreams" thinking about how monsters and fantasies do not defy but "show the law" by their "double consciousness": "They make me feel that every act, every thought, every cause, is bipolar & in the act is contained the counteract. If I strike, I am struck. If I chase, I am pursued. If I push, I am resisted." In the same way, "Cause & effect are two sides of one fact," he wrote in 1840; and again in 1841, "The whole game at which the philosopher busies himself" is, given one side, to find the other. "Life is a pitching of this penny,—heads or tails," he added in revising this passage for "Montaigne." "We never tire of the game, because there is still a slight shudder of astonishment at the exhibition of the other face." Emerson, at least, never tired of it: "Some play at chess, some at cards, some at the stock exchange. I prefer to play at Cause & Effect."[35]

But of course it was far more than a game. Already in 1838 he was growing worried that his circle of sympathizers failed to understand the rules, and so fell victim to the very disease they were trying to heal: "Even the disciples of the new unnamed or misnamed Transcendentalism that now is, vain of the same, do already dogmatize & rail at such as hold it not, & can not see the worth of the antagonism also." Certainly *he* was willing to antagonize, as he showed with Frederick Henry Hedge a year later: "If as Hedge thinks I overlook great facts in stating the absolute laws of the soul; if as he seems to represent it the world is not a dualism, is not a bipolar Unity, but is *two*, is Me and It, then is there the Alien, the Unknown, and all we have believed & chanted out of our deep instinctive hope is a pretty dream."[36]

Here the stakes of the game are clarified. What Emerson is arraying his own forces against is not a monism or a one-sidedness that refuses to flip the penny around, but a dualism that refuses to see that the two sides finally, truly, are two sides of the *same* coin, a "bipolar unity." Belie that, and Emerson foresees his entire system evaporating into a "pretty dream," for Hedge's alternative admits something Emerson has systematically ruled out: "the Alien, the Unknown," the excluded Other that instead of playing back into the bipolar dance will undo it altogether. Taking up the argument again, Emerson defends his philosophy against an unknown assailant: "We are not Manichaeans, not believers in two hostile principles but we think evil arises from disproportion, interruption, mistake of means for end. Is Transcendentalism so bad? And is there a Christian, or a civilian, a lawyer, a naturalist or physician so bold as not to rely at last on Transcendental truths?" Where others were seeing Emerson himself as a sort of Alien or Unknown, he knew himself to be implanted in the heart of the civil state, together with the lawyers, scientists, and doctors. In the same period he erupts in his journal with another complaint: "I resent this intrusion of *alterity*. That which is done, & that which does, is somehow, I know not how, part of me." Whence can alterity come if all that is said and done is, in the most profound sense, a reflection of himself?[37]

And yet self-reflection, too, is arguable. In 1838 Emerson set himself to "Write the Natural History of Reason. Recognize the inextinguishable dualism." His first entry into this field was the essay "History," in which he hastened to show that even polarity has two sides. Most of the essay works out the ramifications of gnomic science applied to civil history: like natural history, civil history, too, is the outgrowth of the many from the same few laws, infinite effects from simple causes. "In the man, could we lay him open, we should see the reason for the last flourish and tendril of his work, as every spine and tint in the seashell preëxist in the secreting organs of the fish." All the facts of history must, then, preexist in the mind as laws. "The creation of a thousand forests is in one acorn, and Egypt, Greece, Rome, Gaul, Britain, America, lie folded already in the first man." Since in this gnomic mirroring the mind that *wrote* history is the same that must *read* it, Emerson's entire trend seems to be that there is no alterity here: "All history becomes subjective; in other words, there is properly no History; only Biography." We are back again to Emerson's quizzical inquiry into the relationship of part and whole: "The Sphinx must solve her own riddle. . . . as the light on my book is yielded by a star a hundred millions of miles distant, as the poise of my body depends on the equilibrium of centrifugal and centripetal forces, so the hours should be instructed by the ages, and the ages explained by the hours." The hours as we live them may not seem to unfold our life so neatly as the seashell or the acorn. "What is our life but an endless flight of winged facts or events!" Emerson exclaims; yet each hour is a question put by the Sphinx, and those who fail her, she will devour: "Those men who cannot answer by a superior wisdom these facts or questions of time, serve them." Here Emerson's answer is to refuse "the dominion of facts" by holding fast to principle. Then the riddle will dissolve and "the facts fall aptly and supple into their places" as servants around their master.[38]

And so Emerson's facts do fall supple into their places, right up to the end of the essay, when abruptly he snatches it all away. "Hear the rats in the wall, see the lizard on the fence, the fungus under foot, the lichen on the log." What, Emerson asks, does he know of *their* worlds? "What does Rome know of rat and lizard?" Just when the gnomic package is sealed and all but delivered, Emerson calls instead for annals that are "broader and deeper," for "the path of science and of letters is not the way into nature."

Nearer to it than "the dissector or the antiquary" stand the "idiot, the Indian, the child, and unschooled farmer's boy."[39]

Taken seriously, Emerson's turn at the end of "History" would give the lie to most of what he had written thus far. A gnomic law that lets the rats and lizards escape is not particularly compelling, for instead of ordering the universe in its entirety, it merely builds an island of order in a sea of chaos, a beautiful artifact with at best a problematic relation to truth. If science is not the path to nature, then exactly what is it? If pursued consistently, this reading would have given Emerson a much less secure, a much more chancy and contingent universe. The grand architecture of science, or moral and physical law, would have to be seen thenceforward as a temporary stay against unmeaning, a social construction that might suffice for Rome, but hardly for other cities on other roads. Only certain roads would lead to the Roman truth.

Perhaps Emerson tossed aside his central insight so blithely in order to open a pathway to his next essay, "Self-Reliance." On the one hand, self-trust is enabled because, as we have seen, the heart of one is the heart of All, and the truth of the heart does prophesy the universe. On the other, the task of "Self-Reliance" is to unhook that private heart from the dictatorship of society, that "joint-stock company," to encourage Emerson's own impulse not to be bound by law but to write on the lintels of his doorpost, "Whim." Anticipating objections, he deflates them with one of his best-known one-liners: "A foolish consistency is the hobgoblin of little minds, adored by little statesmen and philosophers and divines."[40]

And yet, how inconsistent is Emerson, really? Perhaps there are no inconsistencies if the mind is not little but *great*. Even in "Self-Reliance," at the height of his gleeful antinomianism, he reiterates the organic certainty that "the magnetism which all original action exerts is explained when we inquire the reason of self-trust. Who is the Trustee?" A few lines later, he puts his central faith into deeply eloquent phrasing: "We lie in the lap of immense intelligence, which makes us receivers of its truth and organs of its activity." His very faith that the "magnetic" action of that intelligence will stabilize all dangerous excess is what enables his statement about "hobgoblins." As he would say in 1843, "Botany Bay grows up with a clean conscience. . . . the fact of two poles is universal; the fact of two forces centripetal & centrifugal is universal, & each develops the other. . . . Wild liberty develops iron Conscience; want of liberty strengthens decorous & convenient law." In the end, even the criminal refuse of the British Empire is disciplined by polar law. It is hard to see how the wildness of rats and lizards (or for that matter of children and Indians) can be allowed to destabilize the self-sufficient, balancing interplay of centrifugence and centripetence.[41]

That is, what Emerson is doing in "Self-Reliance" is enlarging the groundwork for deploying higher law as, in Thoreau's terms, the wrench that jams the machine of social injustice. Stanley Cavell observes that the word *whim* on the lintel is whimsical only as one writes it; once one crosses the threshold and lives it, whim becomes fatality, "a matter of my life and death." Obedience to the future in one's heart is difficult and frightening: after all, all society conspires against it. Such an act of faith expects to find God beyond the threshold, not rats and lizards; or if they *are* out there, they, too, must be domesticated. In other words, the Emerson who writes "whim" is not laying open his beautifully ordered system of bipolar unity to deep challenge by irrational whimsy or the radically wild. Given the importance of order to Emerson, though—especially as the decades wore on—it is worth noting that he is willing to flirt, even briefly, with that alien Other that his bipolar system is erected to exclude. As a very young man he consciously made the choice to exclude chance; part of Emerson's integrity and attractiveness as an intellectual is that he never entirely forgot that the exclusion was an act of choice, and that chance was still out there—beyond the pale to be sure, but there, in the wilderness at the edge of the civil domain. Meanwhile the game goes on: "So use all that is called Fortune. Most men gamble with her, and gain all, and lose all, as her wheel rolls." Leave aside those unlawful winnings, "and deal with Cause and Effect, the chancellors of God. In the Will work and acquire, and thou hast chained the wheel of Chance, and shalt sit hereafter out of fear from her rotations."[42]

If, as Martin Eger suggests, science is successful partly because of the metaphors it supports, much of the success of polarity for Emerson was its superb applicability to the doctrine of compensation, which polarity refined, extended, and deepened. The issue was a profoundly personal one for Emerson. As a young man he had spent several grim months shadowed by the apparent likelihood that he was losing his eyesight—a poignant gloss on the line in *Nature* in which Emerson imagines that no disgrace or calamity can befall him "which nature cannot repair," so long as it spares him his eyes. When early in January 1826 he can at last return to his journal "with mended eyes," his opening statement is the comforting thought that "Compensation has been woven to want, loss to gain, good to evil." In his own case, his brush with blindness has opened his inner sight, gaining him greater knowledge and greater courage; and what he has learned is that "all things are double one against another. . . . The whole of what we know is a system of Compensations. . . . Every suffering is rewarded; every sacrifice is made up; every debt is paid." Compensation is nothing less than "the enginery of the moral universe." When it comes time to preach his first sermon, a little over six months later, it is to this insight he turns: if the moral laws of this world have any relation to the laws of the next, we may be sure that "the riches of the future are dealt out on a system of compensations." By 1831 the fundamental reciprocity of the universe finds its expression in Newton's law of gravity: "I believe one law prevails throughout the moral universe. . . . The Earth not only falls to the Sun, but the Sun falls to the Earth. All motion is reciprocal and so all influence spiritual influence." Compensation has thus moved from personal insight, to moral law, to a universal truth that relates moral and physical law through a single system of reciprocity.[43]

Polarity extends this convergence still further, by suggesting not only a logic to calamity—the loss of eyesight, or a brother or a wife—but a much richer language for moral reckoning. In turn, the concept of compensation activated the scientific doctrine of polarity, animating it with a moral valence that not only connected the realms of mind and matter but put the entire system into motion as cause and effect, action and reaction, stroke and return. The opening of the main body of "Compensation" presents the completed system:

> Polarity, or action and reaction, we meet in every part of nature; in darkness and light; in heat and cold; in the ebb and flow of waters; in male and female; in the inspiration and expiration of plants and animals; in the equation of quantity and quality in the fluids of the animal body; in the systole and diastole of the heart; in the undulations of fluids, and of sound; in the centrifugal and centripetal gravity; in electricity, galvinism, and chemical affinity. Superinduce magnetism at one end of a needle; the opposite magnetism takes place at the other end. If the south attracts, the north repels. To empty here, you must condense there. An inevitable dualism bisects nature, so that each thing is a half, and suggests another thing to make it whole: as spirit, matter; man, woman; odd, even; subjective, objective; in, out; upper, under; motion, rest; yea, nay.
>
> Whilst the world is thus dual, so is every one of its parts. The entire system of things gets represented in every particle.[44]

As a steel filing caught up in a magnetic field becomes itself a miniature magnet, so does every bit of the universe reproduce the system of the whole, placing polarity at the heart of Emerson's gnomic universe. At every point and across all scale levels nature "keeps her balance true": organic nature, mechanical forces, the individual in society, politics, cities and nations: "the universe is represented in every one of its particles. Every thing in nature contains all the powers of nature. . . . The world globes itself in a drop of dew." As fugitive as the dewdrop, as eternal as the pulse of water through the veins of the Earth: "Thus is the universe alive. All things are moral."[45]

To be alive is to be thrust, willy-nilly, into this relentless system of compensation, which up to this point in the essay has seemed to cradle us in goodness. Soon, however, it shows a sterner side: to be alive is to be, alas, "not quite invulnerable," like Achilles, whose heel had not been washed by the sacred waters. Just so, "There is a crack in every thing God has made." Not even the holiday of "wild poesy" can "shake itself free of the old laws,—this back-stroke, this kick of the gun, certifying that the law is fatal;

that in Nature, nothing can be given, all things are sold." Seen from sufficient distance, nature's all-keeping balance looks like a gentle embrace, but
close up this is an unforgiving, even "vindictive," transactional universe in
which nothing is got for nothing. Every chain has two ends. Put one end
around the neck of a slave, and the fatality of the law will fasten the other
end "around your own." Even in projects of beauty, the law of nature over-
masters our will: "We aim at a petty end quite aside from the public good,
but our act arranges itself by irresistible magnetism in a line with the poles
of the world." Emerson's moral toughness here looks ahead to the merciless
universe of "Fate," in which the overmastered will reacts by building altars
to the "Beautiful Necessity," turning obedience into that tougher power
geared to ensuring that the whole universe is pulling *your* end of the
chain.[46]

The toughness that can repay a spell of blindness with gratitude returns
at the end of "Compensation," when Emerson reveals how much he needs a
universe in which polar compensation exchanges loss for gain. Physical and
moral law join to offer a "natural history of calamity," in which changes
advertise "a nature whose law is growth." The soul, like the nautilus, crawls
"out of its beautiful but strong case, because it no longer admits of its
growth, and slowly forms a new house." Insofar as we are lazy or resistant,
"this growth comes by shocks"—shocks like "a fever, a mutilation, a cruel
disappointment, a loss of wealth, a loss of friends," which at the time seem
unpaid and unpayable. "But the sure years reveal the deep remedial force
that underlies all facts. The death of a dear friend, wife, brother, lover,
which seemed nothing but privation, somewhat later assumes the aspect of
a guide or genius," breaking up one's old life and permitting the formation
of a new one. Perhaps thinking of himself—he who has lost his father, his
wife, his career, and two beloved and talented but overshadowing brothers—he ends the essay by speculating that "the man or woman who would
have remained a sunny garden flower, with no room for its roots and too
much sunshine for its head, by the falling of the walls and the neglect of the
gardener, is made the banian of the forest, yielding shade and fruit to wide
neighborhoods of men." Emerson is not about to offer his public anything
he has not already bought with his own blood.[47]

Yet Emerson's faith in the compensations of calamity barely survived the
sudden and shocking death of his young son Waldo of scarletina, on January 27, 1842. Fate dealt the Emerson circle a double blow: just two weeks
earlier, Henry Thoreau (who was then living with the Emersons) had lost
his beloved and only brother, John, to lockjaw, an equally shocking, sudden, and senseless loss.

Ten years before, after the death of Ellen, Emerson had turned to the
truth of the universe for comfort. On February 4, a little over a week after
Waldo's death, Emerson once again strikes and stands on "Truth": "Twice
today it has seemed to me that Truth is our only armor in all passages of life

& death. Wit is cheap & anger is cheap, but nothing is gained by them. But if you cannot argue or explain your self to the other party, cleave to the truth against me, against thee, and you gain a station from which you can never be dislodged. . . . I will speak the Truth also in my secret heart, or, *think the truth* against what is called God." Tradition, he continues, may force us to deny what we hold sacred, but he must *resist* tradition "& say, 'Truth against the Universe.' " Having reinscribed the standard under which he will continue his forward march, Emerson considers just where it is, in this newly diminished world, that one carries this standard: "Does it not seem as if the middle region of our being were the only zone of life & thought? We may ascend into the region of pure geometry, lifeless science, or descend into that of sensation. Between these extremes is the equator of life, of thought, of spirits, of poetry; a narrow belt." Perhaps most movingly, shortly after this passage he notes one of the middle region's most bitter ironies:

> The chrysalis which he brought in with care & tenderness & gave to his Mother to keep is still alive and he most beautiful of the children of men is not here.
> I comprehend nothing of this fact but its bitterness. Explanation I have none, consolation none that rises out of the fact itself; only diversion; only oblivion of this & pursuit of new objects.[48]

He did have many diversions, many new objects to pursue. Most immediately, Margaret Fuller's resignation as editor of *The Dial* left a vacuum into which Emerson reluctantly stepped; he would see it through another eight issues, until it ceased publication in April 1844. One of his first acts as editor, after noting how reading in natural history "soothes & heals us," was to assign Thoreau the task of reviewing a stack of natural history books, a task that helped Thoreau climb out of his own grief and reconnect with the outer world. Emerson himself continued to lecture and write, completing his second volume of *Essays* for publication in October 1844.[49]

"Experience," the centerpiece of that book, nearly undermines the confident assertions of "Compensation," for now Emerson found that he lacked the facility to turn this newest loss into an opportunity for growth. As he writes, "This onward trick of nature is too strong for us: *Pero si muove*"— but still it moves; Galileo's words, uttered even while yielding to the law of the church. As Galileo insisted, the laws of the cosmos override human desire. The Earth *does* move, though all humanity cry out against it, even as Emerson cried out against the chrysalis that outlived his son. The brutality of this fact had left him feeling numbed and alienated, and in "Experience" he acknowledges (with an equally brutal honesty) his inability to mourn Waldo's death, to incorporate it into his developing selfhood: "The only thing grief has taught me, is to know how shallow it is. . . . Grief too will made us idealists. In the death of my son, now more than two years ago, I

seem to have lost a beautiful estate,—no more. I cannot get it nearer to me."
This calamity, he concludes, "does not touch me: something which I fancied
was a part of me, which could not be torn away without tearing me, nor en-
larged without enriching me, falls off from me, and leaves no scar. It was
caducous. I grieve that grief can teach me nothing, nor carry me one step
into real nature." The "crack" in nature entertained so wittily in "Compensation" has yawned open and swallowed him up; it takes the whole of the
essay for Emerson to identify this grim new mood and teach himself how to
turn this, too, to advantage.[50]

"The Method of Nature" had celebrated nature's endlessness, before
which human limitations can only surrender to transcendent power, passion simmering to truth. The organic body, frail and traitorous, dissolves
into gnomic power. Nothing is safe from these widening circles of power—
certainly not science, which models the ascension through facts to ever-
higher laws; "There is not a piece of science, but its flank may be turned to-
morrow," notes Emerson. Soon the entire familiar world has been
destabilized and begins to pitch and roll: "All that we reckoned settled,
shakes and rattles; and literatures, cities, climates, religions, leave their
foundations, and dance before our eyes." Any mode of thought standing
outside the old circle will "afford us a platform whence we may command
a view of our present life, a purchase by which we may move it"—literature, wild nature, "the din of affairs," a high religion, or even science will all
serve. The point is to step outside the old circle and shake loose all within it,
which Emerson avows is his aim: not "to settle anything as true or false,"
but to send all such polar categories back into circulation: "I unsettle all
things. No facts are to me sacred; none are profane; I simply experiment, an
endless seeker, with no Past at my back."[51]

Now, unsettled and unmoored, his own flank turned, Emerson must
learn how to live in the universe he has created. "Experience" complicates,
even critiques, the gnomic science characteristic of his earlier writings.
What once seemed the key to the universe, unlocking endless possibility,
now swings shut like the jaws of a trap—although, as Emerson reminds his
readers at the end, no essay, and certainly not this one, should be taken as a
definitive statement of achieved truth. "I am a fragment, and this is a fragment of me," he cautions. He has his heart set here on "honesty in this
chapter," and in his current mood it seems to him that a full understanding
of the order of the universe is beyond us.[52]

"Where do we find ourselves?" Emerson opens. We are no longer ascending a hopeful ladder, but stranded on a mysterious staircase: "there are
stairs below us, which we seem to have ascended; there are stairs above us,
many a one, which go upward and out of sight." Instead of grasping the
poles of the universe, we are lodged in a series somewhere between extremes "which we do not know" and in which we do not believe. No longer
asking after the end of nature, Emerson is now asking, as David Van Leer

says, "an altogether more primitive question": "Where are we?" Emerson echoes the Sphinx's complaint that we have been drugged by Lethe, and "Sleep lingers all our lifetime about our eyes." Unmoored and adrift in what he once, in "Circles," had celebrated as nature's "resistless menstruum," Emerson seems no longer free but trapped. In this state, "Dream delivers us to dream, and there is no end to illusion. Life is a train of moods like a string of beads," and those beads are strung on the "iron wire" of temperament. The transparent eyeball that once saw all, now sees nothing. We are shut "in a prison of glass which we cannot see."[53]

The death of Waldo in "Experience," as Van Leer argues, is structurally similar to Nature's transparent eyeball—"the introductory moment that the rest of the essay teaches us to understand and accept"—us, and Emerson as well. Grief, an experience, a phenomenon, cannot touch the world of the real, and, for all its momentous consequences, it changes nothing: in a gnomic world, change is but a trick of scale, an illusion, and nothing really dies. Whatever Emerson felt as a human being (and the evidence is that he grieved as fully as any of us), the logic of his prose performance demands not a personal confession of emotion but an analysis of what death means, and Emerson is, indeed, being brutally honest with himself: in the world he has constructed, death means nothing. It cannot touch him. "The price you pay for invulnerability," Barbara Packer observes, "is invulnerability."[54]

Gnomic science was build on Coleridgean method, by which the mind looks order into the world. Such a method assumes the mind of God, exempt from experience; but the terms of self-culture assume that the mind comes into existence at all only by being plunged into the throes of experience. Man as a method looks order into the world, and sees only the world he animates with his vision. Once again, "Experience" reveals the cost of this power: we live in a world we cannot know because our "moods" reflect the only world they can know. The reciprocity of vision and world shuts us in our own self-made crystal ball.[55]

In this mood, what imprisons us is the gnomic nature of the universe. "On the platform of physics, we cannot resist the contracting influences of so-called science." Emerson's immediate target is phrenology, which read the shape of the spirit in the form of the skull, a reductionism that always repelled Emerson; but the scope of his critique is wider: "I see not, if one be once caught in this trap of so-called sciences, any escape for the man from the links of the chain of physical necessity. Given such an embryo, such a history must follow." Against "the chain of physical necessity," by which the individual must unfold from preexisting and predetermined law, Emerson opposes a "creative power" that jams its foot in the trapdoor so that it never completely closes. What has changed? Before, power was intrinsic to that embryonic unfolding; now, a bitter contest is joined. The role of science becomes, not just to offer encouragement to a process unfolding naturally, but to oppose a brute reality against the force of desire.[56]

Emerson's inability to be touched by Waldo's death recalls to him the atomic theory of the eighteenth-century physicist Ruggiero Giuseppe Boscovich: "Was it Boscovich who found out that bodies never come in contact?" Or at best, as recent experiments in static electricity had shown, they come in contact in the most limited way: "Two human beings are like globes, which can touch only in a point, and, whilst they remain in contact, all other points of each of the spheres are inert." Science, then, offers no help to the bleak reality that the world eludes our touch, refuses our longing for contact: "I take this evanescence and lubricity of all objects, which lets them slip through our fingers then when we clutch hardest, to be the most unhandsome part of our condition." Nature will not be grasped, and to close the gap between mind and matter by force is only to open it all the wider. In this version of polarity, opposites do not attract but resist each other, the poles misaligned.[57]

The effect of pushing gnomic science to this limit is, for Emerson, to realize that life is not lived by its reductive logic. The organic body rebels and declares its own possibilities of expansion. In David Robinson's words, "Emerson locates an axis on which to turn the doctrine of temperament against itself," countering determinism "by affirming the unknown as a source of the possible," converting the limits of knowledge to its liberation. "The very *problem* of not knowing . . . has become the *solution* of not knowing"—and, with this insight, the mood of the essay changes. What emerges is a new voice, a solution, if not resolution: criticism is useless. "Life is not dialectics. . . . Do not craze yourself with thinking, but go about your business anywhere." If we cannot grasp the poles of the universe, then we must set up housekeeping in the "midworld"—the "highway," the "middle region," the "temperate zone"—which, after all, "is best."[58]

It is the solution he had reached in February 1842, so soon after Waldo's death, that actual life must lie in the "middle region" or "equator" between the poles of "pure geometry, lifeless science," and "sensation." Ultimately, as science itself had shown, there was nowhere else to live: "Then the new molecular philosophy shows astronomical interspaces betwixt atom and atom, shows that the world is all outside: it has no inside." Yet as Heraclitus said, "Even here the gods are present"—or, as Emerson said, "And yet is God the native of these bleak rocks." Again, limit is leveraged into power, possibility: "We must hold hard to this poverty, however scandalous, and by more vigorous self-recoveries, after the sallies of action, possess our axis more fully." Robinson sees in this crucial turn not just a renewed call for the old "self-trust," but a somber realization that "the price of self-affirmation" is "the irreducible fact of an illusory world. . . . Illusion thus becomes the fundamental ethical challenge of his later work." Again, Emerson turns lemons to lemonade. If the world is "all outside," as the molecular philosophers proclaim, then Emerson's new goal will be performance rather than principle, artistry rather than truth. Emerson emerges with what Robinson terms "a tour de force of triumphant pragmatism": "We live amid surfaces, and the true art of life is to skate well on them." Idealism's "noble doubt" is set aside: "Let us treat the men and women well," Emerson quips; "treat them as if they were real: perhaps they are."[59]

Soon after recording his feeling, in June 1842, "some efficient faith again in the repairs of the Universe," Emerson records a snatch of conversation with Sampson Reed: " 'Other world?' I reply, 'there is no other world; here or nowhere is the whole fact; all the Universe over, there is but one thing,— this old double, Creator-creature, mind-matter, right-wrong.' " Here, or nowhere: as he writes in "Experience," down here in the midworld where we all do our living together, a new entity emerges: the collective. Another new note that enters the journal after the death of Waldo is Emerson's interest in "Masses," which also disrupt the linear clarity of gnomic science. As Emerson observes, "A drop of water cannot exhibit all the phenomena of the sea; a drop cannot exhibit a storm." Instead of a drop of dew, now it takes an entire ocean to globe the world. Similarly, in "Experience" it turns out that men are but partial after all, for no one man can ever "pass or exceed" that certain idea he represents. So, unlike the hope advanced in "The American Scholar," now it must take all men to get One Man: "Of course, it needs the whole society, to give the symmetry we seek. The parti-colored wheel must revolve very fast to appear white." Just so, the individual can no longer hope to live absolutely in one stunning moment, but must gather himself up in the collectivity of moments lived—a disappointment when we observe how "unprofitable" all our actual days are while they pass; but on the other hand, out of that seeming oblivion something new and marvelous emerges. "The results of life are uncalculated and uncalculable. The years teach much which the days never know." The lesson is, strikingly, that "the individual is always mistaken." Among all the designs, alliances, quarrels, and blunders, "something is done; all are a little advanced, but," he reiterates, "the individual is always mistaken."[60]

To find oneself at the midpoint of extremes vanishing in the distance is still to be conditioned by polarity, only within a narrower range. Man goes from being, in the Greek ideal, a "golden mean," to a "golden impossibility. The line he must walk is a hair's breadth." To walk that tightrope line is to focus the dual energies of polarity within the tight beam of the moment lived. There is no machinery for managing this, no "perfect calculation of the kingdom of known cause and effect." Rather, "Nature hates calculators; her methods are saltatory and impulsive. Man lives by pulses; our organic movements are such; and the chemical and ethereal agents are undulatory and alternate; and the mind goes antagonizing on, and never prospers but by fits. We thrive by casualties."[61]

Yet these pulses or undulations or fits or casualties are emphatically *not* the basis for an apologetic for chance. Emerson reins in the implications of his language: "The ancients," he acknowledges, "struck with this irreducibleness of the elements of human life to calculation, exalted Chance into a divinity"; but for Emerson, this is to mistake the single spark for the principle of fire. Because an underlying order cannot be seen hardly means it is not there. For evidence, Emerson turns this time to organic nature: "In the growth of the embryo, Sir Everard Home, I think, noticed that the evolution was not from one central point, but coactive from three or more points. Life has no memory." What Emerson means is that no one point knows or remembers what the other two are doing; rather, all three are "ejaculated from a deeper cause" of which they are unconscious. Similarly, we who are "now skeptical or without unity" because immersed in "forms or effects" apparently at odds with each other must be patient. "Bear with these distractions, with this coetaneous growth," he advises; "they will one day be *members*, and obey one will. . . . Underneath the inharmonious and trivial particulars, is a musical perfection, the Ideal journeying always with us, the heaven without rent or seam."[62]

The problem arises, unsurprisingly, with the nature of man. Unlike the rest of organic nature, human beings are aware that they exist, and this self-consciousness divides humanity from all nature. Thus, in one of Emerson's most depressive utterances: "It is very unhappy, but too late to be helped, the discovery we have made, that we exist. That discovery is called the Fall of Man. Ever afterwards, we suspect our instruments. We have learned that we do not see directly, but mediately, and that we have no means of correcting these colored and distorting lenses which we are, or of computing the amount of their errors. Perhaps these subject-lenses have a creative power; perhaps there are no objects." The steep inequality between subject, "the receiver of Godhead," and object, which after all may not even exist, means that "thus inevitably does the universe wear our color, and every object fall successively into the subject itself." This has been the burden of the essay from the start, but now this limitation, too, becomes a powerful instrument: "The partial action of each strong mind in one direction, is a telescope for the objects on which it is pointed." And the corrective for this limitation is, again, collectivity: "But every other part of knowledge is to be pushed to the same extravagance, ere the soul attains her due sphericity." If consciousness, the medium or instrument through which we see, distorts, then the combined vision of many such instruments will approach a full and rounded view. Meanwhile, despite the glancing suggestion that the universe may lack "objects" altogether and so collapse back into total subjectivity, in the end Emerson reclaims the physical object, the outer world: "A subject and an object,—it takes so much to make the galvanic circuit complete, but magnitude adds nothing."[63]

From the brink of monism, polarity pulls us back to wholeness: only now the poles of the universe are not located at the extremes beyond our vision, as on an endless staircase, but are inscribed here and now within the act of perception itself—*seeing*, or knowledge, defined, as Coleridge had said, as the coincidence of subject and object at the midpoint or "equator" of the world, a binocular vision that restores us from blindness to "perfect sight." By essay's end, Emerson has regained his formula for redemption, for successful living even, or especially, within the constraints of a world stranded between unapproachable extremes. Thus the conclusion of "Experience" rewrites the end of *Nature:* "Build your own world"—chastened, but rejuvenated nonetheless. Amidst the household minutiae of the midworld, the "sanity" of solitude will give every man "revelations, which in his passage into new worlds he will carry with him. Never mind the ridicule, never mind the defeat: up again, old heart!—it seems to say,—there is victory yet for all justice; and the true romance which the world exists to realize, will be the transformation of genius into practical power."[64]

The "crack" that opened in "Compensation" does not, then, split the universe apart, for it is healed by the discipline of knowledge, a proactive engagement in the world that may, thanks to the contingencies of mundane life, be sporadic and occasional but that even so will still save us from the skepticism and unmeaning that almost seemed the lesson of "Experience." There, the polar vision that seemed about to crack the universe in two is retooled as the instrument that sutures the very cut it has made. If consciousness has expelled man from the Garden, knowledge will restore him to his once and future kingdom.

After the meditative dazzle of "Experience," the essay "Nature" whisks us back from the yeasty and polluted midworld to an earlier age, that of the majestic harmony of a nature unsullied by history, church, or state, in whose beauty we "nestle" affectionately. The primary mission of this new return to an old subject (this essay was intended originally as the conclusion for *Essays, First Series*) is to carry over the bipolar distinction made by Schelling and Coleridge between *natura naturata* and *natura naturans,* nature passive and nature active—object and law. The former is the nature Emerson opens his window to in the morning: physical, existing, material nature. For all its wonder, he knows his reader will quickly tire of nature description—a symptom, after all, of our malaise: "If there were good men, there would never be this rapture in nature. If the king is in the palace, nobody looks at the walls." So Emerson turns quickly to *natura naturans,* Kant's true science, "efficient nature . . . the quick cause, before which all forms flee as the driven snows, itself secret, its works driven before it in flocks and multitudes." This is nature *active,* the creative force of which the thing created is, in Schelling's words, "the mere body or symbol." Or, as Emerson says, far more vividly, "It publishes itself in creatures, reaching from particles and spicula, through transformation on transformation to the highest symmetries, arriving at consummate results without a shock or a leap."[65]

Applying Schelling to his recent reading in Lyell's *Principles of Geology,* Emerson explains: "All changes pass without violence, by reason of the two cardinal conditions of boundless space and boundless time. Geology has initiated us into the secularity of nature, and taught us to disuse our dame-school measures, and exchange our Mosaic and Ptolemaic schemes for her large style." The transformative series from rocks to lichens to the trilobite and quadruped and finally "to Plato, and the preaching of the immortality of the soul" seems "inconceivably" long, "Yet all must come, as surely as the first atom has two sides." Polarity appears once again as the generative law of nature with its gnomic lesson: "Motion or change, and identity or rest, are the first and second secrets of nature: Motion and Rest. The whole code of her laws may be written on the thumbnail, or the signet of a ring." Given matter and the "aboriginal push," the harmony of "the centrifugal and centripetal forces" is generated, and the action of "that famous aboriginal push propogates itself through all the balls of the system, and through every atom of every ball, through all the races of creatures, and through the history and performances of every individual."[66]

Finally, though, in a turn toward the bewildered opening of "Experience," that aboriginal push and its onward course leaves us dissatisfied and unhappy. Nature mocks us, "leads us on and on, but arrives nowhere, keeps no faith with us." The very woods and waters entice and flatter us, but "disappointment is felt in every landscape. . . . Nature is still elsewhere." Where *is* nature? Not in the woods, nature passive, but in nature creative. Once we look for nature not in the woods but in the aboriginal creative principle, we, the king restored to his palace, will not need but *be* nature: we shall find "the fathomless powers of gravity and chemistry, and, over them, of life, preëxisting within us in their highest form."[67]

"Every man is wanted," Emerson remarks, "and no man is wanted much." This line from "Nominalist and Realist" brings the reader back fully into the grim universe of "Experience," and the difference is instructive. In contrast with "Nature," "Nominalist and Realist" sets up a bipolar unity not in order to empower one side over the other, but in order to set both sides in motion around each other. The classic debate referred to in his title had set Realists, who believed that general ideas "really" exist (that is, as preexisting essences that enable us to recognize and know the objects of the world), against Nominalists, who held that universals were only names, convenient labels invented by philosophers, with no "real" existence. Emerson, who called general ideas "our gods," had clear loyalties to the Realists, and to make their case persuasive he turned to one of his favorite metaphors:

The magnetism which arranges tribes and races in one polarity, is alone to be respected; the men are steel-filings. Yet we unjustly select a particle, and say, "O steel-filing number one! what heart-drawings I feel to thee! what prodigious virtues are these of thine! how constitutional to thee, and incommunicable!" Whilst we speak, the loadstone is withdrawn; down falls our filing in a heap with the rest, and we continue our mummery to the wretched shaving. Let us go for universals, for the magnetism, not for the needles.[68]

As Van Leer observes, this passage burlesques Emerson's earlier claim to have found general power in the individual genius. It seems another version of the argument in "Nature": let us go for the creative force, for the soul of the workman and not the work itself. Yet here Emerson recognizes our divided loyalties, our "mummery" to our fellow shavings: "We are amphibious creatures, weaponed for two elements, having two sets of faculties, the particular and the catholic." It is one thing to settle the question "in our cool libraries" that we may "flout the surfaces." "But this is flat rebellion. Nature will not be Buddhist: she resents generalizing, and insults the philosopher in every moment with a million of fresh particulars." Even as the insulted philosopher reacts he only distributes himself into his "class and section. You have not got rid of parts by denying them, but are the more partial. You are one thing, but nature is *one thing and the other thing,* in the same moment." This is the nature who ordains that we cannot live by "general views" but must "fetch fire and water, run about all day among the shops and markets, and get our clothes and shoes made and mended," and if the victim of these details is lucky, he may once in a fortnight "arrive perhaps at a rational moment."[69]

Again, the answer lies in the collective. Nature solves the problem by distributing her powers into the multitude, setting up a great rotation "which whirls every leaf and pebble to the meridian" such that we all take turns at the top; " 'Your turn now, my turn next,' is the rule of the game. The universality being hindered in its primary form, comes in the secondary form of *all sides:* the points come in succession to the meridian, and by the speed of rotation, a new whole is formed." Man is again placed in the middle, celebrating a marriage that has by now spun out of control: "The end and the means, the gamester and the game,—life is made up of the intermixture and reaction of these two amicable powers, whose marriage appears beforehand monstrous, as each denies and tends to abolish the other." We will never reconcile such contradictions or make a sentence that "will hold the whole truth, and the only way in which we can be just, is by giving ourselves the lie"—flipping the coin of every statement, just as he did with Sampson Reed. "Speech is better than silence; silence is better than speech;—All things are in contact; every atom has a sphere of repulsion;—Things are, and are not, at the same time;—and the like. All the universe over, there is but one thing, this old Two-Face, creator-creature, mind-matter, right-wrong, of which any proposition may be affirmed or denied." The upshot is that every man is both part and whole, contributing his fragmented individuality to the great circle, and working out on his own single pulse the universal problem. Emerson has seen his deepest metaphors through to the end. We are both organic *and* gnomic, and each simultaneously, "*one thing and the other thing* in the same moment," living along the Möbius loop of a duality with only one side, single nature with a double face, a monstrous marriage that cannot be escaped.[70]

Magnetism, Emerson's favorite figure for this two-faced unity, is perhaps his oldest and longest-lived scientific metaphor. The shifts in its use track the shifts in his thought. In summer of 1827 he recorded seeing "a skillful experimenter lay a magnet among filings of steel & the force of that subtle fluid entering into each fragment arranged them all in mathematical lines & each metallic atom became in its turn a magnet communicating all the force it received of the loadstone." A few months later, while drafting a letter to Mary Moody Emerson, the image received its spiritual reading: "The soul has a divine power of assimilating all its acquisitions to its own nature. If it is weak & little, events over bear it, and give it their own hue & complexion. If it be great it converts them with instantaneous magic to its own predominant character. It converts calamity to knowledge; knowledge to power; hope to happiness." Another application surfaced the following year as one of Emerson's own aphorisms: "A wise man in certain society is a magnet among shavings." As Eric Wilson details, in 1828 Emerson began reading in Sir Humphry Davy, where he found the exciting notion that all matter might consist of corpuscles endowed with positive and negative poles. In 1829 he began to work out in his sermons a more thorough reading. In Sermon 43, for instance, he explained the spiritual analogy to his congregation:

> Many of you have witnessed the common experiment of the loadstone. If you introduce a magnet into a heap of steel-filings the rubbish becomes instantly instinct with life and order. Every particle at once finds its own place, is found to have a north and a south pole, and all arrange themselves in regular curves about the magnet. The same is true of the earth itself in the common theory of magnetism; the globe is a great magnet. . . . The mind is that mass of rubbish . . . until its hidden virtue is called forth when God is revealed.

In Sermon 48, he added a new dimension: one needle in one locale may point to "some local disturbing cause," but innumerable trials "in every degree of latitude and of longitude" will show the true "magnetic axis of the globe." Truth, then, became a kind of collective consensus, the averaging of all the needles.[71]

The image surfaced again in *Nature,* also as a way to relate the individual to the truth of nature: "The ruin or the blank, that we see when we look at nature, is in our own eye. The axis of vision is not coincident with the axis of things, and so they appear not transparent but opake." Here magnetic polarity coincided with contemporary Newtonian optics, which ex-

plained the puzzling ability of transparent objects to both transmit and reflect light. In Herschel's words, one should regard "every particle of light as
a sort of little magnet revolving rapidly about its own centre while it advances in its course, and thus alternately presenting its attractive and repulsive pole." When the light particle arrives with its repulsive pole foremost, it
is repelled and reflected; when the attractive pole is foremost, it enters the
surface, rendering the object transparent. Emerson read Herschel in 1831,
and as Barbara Packer notes, Emerson's references to transparency and
opacity proliferate thereafter. Although Packer suggests that "we can only
guess why these images appealed so strongly to Emerson," it seems clear
that aligning the polarity of light with the polarity of matter, as Herschel
suggested, allowed Emerson to dissolve matter into law and thence spirit in
just the way of Newton—an imaginative breakthrough behind all Emerson's metaphors of transparency. Now transparency could apply not only
to nature, but to genius: in October 1836 Emerson noted that "as we advance, shall every man of genius turn to us the axis of his mind; then shall
he be transparent."[72]

By 1837 Emerson was ready to generalize magnetism still further, in the
passage that became the thesis of "Compensation": "Polarity is a law of all
being. . . . Light, shade; heat, cold; centifugal, centripetal; action, reaction.
If the mind idealizes at one end perfect goodness into God coexistently it
abhors at the other end, a Devil." In "Self-Reliance," the metaphor
grounded the action of the self-reliant individual: "The magnetism which
all original action exerts is explained when we inquire the reason of self-
trust. Who is the Trustee?" Similarly, in "Spiritual Laws," magnetism related the individual to the whole through the Coleridgean concept of
method: "A man is a method, a progressive arrangement; a selecting principle . . . like the loadstone amongst splinters of steel." In this context, then,
the shift in emphasis by the mid-1840s is startling: "The magnetism which
arranges tribes and races in one polarity, is alone to be respected," as Emerson wrote in "Nominalist and Realist"; "the men are steel filings. . . . Let us
go for universals; for the magnetism, not for the needles." Or, as Emerson
repeated to himself in 1845, "Believe in magnetism, not in needles, in the
unwearied and unweariable power of Destiny."[73]

Taking still another tack with the metaphor, Emerson also used it as a
device to consider memory: "Alas! you have lost something for everything
you have gained, & cannot grow. You are a magnet drawn through steel
shavings: it gains new particles all the way, but one falls off for every one
that adheres." However, late in life when his memory really was failing, he
found a way to turn the metaphor in a more reassuring direction: "*Memory.* A man would think twice about learning a new science or reading a
new paragraph, if he believed that the magnetism was only a constant
amount, & that he lost a word therefore for every word he gained. But the
experience is not quite so bad." Every new word is a "lamp lighting up re-

lated words," and so with "each fact in a new science," each one adding "transparency to the whole mass."[74]

Ultimately, Emerson decided that the magnetic metaphor could be used not only to subdue the individual "steel filing" to the currents of destiny, but to free him from the tyranny of the whole: "Polarity. Every nature has its own. It was found, that, if iron ranged itself north & south, nickel or other substance ranged itelf east & west; & Faraday expected to find that each Chemic element might yet be found to have its own determination or pole. And every soul has a bias or polarity of its own, & each new: Every one a magnet with a new north." As magnetism was one of Emerson's first fascinations, it was also one of his last. One of the final entries in his journal reads: "The magnet is the mystery which I would fain have explained to me, though I doubt if there be any teachers. It is the wonder of the child & not less of the philosopher."[75]

The range and flexibility of the magnetic metaphor, its symbolic resonance, and particularly the way it could be used to relate the individual mote to the social field made it a useful structuring device for the consideration of character, "this moral order seen through the medium of an individual nature." Indeed, an individual in perfect alignment with the moral order cannot *be* seen, for his own axis will coincide with the axis of things: "A healthy soul stands united with the Just and the True, as the magnet arranges itself with the pole, so that he stands to all beholders like a transparent object betwixt them and the sun, and whoso journeys towards the sun, journeys towards that person." Emerson then arrays the powers of this healthy soul along the magnetic poles of the Earth: "Everything in nature is bipolar, or has a positive and negative pole. There is a male and a female, a spirit and a fact, a north and a south. Spirit is the positive, the event is the negative. Will is the north, action the south pole. Character may be ranked as having its natural place in the north. It shares the magnetic currents of the system. The feeble souls are drawn to the south or negative pole." As the metaphor insists, character does not—cannot—exist in isolation, but only in a social context.[76]

In "Politics" Emerson used this metaphor to show how society, including "our republic," is protected from the "waves of chance" by the stabilizing, mutually self-developing cycle of polarity: "The fact of two poles, of two forces, centripetal and centrifugal, is universal, and each force by its own activity develops the other. Wild liberty develops iron conscience. Want of liberty, by strengthening law and decorum, stupefies conscience." Individual character becomes a mote in this social field, both producing and produced by the energies swirling around and through it. Within this field all persons arrange their world around themselves, according to the extent of their magnetic power, in a series of subsystems, tributary to the ascending hierarchy of systems that ultimately embraces the entire globe. Yet our individual experience is not global but local: "I am always environed by myself,"

Emerson comments in "Character." Thus we no longer "build our own world" in splendid isolation, but subject to the constraints and opportunities in which we are embedded; and the world we build interdigitates with all those by which we are neighbored, in a complex play of reinforcements and resistances.[77]

The magnetic metaphor also became an invaluable structuring device for that most extended of Emerson's character studies, *Representative Men.* Soon after the publication of *Essays, Second Series,* Emerson began writing *Representative Men* as a new lecture series, and it was this series that carried him through his triumphal lecture tour of England in 1847–48. He revised the lectures over the next three years, and they were published in 1850. The book opens with a magnetic joke: great men are hard to find. "But if there were any magnet" that would point to them, Emerson would "sell all, and buy it, and put myself on the road today." The difficulty is, how may we allow ourselves to be "attracted" or "drawn" to great men, without being wholly captured and imprisoned within their field of force? The risk is worth taking, for great men are a kind of "collyrium" for clouded vision, who "clear our eyes from egotism." Yet their "attractions warp us from our place," turning us into "underlings and intellectual suicides." Fortunately, nature's polarity defends us from this danger of overinfluence. "The centripetence augments the centrifugence. We balance one man with his opposite, and the health of the state depends on the see-saw." So although on the one hand they are very "attractive," nature sees to it that "the more we are drawn, the more we are repelled," allowing us to grow, but protecting our "selfdefended" boundaries.[78]

The term *representative* occurs in Schelling as the notion that the self can liberate itself by detaching from itself an object, which it can then contemplate freely, "re-presenting" it to mind. In the same way, Emerson works to detach himself from the great men whose attractions draw him strongly, liberating himself from them by re-presenting them as objects of contemplation. Emerson, in effect, analyzes the dynamic relationship he has with these "representative" figures: the attraction to them, the reaction or repulsion from them. In all, he shows how such figures may be *used* (as he says in his opening chapter title, "The Uses of Great Men") to achieve our own latent greatness, through educating or "educing" responsive aspects of our own being. It is in this *in*direct sense that great men can serve us, not through direct gifts: "Man is endogenous, and education is his unfolding." Great men exist to draw out some latent power of nature, to "convert" some "raw material . . . to human use." Thus, "Each man is, by secret liking, connected with some district of nature whose agent and interpreter he is, as Linnaeus, of plants; Huber, of bees; Fries of lichens." Plants, bees, lichens are "still hid and expectant," sealed up by enchantment, until their destined human deliverer disenchants them, and they "walk forth to the day in human shape." Just so, "A magnet must be made man in some Gilbert, or Swedenborg, or

Oersted, before the general mind can come to entertain its powers." Each man can draw out or "represent" that material part of nature only because the material has its "celestial" or spiritual side, with which the human agent identifies, and which allows him to be not just "representative but participant": the scientist can know nature because he *is* nature, and conversely, nature *is* the scientist. "Shall we say that quartz mountains will pulverize into innumerable Werners, Von Buchs, and Beaumonts, and the laboratory of the atmosphere holds in solution I know not what Berzeliuses and Davys?" Person by person, nature takes human shape and walks forth into the world: "Thus we sit by the fire, and take hold on the poles of the earth."[79]

Thus far, Emerson's "representative men" have all been scientists, for it is they who model the method by which nature and man realize each in the other: the coincidence of subject and object, the binocular focus that produces knowledge. By this point in history, "Life is girt all round with a zodiac of sciences, the contribution of men who have perished to add their point of light to our sky." Emerson's task, his own point of light, will be to show how the zodiac of great men can itself be approached, each single point of light an object to our subject, drawing out some power in ourselves. Since we are surrounded not by a few such stars but a whole "zodiack," we are defended from falling under the spell of any one. Instead we approach them one by one in a rotation, rounding off the world point by point, using each one and then moving on. As Emerson repeats, parenthetically, "(The needles are nothing; the magnetism is all.)."[80]

Liberated in this way, Emerson himself can freely critique even the greatest luminaries in his own personal firmament: Plato the philosopher, for all his virtue, lacks "a system." Swedenborg the mystic, ironically, has too much system: he "denotes classes of souls as a botanist disposes of a carex, and visits doleful hells, as a stratum of chalk or hornblende!" Montaigne the skeptic suffers from the skeptic's inevitable superficiality. Shakespeare was poet, but not priest; Napoleon had intellect without conscience; Goethe served culture but not truth.[81]

As the binary structuring of the critiques suggests, the individual essays also tend to be patterned across polar structures. So, for instance, in "Plato," while the universe fractures across the great divide of polar principles—unity and diversity, being and intellect, necessity and freedom, rest and motion, power and distribution, strength and pleasure, consciousness and definition, genius and talent, earnestness and knowledge, possession and trade, caste and culture, king and democracy, "pure science" and "executive deity"—it is Plato's contribution to have *joined* these opposing "gods of the mind," enhancing the power of each and so realizing "a balanced soul," who could see "two sides of a thing." Like Emerson, Plato was a virtuoso at the game of cause and effect, of flipping "the upper and the under side of the medal of Jove." Like Emerson, too, Plato "cannot forgive in himself a partiality, but is resolved that the two poles of thought shall appear in his statement. His argument and his sentence are selfpoised and spherical. The two poles appear, yes, and become two hands to grasp and appropriate their own." The principle becomes general: "Every great artist has been such by synthesis," spinning "a thread of two strands." Similarly, Swedenborg's theory of correspondence spun together matter and spirit, not just as a poet who plays with allegory as a toy, but as a man of science who puts "the fact into a detached and scientific statement," even as "the magnet was known for ages, as a toy" before science offered it up for use. Swedenborg took up "such rightness of position, that the poles of the eye should coincide with the axis of the world," but sadly, he could not maintain his equilibrium, and so became an "example of a deranged balance." His very grounding in science became the source of his failure, and he narrowed nature's protean forms into a fixed and rigid theology. As a result, "The universe in his poem suffers under a magnetic sleep, and only reflects the mind of the magnetizer"—instead of ecstatic science, mere hypnosis.[82]

Montaigne offers another version of synthesis. He is the "third party" who occupies the "middle ground" between the "abstractionist" and the "materialist"—namely, "the skeptic," who "finds both wrong by being in extremes. He labours to plant his feet, to be the beam of the balance." Shakespeare represents the man whose axis of vision is fully aligned with the axis of the world, "suffering the spirit of the hour to pass unobstructed through the mind," but thereby presenting a paradox: as a channel open to all others, the man himself is curiously invisible, so transparent that he seems to have no biography. Napoleon, by contrast, seems to be *all* biography, the "best known" eminence of the nineteenth century, whose tremendous power magnetized all the shavings around him into miniatures of himself: "if Napoleon is France, if Napoleon is Europe, it is because the people whom he sways are little Napoleons."[83]

Finally, Goethe's role as "the writer" is to be the man through whom nature "reports" or "self-registers," a member of "the class of scholars or writers, namely, who see connexion, where the multitude see fragments, and who are impelled to exhibit the facts in ideal order, and so to supply the axis on which the frame of things turns." If the century had become "one great Exploring Expedition, accumulating a glut of facts and fruits too fast for any hitherto existing savans to classify, this man's mind had ample chambers for the distribution of all. He had a power to unite the detached atoms again by their own law. He has clothed our modern existence with poetry." Leaving behind "French tabulation and dissection," telescope and microscope, Goethe contributed a poet's vision to nature, providing thereby the key: to botany, the insight that the leaf is the unit on which all plants are built; to osteology, the parallel insight that the vertebra is the unit of the skeleton; to optics, the dismissal of Newton's prism of seven colors, in favor of the (more pleasingly polar) theory that "every color was the mixture of light and darkness in new proportions."[84]

But what finally marks Goethe for Emerson is not nature, but culture. Goethe is incapable of "selfsurrender" to pure truth, to the "moral sentiment." He devotes himself instead "to truth for the sake of culture," becoming, indeed, "the type of culture, the amateur of all arts and sciences and events." His autobiography familiarized a new generation with the idea "that a man exists for Culture; not for what he can accomplish, but for what can be accomplished in him." The consequence was that Goethe's actual accomplishments were limited. "He is fragmentary; a writer of occasional poems, and of an encyclopaedia of sentences." Emerson ends his book with an admonition: the aim of culture may be high, but "the idea of absolute eternal truth without reference to my own enlargment by it, is higher." Goethe's great value lay in his ability to distribute and make "subservient" the "mountainous miscellany" of facts that threatened to glut the nineteenth-century mind and, with Napoleon, to represent nature's impatience with dead convention. But Emerson's own ambition was indeed "higher": not to "distribute" worldly facts but to mediate between Earth and absolute eternal truth, to sit at the fireplace and grasp the poles of the universe. "We too must write Bibles," he ends, "to unite again the heavenly and the earthly world." The culture Emerson aimed at would accomplish more than worldly refinement. It would align the axis of the individual with the poles of the world. It would be, in short, not a culture of self, but a culture of truth.[85]

CHAPTER FIVE

Truth against the World

FROM VESICLE TO COSMOS: THEORIES OF EVOLUTION

In 1844 Emerson began drafting his first major antislavery address by thinking about, of all things, geology. "The use of geology has been to wont the mind to a new chronology." It has broken up our "little-dame school measures," "our European & Mosaic & Ptolemaic schemes for the grand style of nature & fact." Over "weary patient periods" rock is formed and broken; the first lichens are succeeded by trilobites, quadrupeds, and at last man. "All duly arrive, & then race after race. . . . all must come, as surely as the first atom has two sides." Yet "nothing interests us of these or ought to." Instead of looking into our geological past, we should look forward, into our future, "that evergoing progression": "Who cares for the crimes of the past, for oppressing whites or oppressed blacks, any more than for bad dreams? These fangs & eaters & food are all in the harmony of nature: & there too is the germ protected, unfolding gigantic leaf after leaf."[1]

For Emerson, the problem of race could not be approached apart from theories of development and their implications for types of mankind. Emerson's point here is that in this unfolding, nature saves only "what is worth saving & it saves not by compassion but by power. It saves men through themselves." No police guards the lion. He guards himself with "teeth & claws," as does the bird with wings, and "flies & mites" with spawning numbers—and so with men: "Ideas only save races. If the black man is feeble & . . . not on a par with the best race, the black man must serve & be sold & exterminated." But if—as Emerson will argue—the black man does carry "in his bosom an indispensable element of a new & coming civilization," then "no wrong nor strength nor circumstance can hurt him": "if you have man, black or white is an insignificance. Why at night all men are black." In sum, only the black man can save himself (although, Emerson adds shrewdly, when he does so the white man will step in and take the credit). However, "If the negro is a fool all the white men in the world cannot save him though they should die." If gnomic determinism shuts us in, no effort of will can free us.[2]

This disturbing passage illustrates several important aspects of Emerson's turn of mind in 1844. First, the question of race (and therefore slavery) was tightly connected with science and, more particularly, with geology and the development of races and species. Second, there is an oddly "Darwinian" flavor (using the term in its corrupted sense) to the developmental unfolding he envisions, wherein only the strongest, most fleet, most prolific, or smartest survive. Third, applied to humanity, this survivalist view creates a troubling ambiguity. It is *nature* that has destined the black race for either success or failure, as if the white race bears no responsibility for blacks' oppression (dismissed as a "crime of the past") or for their freedom.[3]

Yet clearly, this would not do. In the address as he actually delivered it that August, "An Address . . . on . . . the Emancipation of the Negroes in the British West Indies," Emerson deleted the skeptical qualification that if the Negro was a "fool" he was beyond salvation, and emphasized instead that now was the time for the black race to "emerge." Then he added the important argument that "the civility of no race can be perfect whilst another race is degraded. . . . man is one, and . . . you cannot injure any member, without a sympathetic injury to all the members." But his public statement masked another ambiguity, also tied to science: If man was one, what were the races of man? Were they species? Were species immutable? Were the races all descended from one origin, or would science show that races were immutable and hence confront Genesis with the hard truth that man does not have a common origin? If human races *were* separate species, how many were there? What were their characteristics, and how were they to be related and ranked? Were nations races? How did physical law apply to social groups—or did it?[4]

Science was not a refuge from questions of class, gender, or race, but a crucial means of addressing them, especially in the superheated climate of the antebellum United States. This was all the more true for that class of public intellectuals who were striving to marry the physical and moral universes. Hence Emerson was keenly interested in three of the liveliest scientific arenas of his day: theories of transmutation, or evolution; race and species; and statistics and probability. All three suggested how the masses could be organized into a true social science, and all three hinged on controversies that confronted the Emersonian individual with dilemmas of chance, choice, and determinism.

When Charles Darwin sat down in the late 1850s to draft *The Origin of Species,* evolution was already an old idea. It meant "unfolding," as in Emerson's sense that the world unfolded out of the first atom, and it carried overtones of destiny, of inevitability, of providential design. Darwin's task was not to invent evolution, but to reinvent it so as to demolish design and to enthrone chance. When Emerson came to read Darwin, which he did in 1860, he saw nothing he had not seen before—a fact that reveals little about Darwin and a great deal about Emerson.

The long journal passage from 1844 shows just how far Emerson had come by then in his thinking about evolution. Since the 1820s it had been an article of faith for him that the planet had undergone a nearly endless series of sweeping changes over an incalculable period of time, each stage characterized by an array of life forms, all moving ever closer to the assemblage of flora and fauna existing in the nineteenth century. His reading in Charles Lyell and Georges Cuvier, not to mention the marvelous displays at the Paris Muséum d'Histoire Naturelle, had acquainted him long since with the idea that life had undergone successive transformations. Yet Emerson did not as yet have a mechanism to account for the changes other than a vague sense that the changes composed an "onward & onward" progression, together with the leftover assumption that the progression bore the stamp of divine design. Furthermore, as William Rossi observes, Emerson's contemporaries were quite sensitive to the distinction between "progressionism," which saw the increasing complexity of life "as a goal-directed progress toward humanity and the unfolding of a divine plan," and "transmutationism," which interpreted the fossil record "as a history of genealogical descent down to humanity." Emerson, like most of his contemporaries, favored the former while finding the materialist implications of the latter repulsive.[5]

At least one mechanism had been proposed, but it came from the suspect transmutationist camp. His 1836 lecture "The Humanity of Science" reveals Emerson's enthusiasm for the transformationist ideas of Goethe, who had shown that every part of the plant was a transformed leaf, as every part of the vertebrate skeleton was a transformed vertebra. Emerson also showed some familiarity, but relatively limited enthusiasm, for Jean-Baptiste Lamarck. According to Emerson, Lamarck's system "aims to find a monad of organic life which shall be common to every animal, and which becomes an animalcule, a poplar-worm, a mastiff, or a man, according to circumstances. It says to the caterpillar, 'How dost thou, Brother! Please God, you shall yet be a philosopher.' " Instead of endorsing the idea, however, Emerson offers Lamarck as one of the "extreme examples" of the impatience of the human mind to unify.[6]

Had Emerson actually read Lamarck, he might well have been more receptive, for in his landmark book *Zoological Philosophy* (1809), Lamarck is a likeable, modest, and enthusiastic teacher delighted with the power of nature to generate ever-new forms. However, Emerson's knowledge of Lamarck came at second hand, from Lyell's *Principles of Geology*. Lyell spent the first two chapters of his second volume demolishing the threat posed by transformationist ideas, of which Lamarck's were the most odious. Lamarck's theory rested on the view that species were not fixed but variable, so much so that they actually blended together. As circumstances changed, organisms responded by altering their actions and habits, developing new organs in the process, and so leading to *progressive improvement*. But, as Lyell hastened to add, Lamarck had no data. His "strange conclusions" were riddled with defective evidence and fallacious reasoning. Naive beginners were attracted to Lamarck because to them it might actually seem possible that species were variable—after all, some even interbred, so perhaps they *did* blend together. However, a more informed view recognized that species were essentially *in*variable. Wild dogs do not revert to wolves; in fact, evidence from Egyptian burials proved that even the most variable of species, domestic dogs and cats, were then exactly the same as now. Lamarck's conjectures looked ahead to the "future perfectability of man," but Lyell found such an outlook a poor compensation for renouncing "the high genealogy of his species."[7]

Lamarck was notorious in England as a French radical and materialist who believed that education would level all inequalities and erase all social hierarchies. Lyell's hatchet job was typical of mainstream British science, which was orthodox, conservative, progressivist, and to a surprising extent dedicated to proving Lamarck wrong. The intellectuals Emerson had met in England, and the science he awoke to there, were largely in this conservative tradition. On two counts, Emerson's small jest about Lamarck's caterpillar reveals a misunderstanding that he had no chance to correct. A more radical sympathizer might have had the *caterpillar* say: "Please God, *I* shall yet be a philosopher!" Second, Lamarck's caterpillars knew that pleasing God had nothing to do with it; God had given them the power to please themselves.[8]

Thus, although Lamarck did provide a plausible mechanism for transformation, it was not one available to Emerson, and for some time transformation remained an attractive but vague concept. All this changed rapidly, however, in 1845, when he read the anonymous best-seller *Vestiges of the Natural History of Creation* by the Scottish journalist Robert Chambers. *Vestiges* was attacked vehemently by the scientific press, for, as Rossi says, it rewrote the progressivist narrative as a transmutationist one. Worse, it was written for a lay audience vulnerable—just as Lyell had feared—to the seductions of developmental cosmologies. By 1860, eleven editions had sold 23,350 copies, and *Vestiges* had been reviewed over eighty times. Emerson owned both the second American edition and Chambers's lengthy rejoinder to his critics, *Explanations*. The key concept Emerson derived from Chambers is encapsulated in his journal: "We owe to every book that interests us one or two words. Thus to 'Vestiges of Creation' we owe 'arrested development.' " The term became a crucial one in Emerson's thinking, as indicated by his summary in "Poetry and Imagination": "The electric word pronounced by John Hunter a hundred years ago, *arrested and progressive development,* indicating the way upward from the invisible protoplasm to the highest organisms, gave the poetic key to Natural Science, of which the theories of Geoffroy St. Hilaire, of Oken, of Goethe, of Agassiz and Owen and Darwin in zoölogy and botany, are the fruits,—a hint whose power is not yet exhausted, showing unity and perfect order in physics." But as Emerson's biographer Gay Wilson Allen concluded, the expression "arrested and progressive development" was neither Hunter's nor Chambers's, but Emerson's own.[9]

What Emerson's words named was the method by which organisms developed from lower to higher, a method Emerson synthesized from much of his reading but adapted primarily from Chambers: the final form of an organism was determined by the stage at which development was arrested in the embryo. As Emerson put it in his journal, "The trilobium, which is the eldest of fossil animals, reappears now in the embryonic changes of crab & lobster. It seems there is a state of melioration, pending which, the development towards man can go on; which usually is arrested." As Chambers wrote, the "fundamental unity" of organic form shows all organisms, "from the humblest lichen up to the highest mammifer," to be part of one single system, which must be the result of "one law or decree of the Almighty." Each organism, as it germinates, passes "through a series of changes resembling the *permanent forms* of the various orders of animals inferior to it in the scale." Thus, the insect larva is a worm, the lowest form in the same class; the tadpole is the frog in a fish stage, and the human being, in the nine months of gestation, "passes through conditions generally resembling a fish, a reptile, a bird, and the lower mammalia, before it attains its specific maturity." Change occurs when, in response to external conditions, the "embryotic progress" is arrested or advanced. That is, the law of development is fully reversible. "Give good conditions, it advances; bad ones, it recedes." Under bad conditions, the human fetus may be born with a three-chambered heart, regressed to the reptile stage; good conditions could make "a fish mother develop a reptile heart, or a reptile mother develop a mammal one." Despite the equivocation, the general trend is ever upward, and as successively higher species are produced, their vacated place is taken by the species next in line. The whole series is continually starting anew at the bottom with the formation of germinal vesicles out of inorganic matter.[10]

Chambers's system may look superficially like Lamarck's, but Chambers was careful to delineate the essential distinctions. Whereas Lamarck's theory "deservedly incurred much ridicule," he, Chambers, followed "the laws of organic development": "I take existing natural means, and shew them to have been capable of producing all the existing organisms, with the simple and easily conceivable aid of a higher generative law, which we perhaps still see operating upon a limited scale. I also go beyond the French philosopher to a very important point, the original Divine conception of all the forms of being which these natural laws were only instruments in working out and realizing." Lamarck's theory, in other words, was lawless and anarchic, whereas Chambers's theory showed the law-bound and purposive unfolding of life from vesicle to man along a preordained path. This path had the consent, and even the assistance, of the entire universe—as Emerson recognized:

We want a higher logic to put us in training for the laws of creation. How does the step forward from one species to a higher species of an existing genus take place? The ass is not the parent of the horse, no fish begets a bird. But the concurrence of new conditions necessitates a new object in which those conditions meet & flower. When the hour is struck in on-ward nature announcing that all is ready for the birth of higher form & nobler function not one pair of parents, but the whole consenting system thrills, yearns, and produces.[11]

As Philip Nicoloff writes, Chambers allowed Emerson to "persist in his grand assumption" that the ends of the universe were complete in law and implicit since the beginning; organic progress could indeed be seen "as an 'unswaddling,' and not as an arbitrary and meandering ascent." In effect, by showing evolution to be a lawful or gnomic unfolding of latent potential, Chambers made evolution safe for the Anglo-American middle classes. As Nicoloff continues, Emerson adopted Chambers's theory and vocabulary "wholeheartedly": "As his mature historical theory reached its final devel-opment, he was to employ consistently the scientific principles of arresta-tion, progression, latency, environmental selection, embryological recapitu-lation, and racial competition as emblems of the human condition." Emerson's first reaction to *Vestiges* had been to praise "everything" in it as good—everything "except the theology, which is civil, timid, and dull." Chambers shows how flexible and adaptable the framework of natural the-ology could be, and how useful in rendering exotic ideas palatable to the public. Emerson's reaction shows how far he had moved beyond the need for such a framework, how far he had come since his early science lectures.[12]

Just how much further Emerson was willing to go is shown by his con-tinued enthusiasm for the most extreme of the transcendental biologists, all of whom, in his reading, variously confirmed and extended his own synthe-sis of "arrested and progressive development"—even Louis Agassiz, who didn't believe in development at all. One of the most important texts in this tradition after Chambers was Johann Bernhard Stallo's *General Principles of the Natural Philosophy of Nature* (1848), which Emerson read in 1849. Stallo had immigrated to the United States from Germany at age sixteen and, after teaching languages in Cincinnati, had spent a brief period as a professor of chemistry and physics at St. John's College in New York, where he wrote *General Principles* (a book that he later disavowed). *General Prin-ciples* opened with a long essay synthesizing German philosophy for Amer-ican audiences, followed by lengthy explications of Kant, Fichte, Schelling, Oken, and Hegel. Of them all, Emerson was most immediately attracted to Lorenz Oken, a German biologist who was thought at the time to be a greater philosopher than Schelling because of his wider knowledge of natu-ral science. Oken saw the body of man as, in Emerson's term, the "master key" to all nature, which consisted of the organs of man distributed throughout the organic world: nature was, literally, man writ large. The world was germinated and sustained by the dual force of polarity, "+ −," which for Oken meant that the origin of all things was "*gender,* the sexual antithesis." In this world, all things were alive, nothing was dead; and since polarity returned to itself, "all life is self-returning, circular," the balance of two processes, one individualizing and vivifying, the other universal and mortifying.[13]

Oken had arrived independently at the notion Emerson celebrated in Goethe, that the vertebra was the unit of the skeleton. Emerson quoted with delight Oken's saying "The limbs are emancipated ribs." Shortly after reading Stallo, Emerson played with the idea in his journal:

> Nature is likest herself. At the end of the spine she projects two little spines for arms; at the other, two more for legs; at the end of these, she repeats the gift, each time a little modified to suit the want. Nature is likest herself. In the brain, which she prepared by bending over a spine, she recites her lesson once more, the well known tune in a higher key. Here are male & female faculties of mind, here is marriage, here is fruit. There is also constant relation to the lower & a stern Nemesis binds the farthest series in her divine grasp.

The connection with "Nemesis" is also behind such later remarks as "The Menagerie of forms & powers of the spine is a book of fate." Emerson was also taken with Oken's saying that "The development of all individual forms will be spiral"; and that "Animals are but foetal forms of man," which confirmed Chambers's idea that organic succession was mirrored by fetal development.[14]

Stallo was not the only European immigrant working to introduce German *Naturphilosophie* to American readers. Far more important still to Emerson was the arrival in Boston of the Swiss naturalist Louis Agassiz, who was famous in Europe for his definitive work on fossil fish and his groundbreaking and controversial "Ice Age" theory that glaciers had once scoured the surface of the Earth. Agassiz's enormous ambition had in 1846 precipitated a personal and financial crisis. His wife, tired of his overwork, took their children and left him, and his absurdly overextended finances led to bankruptcy, forcing him to close his research and publishing center. Help came from two powerful friends: his mentor Alexander von Humboldt engineered a sizable grant from the king of Prussia for a trip to the United States, and in England Charles Lyell put Agassiz in touch with Boston's John Armory Lowell, who offered to fund a lecture series at the Lowell Institute. Agassiz arrived in Boston in October 1846, and his lecture series, "The Plan of Creation in the Animal Kingdom," was a huge success: tickets were free, but demand was so great that they had to be distributed by lottery. The audience ranged upward of five thousand, and even so, the lectures had to be repeated before a second audience.

Agassiz was an enormously charming and charismatic personality, lionized wherever he went. He quickly decided to make Boston his headquarters and began importing his disbanded staff from Switzerland. In 1847, when Lowell's friend Abbot Lawrence, a cotton manufacturer, announced he was donating $50,000 to found a scientific school at Harvard, the university immediately created a professorship for Agassiz, and soon he was installed at the new Lawrence Scientific School. America quickly became his home, and the most extravagant offers from Europe never enticed him to leave it. For the next twenty-five years he would dominate American science. Agassiz organized massive collecting networks (Thoreau sent him specimens from Walden Pond); mounted expeditions to Lake Superior and to South America; flooded the market with lectures, scientific papers, books, and popular articles; helped to found the American Association for the Advancement of Science in 1848 and the far more exclusive National Academy of Sciences, "the official scientific advisory body of the federal government," signed into law by President Lincoln in 1863; founded the extraordinary Museum of Comparative Zoology at Harvard, and professionalized science and science education in the United States. Agassiz's impact on American culture was immediate, profound, and far-reaching. Furthermore, by the mid-1850s he had become one of Emerson's closest friends. Agassiz's marriage, shortly after the death of his first wife, to James Elliot Cabot's sister Elizabeth had consolidated Agassiz's ties with Boston's finest families; and when, to raise money for his publications, Lizzie Agassiz started a school for girls in their home, Waldo and Lidian enrolled their eldest daughter, Ellen.[15]

Emerson's ties with Agassiz were intellectual as well as personal. Agassiz's central insight, borrowed from his teacher Cuvier, was that the animal kingdom was divided into "four great departments," arranged in a circle, with vertebrates, the most complicated, at the top (headed by man), molluscs and articulates to right and left, with radiates, as the simplest, at the bottom. All four departments had been present since the Creation as God's four fundamental ideas, and in successive ages all life had been swept away and replaced according to God's cosmic plan—thus the importance of glaciers, which Agassiz called "God's great plough," in preparing the global stage for life in its present form. Each successive creation added ever-higher and more-complicated forms. The full scale could be read in four correspondent series: the series of "Rank," or structural complication among all living adult animals; of "Growth," or the successive stages of embryological development; of "Time," or the successive introduction of animals on Earth; and of "Geographical Distribution" across the globe in fixed zoological provinces. This elegant fourfold correspondence made it clear that the connection between animals could not be material, but must be intellectual. As Agassiz wrote in his popular textbook *Methods of Study in Natural History* (1863; Emerson owned the first edition):

These correspondences are correspondences of thought,—of a thought that is always the same, whether it is expressed in the history of the type through all time, or in the life of the individuals that represent the type at the present moment, or in the growth of the germ of every being born into that type today. In other words, the same thought that spans the whole succession of geological ages controls the structural relations of all living beings as well as their distribution over the surface of the earth, and is repeated within the narrow compass of the smallest egg in which any being begins its growth.

That is, not only were individual species of animals thoughts of God embodied, but thought, realized as law, controlled the development of nature at every moment, in all places, and across all eras of geological time.[16] Agassiz's own nemesis was the developmental hypothesis, which he abhorred as the "material" connection that his entire intellectual system was designed to repudiate. His most powerful ally was the English paleontologist Richard Owen, whose theory of vertebrate archetypes dominated mainstream British science. Agassiz really had no serious opponents (except perhaps that irritating botanist next door, Harvard's Asa Gray) until Darwin published On the Origin of Species in 1859. Thenceforth much of Agassiz's attention was diverted to showing the world the error of Darwin's ways. During one stage of his campaign, Agassiz led his students (including a young William James) on an expedition to Brazil in 1865–66, during which he collected staggering numbers of freshwater fishes across as many Amazon river systems as he could reach, all to the end of disproving evolution by showing that God had created each of the myriad of species in and for its designated place. He also collected evidence that proved—to him, at least, if to no one else—that glaciers had scoured even the equatorial tropics. Agassiz's highly visible public campaign against Darwin's ideas called out a reluctant Asa Gray, Darwin's major American ally, to a series of debates, and the prominence these debates gave to an arcane scientific dispute elevated it into a major public controversy, one that, ironically, simmers to this day in the United States, where Agassiz's creationist arguments still buttress a lively movement against teaching Darwin in the public schools.

Emerson valued whatever in Agassiz corresponded with his own ideas, and the rest he simply ignored. He himself was fully comfortable with transformationist theories, and found the stodginess of those who had not yet caught on amusing. For instance, while he was visiting Agassiz's friend Richard Owen in England in 1848, Owen gave Emerson a pass to his course of lectures before the Royal College of Surgeons, showed Emerson around its newly renovated Hunterian Museum (of which Owen was now the curator), and escorted Emerson to the studio of the great artist Joseph Mallord William Turner (who, unfortunately, had stepped out). Emerson attended as many of Owen's lectures as he could, and was duly impressed with the man's excellence as a lecturer, his "surgical smile," and his "air of virility." But Emerson did not like to hear Owen "abusing without mercy . . . these poor transmutationists." He saw no need for Owen to "run counter to his own genius, & attack the 'transmutationists'; for it is they who obey the idea which makes him great." Of Agassiz, for whom Emerson felt real affection, the worst he could say was that he violated one of Emerson's own rules: "omit all the negative propositions. I fear, Agassiz takes quite too much time & space in denying popular science. He should electrify us by perpetual affirmations unexplained."[17]

What really impressed Emerson was Agassiz's discovery that embryological development recapitulated the unfolding of organic forms from age to age. He found in both Agassiz and Oken scientific authority for one of his dearest ideas, the rhyming partnership of poetry and science: "The iterations of rhymes of nature are already an idea or principle of science, & a guide. . . . What rhymes are these which Oken or Agassiz show, in making the head only a new man on the shoulders of the old, the spine doubled over & putting out once more its hands & feet," leading the anatomist on to "the anatomy of the Understanding, which is the material body of the mind, whilst Reason is its soul; and the law of Generation is constant, & repeats on the higher plane of intellect every fact in the animal." What also struck Emerson was Agassiz's raging success in bringing ideas once ridiculed into the popular and scientific imagination. "England cannot receive Oken but nibbles, gnaws, accommodates by Owen and Chambers," he grumbled. On the other hand, America was more fortunate—and Agassiz more shrewd: Schelling's science was "a forlorn hope," Oken was "ridiculous" and Hegel "nonsense," yet their science resounded to the people "in melody from songs of Goethe" and from Geoffroy Saint-Hilaire. Then "Agassiz brought it to America, & tried it in popular lectures on the towns. It succeeded to admiration, the lecturer having of course the prudence to disown these bad names of his authors. The idea was that the form or type became transparent in the actual forms of successive ages as presented in geology."[18]

The upshot was that in the concept of "arrested and progressive development," Emerson had both found and forged a synthesis of both the most widely accepted and the most controversial ideas in contemporary science, and he found his synthesis both deeply satisfying and extremely useful. This loose and flexible system embraced brute matter and intellectual law, animated it through polarity, and sent it climbing ever higher. No longer stalled on the bewildering staircase of "Experience," Emerson could look ahead to continued ascent up the ladder of destiny, a pathway written into Creation and certain not to fail.

Emerson saw confirmation of this upward destiny everywhere he looked in science, even when it was not actually there. After finishing Chambers in 1845, Emerson took up a far more authoritative and scientific treatment of similar questions, the first volume of Alexander von Humboldt's *Cosmos*. Emerson responded not to Humboldt's complex opening essay on the nature of science, but to the spectacle of Humboldt taking on the cosmos, a single mind grasping the poles of the universe: "The wonderful Humboldt, with his extended centre & expanded wings, marches like an army, gathering all things as he goes. How he reaches from science to science, from law to law, tucking away moons & asteroids & solar systems, in the clauses & parentheses of his encyclopaediacal paragraphs!" Humboldt always stood for Emerson as the hero of modern science, one of the age's great men, whose faculties were "all united by electric chain" and who showed the true power of man. Humboldt himself was wary of transmutationist speculations and harshly critical of the German school of nature philosophers, with whom he had early been associated. His writing was relentlessly, exuberantly empirical, tracing the ramifications of material nature from nebulae and comets to volcanoes, jaguars, and lichens, all bound not by divine Mind but by the searching, inquiring, synthesizing intelligence of the human participant-observer. Humboldt's nature did not unfold from any primordial germ but sprawled across thousands of pages in unaccountable diversity, held together not by generative law but by the lived history of natures and peoples working in time with whatever chance has made available. Thoreau responded to Humboldt with an intensity at least equal to that which Emerson brought to Chambers, Stallo, Oken, and Agassiz, but to the distinctive character of Humboldt's cosmos Emerson himself could never find the key. Instead, he folded Humboldt back into the fatal universe of Chambers and Oken as seen by Emerson: "The sun is as much a creature of fate as any worm which his heat engenders in the mud of earth. Large & small are nothing. Given a vesicle you have the Cosmos."[19]

Similarly, the fuss over Humboldt's disciple Charles Darwin escaped Emerson, who read *The Origin of Species* in 1860 and quietly assimilated Darwin, that contingent and radical materialist, as simply the latest in the same old beloved developmental train: Saint-Hilaire, Oken, Goethe, Agassiz, Owen—and Darwin, who had shown the whole world to be "only a Hunterian museum to exhibit the genesis of mankind." Toward the end of his journal, Emerson jotted this dismissal: "Darwin's 'Origin of Species' was published in 1859, but Stallo, in 1849, writes, 'Animals are but foetal forms of man,' &c." Emerson, having seen it all before, could never see the difference. He had already made his statement, in the little poem he appended to the 1849 republication of *Nature* (as a replacement for the epigraph from Plotinus): "And, striving to be man, the worm / Mounts through all the spires of form."[20]

THE DILEMMA OF RACE

Applied to humankind, the theory of arrested and progressive development meant that race was a biological category, and races could be arrayed across a scale of regressive and progressive tendencies. Most biological theorists of Emerson's day did so, and for the most part so did Emerson, although doing so often made him queasy. He was inclined to accept race as a valid category, as when he ventured in his journal that "the Atlantic is a sieve through which only or chiefly the liberal adventurous sensitive *America-loving* part of each city, clan, family, are brought. It is the light complexion, the blue eyes of Europe that come: the black eyes, the black drop, the Europe of Europe is left."[21]

However, in his most considered public work on the subject, *English Traits* (1856), he waffles. Individuals within a race can be as unlike as wolf and lapdog, and each variety of man "shades down imperceptibly into the next, and you cannot draw the line where a race begins or ends." Scientists are no help—Emerson even uses them to deconstruct the concept: "every writer makes a different count. Blumenbach reckons five races; Humboldt three; and Mr. [Charles] Pickering . . . makes eleven." Yet on the next page Emerson accepts the notion that English power is "due to their race," which, he decides, is "something like that law of physiology," which is—in essence—like father, like son. Thus, "It is race, is it not? that puts the hundreds of millions of India under the dominion of a remote island in the north of Europe. . . . Race is a controlling influence in the Jew, who, for two millenniums, under every climate, has preserved the same character and employments. Race in the negro is of appalling importance." Yet if race is an agent of fate, other forces resist it: the power of civilization "eats away" at it; social beliefs "counteract" it; finally, even the best argument for race, "the fixity or incontrovertibleness of races as we see them," is a poor one, for our historical span is but "a point" in relation to geological time. All our experience, then, "is of the gradation and resolution of races," of their mixture and blending: witness the English, "a fusion of distant and antagonistic elements," who owe their power to their very hybridity. And yet, pragmatically, Emerson cannot do without the category of race. "The kitchen-clock is more convenient than sidereal time. We must use the popular category, as we do by the Linnaean classification, for convenience, and not as exact and final." In other words, race may be a nominal not a "real" category—but usage makes it real enough, and so he will proceed with his fine-grained analysis of the English racial character.[22]

Philip Nicoloff has established that *English Traits* "provides us with one of the best illustrations of Emerson's use of contemporary science." Emerson's visit to England in 1847–48 had provided not only "the stimulus to write some lectures, and finally a book, on the problem of English power, but also a stimulus to his interest in modern science." David Robinson additionally credits this journey with "the rebirth of his conviction that science might offer a usable interpretive paradigm for the ultimate translation of nature's metaphysical code." Nicoloff lists nine major British scientists visited by Emerson in England, including Lyell, Faraday, and Owen. Emerson himself adds several more, plus the note that two of his three most "signal days" on the trip were spent with scientists: the first, when the great botanist Sir William Hooker showed him "all the riches" of the vast Kew Gardens; the second, when Richard Owen accompanied him through the Hunterian Museum.[23]

Nicoloff notes that through the 1850s Emerson's thoroughgoing reexamination of the scientific basis of his thinking substantially benefited both *English Traits* and *The Conduct of Life*. The second trip to England coincided with Emerson's increasing acceptance of "organic process as a mediator between the spiritual aspirations of the individual man and the eternal law which he could postulate but not yet attain"; man may still have been, in Emerson's words, a "zoological Fourth of July," but now his freedom was only relative. The result in *English Traits* was, according to Nicoloff, a rather bleak cyclical argument: the English race, a superior hybrid of many races, emerged from a state of savagery and war to be knit together by geographical forces. Its golden age synthesized idealism with materialism, as represented by Sir Francis Bacon; but the height could not be sustained, and England descended into Lockean empiricism, followed by two centuries of "spiritual drabness." England now had sunk to the solstice of its power and the senescence of its institutions—university, church, literature, science—and although the old idealism occasionally flickered back into life, the cycle had concluded and could only start anew in another location—say, America. England was finished. "Their mind," Emerson said, "is in a state of arrested development,—a divine cripple like Vulcan; a blind *savant*," trapped in "cramp limitation" and "sleepy routine." England, then, displayed the full cycle of arrested and progressive development, transferred from organic life to civil history, but subject to the same law. It was a sad and noble spectacle.[24]

Back home in America, though, the day was still young, and the question of race considerably more vexed. Perhaps hybridity had been the key to English power, but Emerson did not propose that America's future lay in the hybridization of the white and black races. On the developmental trajectory, the black race was clearly in an arrested mode, and it was not clear if that condition spoke to their permanent inferiority or rather to a power still latent, waiting to emerge. As the debate with himself in 1844 had indicated, the resolution, for Emerson, could go either way. Perhaps the Negro race was "feeble & not important to the existing races," in which case its fate would be extermination. On the other hand, it seemed at least as likely that the race "have been preserved" for a purpose, and that it was now their time to "begin to contend with the white" and be honoured for their unique genius: "Now let them emerge clothed & in their own form." If the latter, then the abolitionist societies could hardly claim credit; they were but "the shadow & witness to that fact." In his finished address, however, the moral burden shifted significantly: if the black race was inferior, then it was all the more incumbent on a civilized society to ameliorate their condition rather than degrade them further.[25]

Either way, slavery was a sin, not only for what it did to the enslaved but also, most certainly, for what it did to the slaveholder. This notion arises in Emerson's journal in 1840: "The negro must be very old & belongs, one would say, to the fossil formations. What right has he to be intruding into the late & civil daylight of this dynasty of the Caucasians & Saxons? It is plain that so inferior a race must perish shortly like the poor Indians." But on reflection Emerson is able, after all, to imagine a redeeming purpose for their intrusion: "Yet pity for these was needed, it seems, for the education of this generation in ethics. Our good world . . . must be stimulated by some-what foreign & monstrous, by the simular man of Ethiopia." This was not just an early sentiment. In 1851 Emerson noted: "The absence of moral feeling in the whiteman is the very calamity I deplore. The captivity of a thousand negroes is nothing to me." The ethics of compensation imagined a mirror world in which one's actions redounded implacably on one's own self; to put a chain around the neck of another was to loop the other end around your own. What one saw, looking into that mirror, was less the hor-ror of the enchained other than the revelation of the enchained self; to free oneself was also to free the other, though that might be just a collateral benefit.[26]

As the discussion of race in *English Traits* indicated, science did not speak with a clear voice on the status of the black race. Some, like Hum-boldt, repudiated a biological basis for racial inequality and argued that all human inequalities were due to civil and environmental history. "We . . . repel the depressing assumption of superior and inferior races of men," Humboldt insisted; "All are in like degree designed for freedom." The nat-uralist Charles Pickering was interested not in constructing a moral scale but in mapping patterns of geographical distribution, migration, and cross-cultural contacts. The British doctor and armchair researcher James Cowles Prichard, with whom Emerson was also familiar, listed so many scores of races as to trivialize racial difference altogether, to the end of showing, like Humboldt, the unity of mankind as "one species and one family." At the other extreme, the American John Knox, the "ingenious anatomist" at the head of Emerson's chapter "Race," dismissed Humboldt's doctrine as a fail-ure and Prichard as timid and feeble. He asserted that "the human mind is free to think, if not on the Rhine or the Thames, at least on the Ohio and the Missouri." "Race," Knox insisted, "is everything: literature, science, art, in a word, civilization, depend on it." Only the transcendental anatomy of Oken, Goethe, and their followers could offer a true "theory of na-ture"—for, as Knox said, "mere details are not philosophy . . . we require *laws*, not *details*," and the law they provided was "the law of the arrest of development."[27]

In this lineup, Emerson was most immediately indebted to Knox, a "rash and unsatisfactory writer" whose conclusions were "unpalatable," but who nevertheless was "charged with pungent and unforgetable [*sic*] truths." Indeed, majority opinion in the nineteenth century increasingly arrayed the races along a biological hierarchy and equated biology with destiny. Surficial variations in skin color and hair texture signaled deep anatomical difference, and difference slid into the inequalities that placed each race along a natural, hence inviolable, hierarchy. This was so because nature exhibited not a simple proliferation but an orderly unfolding of series and progressions and gradations. And unsurprisingly, few authors whom Emerson read entertained the thought that the Caucasian race might not be the summit of Creation.[28]

Emerson's speculations in 1844 lacked only the actual words *arrested* and *progressive* to fit with Chambers's evolutionary theory. Chambers showed how one type might succeed, under favorable conditions, to the next, but the types themselves persisted without changing. Among human beings he identified five: Caucasian, Mongolian, Malayan, Negro, and the "aboriginal American," with varieties explicable as developmental differences from a single origin. Namely, the human brain, having completed the animal transformations from fish to reptile to mammal to human, then underwent a second sequence, from Negro through Malay, American, and Mongolian to the highest and most mature type, Caucasian. However, this sequence of nonwhite races represented stages not of progress but of regress. All had degenerated from the original Caucasian type, and the degree of retrogression was registered in the face: "The Negro exhibits permanently the imperfect brain, projecting lower jaw, and slender bent limbs, of a Caucasian child, some considerable time before the period of its birth. The aboriginal American represents the same child nearer birth. The Mongolian is an arrested infant newly born. And so forth."[29]

Chambers's preferred system of classification (at least in the early editions) was the "MacLeay [or Quinary] System" of William Sharp MacLeay, which arranged all natural groups of animals in circles of five components: typical, subtypical (including an infusion of evil), and three aberrant forms. Chambers, the committed evolutionist, could not accept the conclusion of MacLeay's disciple William Swainson that man stood apart from the rest of animated nature, so Chambers proposed a revision that placed man at the "genuine head" of the highest group of animals, those with hands; thus man stood at the summit of the summit. The five races presumably formed the MacLeay circle of mankind, but Chambers did not develop this implication. Instead he was intrigued by the notion that man might actually *initiate* a *new* circle: "Are there yet to be species superior to us in organization, purer in feeling, more powerful in device and act, and who shall take a rule over us!" If so, "There may then be occasion for a nobler type of humanity, which shall complete the zoological circle on this planet, and realize some of the dreams of the purest spirits of the present race." Emerson was also intrigued: "And how excellent is this MacLeay and Swainson theory of animated circles!" he exclaimed, and added sententiously, "Well & it seems there is room for a better species of the genus Homo. The Caucasian is an arrested undertype." There was, as he suspected all along, always room at the top.[30]

Stallo's treatment of Oken acquainted Emerson with a rival theory of races. Given that nature was man distributed, Oken saw the five races as corresponding with the five senses: African, touch; Australian and Malay, taste; American, smell; Asiatic, hearing; and European-Caucasian, sight. Stallo himself was unconvinced, and advanced another theory that divided the races into *active* (Caucasian) and *passive* (all the other, that is, colored, races). Oken's notion that animals are "foetal"—that is, infant—men jibed with Chambers's presumption that language was also in different stages of development among different races, evidencing here the Chinese, who still spoke in uninflected monosyllables varied only by accent. Thus "the language of this immense nation—the third part of the human race—may be said to be in the condition of infancy"—in short, baby talk.[31]

Had Emerson attended the 1849 lecture series of Agassiz's friend Arnold Guyot or opened his copy of *Earth and Man* for assistance, he would have found that physical geography offered a parallel argument. According to Guyot, the ascending series of organic forms in nature developed under the stimulus of heat, and so reached its apex at the Equator; however, man followed not physical law, but the new law of the *moral* order. Man could not unfold except through education. Human potential could not, then, be realized in the treacherous abundance of the tropics, or in the impoverishing toil of the poles, but only in the temperate regions, the "cradle" of humanity. The regularity, grace, and perfect harmony of the Caucasian race was "the most pure, the most perfect type of humanity." To depart from the temperate zone, the geographical center, was to meet ever more degenerate forms, in a descent that took different shapes on each continent: in Africa, from Arab to Hottentot and, at the bottom, "the miserable Bushman"; in Asia, from Mongolian to South Australian, who exhibited "the last degree to which ugliness can go" in this being created to be lord of the world—and so on. How, Guyot wondered, had man fallen so low? Because as a free and perfectible being, man was "consequently capable also of falling. In the path of development, not to advance is to go back; it is impossible to remain stationary." In this strenuous tightrope universe of contending forces, there was no stable equilibrium, no point of rest. The law was unrelenting. Onward, or downward; swim, or sink.[32]

Of Emerson's immediate circle, the racial doctrines of Louis Agassiz were the most extreme and exerted incalculable influence on American society, if not on Emerson himself. Agassiz had not attended to race as a scientific problem until he arrived in the United States and saw, for the first time, "negroes." As a long-suppressed letter to his mother (as well as her friends and neighbors) reveals, his reaction was visceral, irrational, and extreme:

I can scarcely express to you the painful impression that I received, especially since the feeling that they inspired in me is contrary to all our ideas about the confraternity of the human type and the unique origin of our species. But truth before all. . . . it is impossible for me to repress the feeling that they are not of the same blood as us. In seeing their black faces with their thick lips and grimacing teeth, the wool on their head, their bent knees, their elongated hands, their large curved nails, and especially the livid color of the palm of their hands, I could not take my eyes off their face in order to tell them to stay far away. . . . What unhappiness for the white race—to have tied their existence so closely with that of negroes in certain countries! God preserve us from such a contact!

Up to this point Agassiz had accepted the older doctrine, sanctioned by science and Genesis alike, that humanity had one common origin; but over the next several years he developed a contrary view. The differences among races were so radical that they could be accounted for only by "polygenesis," or a multitude of separate creations. By 1854 Agassiz had concluded that the varieties of man were not just races; they were actually separate species, so separate that they could not successfully interbreed.[33]

Agassiz aired his views first to the American Association for the Advancement of Science in 1850, then in a series of three articles in the *Christian Examiner*, from March 1850 through January 1851. The first of these laid the foundation by establishing the special and separate creation of all animals in distinct and nonoverlapping zoological provinces. The second, "The Diversity of Origin of the Human Races," applied this argument to mankind, who were likewise created in "nations" corresponding to the same zoological provinces. The much-vaunted "Unity of Mankind," Agassiz insisted, was true only in a spiritual sense. There was no unity of origin, there were no "ties of blood." In fact, unity *meant* diversity, for the doctrine of "unity in multeity" (in Coleridge's phrase) or "diversity in unity" (in Agassiz's) "does not mean oneness, or singleness, but a plurality in which there are many points of resemblance, of agreement, of identity"—and this was "the fundamental law of nature." The question science had yet to investigate properly, Agassiz insisted, was not what connected species, but what "keeps them apart."[34]

It was up to science courageously to settle these tough questions. Agassiz's opening paragraph laid the problem squarely on the table: "We have a right to consider the questions growing out of men's physical relations as merely scientific questions, and to investigate them without reference to either politics or religion." Science was not and could not be influenced by social concerns, and to those who protested that his views supported slavery, Agassiz replied that that was not his problem. "Is that a fair objection to a philosophical investigation? Here we have to do only with the question of the origin of men; let the politicians, let those who feel themselves called upon to regulate human society, see what they can do with the results." Nor did Genesis have any authority. As Agassiz reminded his readers, science, like Galileo, must have the courage to confront religious superstition.[35]

The hard scientific truth, then, was that human species were separately created; they had not changed and could not change. "The monuments of Egypt teach us that five thousand years ago the negroes were as different from the white race as they are now." It was, then, the obligation of science "to settle the relative rank among these races, the relative value of the characters peculiar to each, in a scientific point of view." However difficult, "as philosophers it is our duty to look it in the face." What Agassiz saw when he looked the Negro in the face was an apathy and indifference to civilization in striking contrast with the Native American. "The indomitable, courageous, proud Indian,—in how very different a light he stands by the side of the submissive, obsequious, imitative negro, or by the side of the tricky, cunning, and cowardly Mongolian! Are not these facts indications that the different races do not rank upon one level in nature?" Such had been God's plan, and we—including politicians, presumably—would be well advised to conduct human affairs "guided by a full consciousness of the real difference existing between us and them . . . rather than by treating them on terms of equality."[36]

Despite his protestations, Agassiz knew that his arguments added the authority of science to the support of slaveholders. He was quickly captured by the team of Nott and Gliddon, the most extreme of America's racial scientists. The Egyptologist George R. Gliddon (famous for arguing that the ancient Egyptians had been white, not colored) and Dr. Josiah Nott, of Mobile, Alabama, were disciples of the Philadelphia paleontologist Samuel George Morton, famous for measuring human skulls to determine that races were distinct and unchanging, hence real rather than nominal categories. Morton, Nott, and Gliddon had been passionately gathering converts, and in 1846, when Agassiz was staying with Morton in Philadelphia (where he had his memorable encounter with racial otherness), he became their most important catch: not only was he a European scientist with impeccable credentials, but as a European his hands were clean on the slavery issue. Here was a truly objective and scientific voice, and it spoke on their behalf. The campaign of Nott, Gliddon, Morton, and Agassiz advanced to remove the final obstacle: the apparent fertility of "hybrids," which would seem to confound the argument that species were biologically separate. By the time of Morton's sudden death in 1851, the "American School of anthropology" was solidly established and had the scientific field virtually to itself. Agassiz's three articles were widely endorsed, and in the United States only John Bachman, the zoologist in Charleston, South Carolina, widely known for his work with John James Audubon, openly contested their doctrines.[37]

When Nott and Gliddon published their "fighting book," Types of Mankind, in 1854, Agassiz contributed a "Sketch" which emphasized that Morton's definition of species as *primordial organic forms* meant that species of man were equally fixed and unchanging. Agassiz then outlined the eight zoological provinces and the eight corresponding species of mankind, complete with a map and a four-page pullout tableau of differences. Nott himself called attention to Agassiz's crucial role in their campaign: "those grand Truths, for which I have long 'fought and bled,' are at last established by the unanswerable 'Sketch' of our chief naturalist, Prof. AGASSIZ." Agassiz also appeared, even more disturbingly, in Nott's discussion of comparative brain anatomy: "Prof. Agassiz also asserts, that a peculiar conformation characterizes the brain of an adult Negro. Its development never goes beyond that developed in the Caucasian in boyhood; and, besides other singularities, it bears, in several particulars, a marked resemblance to the brain of the orang-outan. The Professor kindly offered to demonstrate those cerebral characters to me, but I was unable, during his stay at Mobile, to procure the brain of a Negro." Types of Mankind sold well and prompted a flurry of reviews, but the controversy was largely spent. Nott and Gliddon brought out a second volume, Indigenous Races, in 1857 (again with a contribution by Agassiz), but shortly afterward Gliddon died of an opium overdose in Honduras. Nott went on to become a Confederate surgeon and a convert to Darwinism, Agassiz to insist to the end that the white and black races were separate and unequal. William Stanton concludes that the power of the American School to do social damage by upholding slavery was blunted by its antibiblical stance, which made it difficult for southerners to accept its "proferred assistance."[38]

Yet the social damage inflicted on American society was profound. The literature of the American School of anthropology is replete with stereotypes that still haunt America today, allied with the "scientific" certainty that biology is forever: a leopard cannot change its spots. As species are fixed and immutable, so must races be, whether conceived as physical way stations through which the human spirit passed (or to which it regressed) on its upward journey, or as fixed and permanent aboriginal entities. The lack of an effective scientific opposition, at least in the United States, meant that Emerson—committed as he was to scientific truth and the certainty that truth was single and coherent,—was left on his own to confront the "facts" as presented by science.

Given this, Emerson's ability to question established science on moral grounds is remarkable. In English Traits he used the confusion of scientific opinion on racial boundaries to deconstruct the idea of race as a "real" category, arguing against majority opinion that races do mix together and intergrade, and even that the strength of a society might arise from such creole mixtures. Ultimately the concept of race was a convenient fabrication—a social construction. By contrast, in his contribution to Types of Mankind Agassiz had insisted that the fixity of races was part of a plan, "determined by the will of the Creator" and uniting "all organized beings into one great organic conception." Otherwise we would "run inevitably into the Lamarkian development theory, so well known in this country through the work entitled 'Vestiges of Creation,' " abandoning us in a world without plan or order.[39]

Emerson's task was to resolve the contradiction highlighted by Agassiz between the apparent fixity of kinds and his own belief in unlimited human possibility. On the one hand, Emerson had to accept as fact that Indians and blacks were less developed, hence a link with the animal kingdom. "*Races.* Nature every little while drops a link. How long before the Indians will be extinct? then the negro? Then we shall say, what a gracious interval of dignity between man & beast!" Also, alas, there was the fact that a race once vanished can never return. Extinction was forever: "You cannot preserve races beyond their term. St Michael pears have died out, and see what geology says to the old strata. Trilobium is no more except in the embryonic forms of crab & lobster." That he accepted these things did not mean he was unconflicted about them, as is suggested by his dry little comment: "The whole circle of animal life, internecine war, a yelp of pain & a grunt of triumph, until, at last, the whole mass is mellowed & refined for higher use,—pleases at a sufficient perspective."[40]

Meanwhile, back in the thick of the midworld, one essential weapon was provided by the very developmental theory Agassiz had rejected: the "melioration" in familiar domestic species—pears, sheep, horses—was our only hint to explain human origin, for they, like all beings, bore a "family likeness" to man: "All natural history from the first fossil points at him. The resemblances approach very near in the satyr to the negro or lowest man, & food, climate, & concurrence of happy stars, a guided fortune, will have at last piloted the poor quadrumanous over the awful bar that separates the fixed beast from the versatile man." Emerson used this passage in *English Traits* to eviscerate the standard and relentlessly repeated argument that Egyptology proved that races and species could not change; five millennia recorded not a jot of difference. On the contrary, said Emerson. "The fixity or unpassableness or inconvertibility of races, as we see them, is a feeble argument, since all the historical period is but a point to the duration in which nature has wrought. Any the least & solitariest fact in our natural history has the worth of a *power* in the opportunity of geologic periods." The correct lesson from geology was the power of deep time to move mountains and transform species. Emerson's application of the deep time he had long ago learned from Lyell operated here to move him surprisingly close to one of Darwin's key arguments, although Emerson's blithe teleology—"All natural history" points at man—would have made Darwin wince. Yet Emerson's most important point here was that we *are*, somehow—through "guided fortune"—helped over the "awful bar" at last, and this applied to everyone, even that "lowest man," the negro.[41]

Emerson's views on race have precipitated considerable scholarly debate. Cornel West maintains that race circumscribed Emerson's sense of the worth and dignity of human personality, and that Emerson's reflections on race are the pillar for his turn toward "history, circumstance, fate, and limitation" in his later years. In those years, according to Nicoloff, Emerson "was progressively spending less and less time bathing [him]self in the blithe currents of universal being and more time scanning the iron pages of geological and biological history," wherein he read the destiny of mankind in its racial seed. West fears that Emerson's views can be appropriated to a defense of white domination, as in fact Emerson's own language can be used to demonstrate; yet Eduardo Cadava counters that Emerson's thought "cannot be reduced to the project of racism," because his texts are too "open, multiple, and fragmented"—parts connect with racism, but parts resist it as well.

I conclude that whether race for Emerson was biological or historical (and it was both), what counted was that it was part of fate, and of "appalling" importance. In his antislavery work, Emerson enacted the same project he described in his essay "Fate," converting fate to power: when passed through the fire of thought, race was not sustained as a stable and limiting category but destabilized and fragmented. What remained firm was the commitment to the highest truth of all, Emerson's "infinitude of the private man," of the God within, truth at the bottom of the heart. So Emerson was able to use the controversial development theory to create two key arguments that would counteract rather than reinforce racial determinism: first, all beings did "meliorate" or progress in time, including the "lowest man"; second, any apparent lack of melioration showed not the fixity of race but the limitation of our perspective in the abyss of geological time.[42]

Thus, that races might not be—*could* not be—fixed by biology was Emerson's saving insight, which he leveraged into powerful arguments for the abolition of slavery. *Pace* Agassiz, Emerson had decided long before that the best use of science *was* to advance the human condition, not to enforce human limitation and inequality. As Robinson points out, the antislavery movement rendered race "morally unacceptable for Emerson as a causative explanation." Passage of the Fugitive Slave Act in 1850 precipitated Emerson's abstract theorizing about race into the need for decisive language. Race might be a convenient device by which to organize a literary work such as *English Traits,* but such a nominal fiction as race had no place in politics, where it was used by real men and women to condemn equally real men and women to slavery.[43]

Len Gougeon has established that Emerson's 1844 antislavery address "would propel him in the direction of active social protest, the natural result of his transcendentalism and a phenomenon that would accelerate throughout the 1840s and 1850s." Similarly, as Albert von Frank makes clear in his study of Emerson's role in the 1854 trial of the fugitive slave Anthony Burns, "Emerson was a force in antislavery because of his idealism, not in spite of it." Emerson believed that political action must be based on truth, and as he said in his 1855 "Lecture on Slavery," "Truth exists, though all men should deny it." Nor in this lecture did Emerson hesitate to draw his standard link between the truth of the physical and moral universes: "The idea of abstract right exists in the human mind, and lays itself out in the equilibrium of nature, in the equalities and periods of our system, in the level of seas, in the action and reaction of forces, that nothing is allowed to exceed or absorb the rest; if it do, it is disease, and is quickly destroyed." In the end, all nature cried out against slavery. At times of profound challenge, Emerson always returned to the lodestone that had pointed the way ever since the death of Ellen: "I will . . . *think the truth* against what is called God . . . & say, 'Truth against the Universe.' " The higher law that had guided him since the crises of his youth would arm him against the corrupted laws of the national government. The same theories of development that others used to rationalize slavery and racial inequality became, in Emerson's hands, new weapons in the fight against racism.[44]

"QUETELET FATE"

Finally, one last piece of the puzzle remained. Granted that development, arrested and progressive, was the method that lifted and shaped great natural groups, from species to nations; amelioration might then advance the race, but only "at a sufficient perspective." As Emerson knew, the wave moved on while the particles that composed it did not. So, what of the individual? How did science link the single soul to the social body, mind and moral law to that inviolable physical law? Chambers had been quite solemn on this question. The workings of the law meant that the individual must suffer. The "Great Ruler of Nature" had established two sets of laws, physical and moral, both absolute and unswerving; "But the two sets of laws are independent of each other. Obedience to each gives only its own proper advantage, and not the advantage proper to the other." In this great scheme "the individual . . . is to the Author of Nature a consideration of inferior moment." In order to perfect the species, the individual must be left to "chance amidst the *mêlée* of the various laws affecting him." The universe was exactly as fair as a lottery.[45]

To the individual thrashing about in its grip, the universe might seem chaotic, but science was teaching that the victim must take the larger view. Chambers observed that the mind, which seemed so "irregular and wayward . . . wild and impulsive!" was really just as chained as any natural phenomenon. "The irregularity is exactly of the same kind as that of the weather." Local wildness, taken in the mass, surrendered to statistical regularity: from year to year one in 650 Frenchmen would commit a crime; from week to week the London police took in the same number of drunks, and post offices in large cities collected from year to year the same number of dead letters. It was this revelation that settled the nature of the elusive connection between mind and matter. "This statistical regularity in moral affairs fully establishes their being under the presidency of law. Man is now seen to be an enigma only as an individual; in the mass he is a mathematical problem." Mental action, "proved to be under law, passes at once into the category of natural things," and the metaphysical nature of mind evaporated. In *Explanations,* Chambers credited this insight to the great work of Adolphe Quetelet, and in 1849 Emerson turned directly to Quetelet himself.[46]

Emerson already had a working knowledge of probability from his reading in the early 1830s of Laplace's *Philosophical Essay on Probabilities,* in which the French astronomer had decreed that the great questions of life were "for the most part only problems of probability"—or, in other words, ignorance: the more we know, the more mere probability approaches to absolute certainty. This was the meaning behind Laplace's famous assertion that an intelligence knowing the complete state of the universe at any one moment could predict all the future: "it would embrace in the same formula the movement of the greatest bodies of the universe and those of the lightest atom; for it, nothing would be uncertain and the future, as the past, would be present to its eyes." This foundational assertion of gnomic science lay behind Emerson's statement, in 1849, that Plato's "celestial geometry" had anticipated the astronomy of Laplace: "These expansions are organic," from the lightest atom to the greatest bodies. Laplace cautioned that the human mind, even in its closest approach to perfection, astronomy, offered "a feeble idea of this intelligence," and he cautioned as well that since each single human mind had but a small piece of the truth, we should "examine critically" our own opinions and indulge the opinions of others. Or as Emerson put it, many needles were needed to detect true north.[47]

As for chance—it didn't exist. What appeared to be chance was only a symptom of human limitations, our inability to see the universe from the viewpoint of Laplace's all-comprehending intelligence. Furthermore, because of our ignorance, Laplace advised, we must be extremely wary of tinkering with social institutions, for we cannot know of the evils change will unleash. "The theory of probability directs us to avoid all change; especially is it necessary to avoid the sudden changes which in the moral world as well as in the physical world never operate without a great loss of vital force." So probability eliminated not only chance, but revolution.[48]

Quetelet extended Laplace's reasoning much further into the social sphere. Originally, like Laplace, an astronomer, Quetelet set his sights on applying the law of errors that guided the deviations characteristic of *astronomical* observation to the observation of *human* variability—which became, in effect, the study of errors around a mean. Quetelet's "social physics," as he called it, was modeled on celestial physics. As the word *physics* implies, Quetelet set himself quite consciously to be the Newton of the social sphere, reducing society's perturbations to scientific law. Increasing awareness of government's inability to control society (dramatized so memorably by the French Revolution) had turned attention to society as the more fundamental reality, leading to the hope that "statistics" (from the German *Statistik*, or "description of the state") would show the path from the confusion of politics to the clarity of rationally guided social reforms, an orderly reign of facts in the hands of enlightened experts. Quetelet entered the field in the early 1830s, just as Emerson was reading Laplace and recommending in his early science lectures that the trust felt in physics be transferred to the moral order. As Theodore Porter explains, "Social physics was an elaborate metaphor that integrated Quetelet's genuine concern for the advancement of scientific knowledge with his desire to turn science to the promotion of sound government and social improvement. Mathematics would bring order to the apparent social chaos." The moral of statistics was hardly that chance ruled—such an interpretation did not acquire currency until century's end—but rather that science could effectively be extended into phenomena previously regarded as deeply resistant to scientific investigation. Even individual actions—like the flip of a coin—were taken to be, not chance, but only too causally complex to predict.[49]

It might seem hard to escape the materialism and determinism implied by Quetelet's statistical science, but in his *Treatise on Man* (1835), Quetelet himself insisted he was neither materialist nor fatalist, only a truth-seeker following the careful examination of facts, whose insights ennobled our vision of the moral universe just as Newton's did the physical universe. The truth Quetelet was drawn to was the truth of *masses:* he who examined the laws of light only in a single drop of water would find "the brilliant phenomenon of the rainbow" unintelligible, and could hardly even imagine its existence unless he happened to observe it. Truth emerged in the mass, not in the individual. Make enough observations, and even the capricious was seen to be constant—crime, for instance: "experience proves that murders are committed annually, not only pretty nearly to the same extent, but even that the instruments employed are in the same proportions." Even criminal acts were not monstrous exceptions to the law, but were in fact necessary consequences of the organization of the social state—which meant that modifying social institutions would ameliorate crime. For *"society prepares crime, and the guilty are only the instruments by which it is executed."* He who was tried and punished was "an expiatory victim for society" whose crime was only the result of circumstances. A Frenchman, for instance, had one chance in 4,462 of being accused of a crime, and, once charged, 61 to 39 of conviction; the numbers from year to year agreed pretty nearly, as if thousands of individuals were driven "in an irresistible manner" to tribunals and condemnations. The hinge for Quetelet, as for Emerson, was the body: "As a member of the social body," man was subject to "the necessity of these causes . . . but as a man, employing all the energy of his intellectual faculties, he in some measure masters these causes." As we have seen before, science showed how the organic subject could turn and become the gnomic master.[50]

Thus, Quetelet's research focused on that "social body" rather than on the "peculiarities distinguishing the individuals composing it." He presumed that the scientist and legislator would find his work of more interest than the literary man and the artist, who would endeavor to understand the very "peculiarities" Quetelet set aside—although, as he added shortly, the knowledge gained from social physics would surely aid even the writer and artist in seizing the characteristic features of those around him, resulting in a more lifelike and expressive art.[51]

At the center of the "social body" was the "social man," who "resembles the centre of gravity in bodies: he is the centre around which oscillate the social elements." In fact this being was a fiction, composed of the averages of all parts of the system, and he would vary with local conditions; yet overall, Quetelet believed that the properties of this "average man" might be the most important contribution of his science, for the average man "is in a nation what the centre of gravity is in a body," that is, the center from which we may apprehend equilibrium and motion. Quetelet's average man was no lowest common denominator, but "the type of perfection," and whatever differed from the average "would constitute deformity or disease." He was, then, a model of harmony and proportion, an ideal of beauty—or at least he became so as society approached perfection.[52]

Quetelet's "average" man translated, then, into the *great* man, who would "represent all which is grand, beautiful, and excellent." Quetelet here borrowed the language of Victor Cousin: " 'The great man . . . is not simply one individual, but he has reference to a general idea, which communicates a superior power to him' " while giving him individuality. He was " 'the harmonious union of particularity and generality' " who represented the " 'general mind of his people.' " Quetelet concluded this extraordinary gloss on Emerson's *Representative Men* and *English Traits* by affirming that a man was great because "he is the best representative of his age," and so was proclaimed its "greatest genius." Science alone was truly progressive, and so science alone could give the measure of the development of human nature. In the end, Quetelet agreed with Cousin: " 'Thus, give me the series of all the known great men, and I will give you the known history of the human race.' " The ultimate consequence of Quetelet's research, at least according to him, was profoundly optimistic. The advance of civilization would contract the wild and monstrous oscillations of suffering and vice into ever-narrower limits. Quetelet looked ahead to "the perfectibility of the human species" in harmony with all the richness of natural diversity.[53]

Quetelet may have asserted that his compelling arguments were, seen truly, liberating not fatalistic; but Emerson was less than fully convinced. Upon reading Quetelet's *Treatise on Man* in 1849, he concluded: "One must study Quetelet to know the limits of human freedom. In 20,000, population, just so many men will marry their grandmothers. Doubtless, in every million, there will be one astronomer, one mathematician, one comic poet, & one mystic." His suspicions were confirmed when he read John Herschel's review article on Quetelet in 1850. The news here was mixed: yes, Quetelet's science did succeed in eliminating chance. But in eliminating chance, the new science also eliminated freedom: "Taken in the mass, and in reference both to the physical and moral laws of his existence, the boasted freedom of man disappears; and hardly an action of his life can be named which usages, conventions, and the stern necessities of his being, do not appear to enjoin on him as inevitable, rather than to leave to the free determination of his choice." By the time Emerson first addressed Quetelet's research in public in *Representative Men,* he was referring to "the terrible tabulations of the French Statists" who reduced every "piece of whim and humour . . . to exact numerical ratios"; yet he was nevertheless able to translate the law of ratios into the revolution led by Swedenborg, the scientist/mystic, who gave science "guidance and form and a beating heart."[54]

By 1860, however, Emerson's tone was grim and militant. The "new science of statistics" had become one of the linchpins of "Fate," "one more fagot of these adamantine bandages," for it ruled that the most casual and extraordinary of events "become matter of fixed calculation." Out of every million there would be one astronomer, and one or two in a dozen millions of "Malays and Mahometans. . . . In a large city, the most casual things, and things whose beauty lies in their casualty, are produced as punctually and to order as the baker's muffin for breakfast. Punch makes exactly one capital joke a week; and the journals contrive to furnish one good piece of news every day." The laws applied alike to "famine, typhus, frost, war, suicide, and effete races, [which] must be reckoned calculable parts of the system of the world." He concluded: "These are pebbles from the mountain, hints of the terms by which our life is walled up, and which show a kind of mechanical exactness, as of a loom or mill, in what we call casual or fortuitous events."[55]

The 1850s had for Emerson been an education in fate. As the decade began, he tried in his journal to work out the "two statements or Bipolarity": "My geometry cannot span the extreme points which I see. I affirm melioration,—which nature teaches, in pears, in the domesticated animals, and in her secular geology, & this development of complex races. I affirm also the self-equality of nature; or, that only that is true which is always true. . . . But I cannot reconcile these two statements. I affirm the sacredness of the individual. . . . I see also the benefits of cities, and the plausibility of phalansteries. But I cannot reconcile these oppositions." Soon afterward he suggested a pathway, if not a resolution: "The question of the Times is to each one a practical question of the Conduct of Life. How shall I live? Plainly we are incompetent to solve the riddle of the Times. Our geometry cannot span the huge orbits of the prevailing Ideas, & behold their return,

& reconcile their opposition. We can only obey our own polarity,—every mind has a polarity,—& that will finally guide us to the sea."[56]

But the sea was no safe haven. On July 19, 1850, Emerson had to record that "Margaret dies on rocks of Fire Island Beach within sight of & within 60 rods of the shore. To the last her country proves inhospitable to her; brave, eloquent, subtle, accomplished, devoted, constant soul!" Emerson gave Margaret Fuller's senseless death by shipwreck a bitter memorial in his essay "Fate": "But Nature is no sentimentalist,—does not cosset or pamper us. We must see that the world is rough and surly, and will not mind drowning a man or a woman; but swallows your ship like a grain of dust." In the late 1840s he had ventured, "Put men to death by principles, & they will not grumble"; but it was hard to see what principle had to do with drowning Margaret Fuller in the fullness of her life, just as she was returning home from Italy.[57]

Emerson wrestled with the issue through the 1850s, as the fight against slavery pressed ever harder on the question of one's proper conduct of life. Barbara Packer points out that "Quetelet's works reached Emerson just at the moment when his own need to think about the freedom or unfreedom of the will was intensified by the political crises of the early 1850s." Once again, science became the means of addressing crucial social pressures. On the surface Emerson's life settled into a solid and secure round of family, writing, lecturing, gardening, and frequent socializing with friends from Alcott and the prickly Thoreau to the upper-crust warmth of the Saturday Club. Yet below the surface, Emerson was waging war. By the time he assembled his new work into his new book, *The Conduct of Life* (1860), the boyish and confident glee of the 1833 lecture, "The Relation of Man to the Globe"—in which the universe seemed to fit like a well-tailored suit—had curdled into a grim realism that knew full well the true and total cost, as well as the ascetic joys, of a world without chance, a world zipped tight into a bipolar unity.[58]

Quetelet had been essential to that view. Emerson's journal recurred to Quetelet several times during the 1850s. "I accept the Quetelet statistics," he stated in 1854; "In a million men, one Homer, & in every million." Packer observes that a Spartan hope is buried here: it may take the million, but Homer will come at last. As Emerson continued, the probabilistic curve, "which looks inexorable when it is predicting suicides and crimes, predicts genius with just as much force"; even the rubbish would contain a little gold. "Most men are rubbish," Emerson asserted in 1855, "& in every man is a good deal of rubbish. What quantities of fribbles, paupers, bed-ridden or bed-riding invalids, thieves, rogues, & beggars, of both sexes, might be advantageously spared! But Quetelet Fate knows better; keeps everything alive, as long as it can live; that is, so long as the smallest thread of public necessity holds it on to the tree Igdrasil." "Quetelet Fate" had affirmed that every particle, every individual, was needed—even if it wasn't needed much.

What else, exactly, did "Quetelet Fate" know? That "we are used as brute atoms, until we think, then we use all the rest." Such was Emerson's answer, explored five years later in his symphonic, even Wagnerian, essay "Fate," the climax of over thirty years of wrestling limitation into opportunity, submission into dominance. The problem was profoundly moral: *How* to live, when we "are incompetent to solve the times"? How to "obey our own polarity," when our "geometry" cannot span and reconcile the extremes of necessity and liberty, individual and world?[59]

The essay puts the answer, self-trust—or obedience to each thought, each moment, trusting in the ultimate emergence of a greater harmony—through the severest test of circumstances. Margaret Fuller's own fate showed the wild and incalculable "ferocity" of nature, but such "shocks and ruins" as shipwreck, plagues, and earthquakes are but incidental fears compared with the "stealthy power" of nature's more subtle laws. Emerson turns first to the lesson learned from Oken: "The menagerie, or forms and powers of the spine, is a book of fate: the bill of the bird, the skull of the snake, determines tyrannically its limits." The body becomes not the gnomic unfolding of the soul but its iron prison: "Every spirit makes its house; but afterwards the house confines the spirit." The bodily chance of black eyes or blue, of facial angle, of poverty or wealth, of birth with "the moral or with the material bias," all "predetermine" our fate and even our politics.[60]

As his turn to Oken suggests, it is science that has presented us with the two sides of this dilemma. "In science, we have to consider two things: power and circumstance. . . . We have two things,—the circumstance, and the life. Once we thought, positive power was all. Now we learn, that negative power, or circumstance, is half. Nature is the tyrannous circumstance." "The book of Nature," the book laid down leaf by leaf through the eons, from granite, to slate, to coal, to mud and vegetation and animals and man and races of man, born to live their term then "come no more again," turns out to be "the book of Fate." To read nature is perforce to read our fate, the "cropping-out" of "reality" in our "planted gardens": the reality of racial limitation; the inevitability of statistical determinism.[61]

But in the essay's famous turning, the same science that decodes the grim book of nature also empowers us to rewrite that book. For "there is more than natural history," Emerson insists. "If Fate follows and limits power, power attends and antagonizes Fate." If organic nature walls us in through the limitations of natural history, our gnomic intellect will seize the law that defines us and turn it into a weapon against those very limitations: we *are* part of nature; so if her weakness is our weakness, so is her strength ours to command. "If the Universe have these savage accidents, our atoms are as savage in resistance." And as we are yet more than atoms, we can do more than merely resist: "The revelation of Thought takes man out of servitude into freedom. . . . If the light come to our eyes, we see; else not. And if truth come to our mind, we suddenly expand to its dimensions, as if we grew to worlds. We are as lawgivers; we speak for Nature; we prophesy and divine."[62]

Men think they serve two gods, one of the house and social circles, art, love, and religion; another of mechanics, steam and climate, trade and politics; but the moral and physical universes are really one after all, and coordinate. "The divine order does not stop where their sight stops." Where sight stops, thought and intellect penetrate:

> Fate, then, is a name for facts not yet passed under the fire of thought;— for causes which are unpenetrated.
> But every jet of chaos which threatens to exterminate us, is convertible by intellect into wholesome force. Fate is unpenetrated causes.

Learn to swim, and the wave that drowns you will carry you. This makes Fuller's death less a blow of fate than a failure of intellect.[63]

Emerson's gnomic figure is wound as tightly as the nautilus, sealing off the organic past to expand into the infinite future: "Behind every individual, closes organization: before him, opens liberty,—the Better, the Best." As the essay turns toward a conclusion, the "intricate, overlapped, interweaved, and endless" structure of stroke and return, shock and recoil, fate and freedom, opens into a "web of relation" in which "person makes event, and event person." In this compensatory universe, the event is contained in the soul even as "eyes are found in light; ears in auricular air; feet on land; fins in water; wings in air; and, each creature where it was meant to be, with a mutual fitness. Every zone has its own *Fauna*. There is adjustment between the animal and its food, its parasite, its enemy. Balances are kept." However confining it may seem, this is a doctrine not of limitation but of liberty, at least to those who know, in Emerson's opening question, *how* to live, how to conduct their lives: "The planet makes itself. The animal cell makes itself;—then, what it wants. . . . As soon as there is life, there is self-direction, and absorbing and using of material. Life is freedom." In other words, nature may seem deterministic, but the polar game of cause and effect wraps us up into the process. We become "golden averages, volitant stabilities, compensated or periodic errours, houses founded on the sea," players in our own fate, hence *self*-determined, however disciplined by the rules whose very constraints promise liberation.[64]

All living beings may be self-directing, but only one living being, man, may be on both sides of the balance at once: may be simultaneously given by the law, and giver of the law. Between the two realms—given and giver, nature and intellect—yawns the gap that man is meant to fill. As Emerson wrote in "The Method of Nature," "A link was wanting between two craving parts of nature, and he was hurled into being as the bridge over that yawning need, the mediator betwixt two else unmarriageable facts." Poised on the hinge between "two craving parts of nature," man becomes the ultimate hybrid, the offspring of God and nature, spirit and matter—a "golden impossibility" living out the absurd contradiction between ideal and real. The human can be purified as divine, man as a spiritual being springing out of brute nature; or as brute, man as a fallen atavism groping toward a vanishing transcendence. This stark choice sets the conditions for Emerson's figure of Man the Intermediary, the tightrope walker trapped between the stars and the mud:

> Man is . . . a stupendous antagonism, a dragging together of the poles of the Universe. He betrays his relation to what is below him,—thick-skulled, small-brained, fishy, quadrumanous,—quadruped ill-disguised, hardly escaped into biped, and has paid for the new powers by loss of some of the old ones. But the lightning which explodes and fashions planets, maker of planets and suns, is in him. On one side, elemental order, sandstone and granite, rock-ledges, peat-bog, forest, sea and shore; and, on the other part, thought, the spirit which composes and decomposes nature,—here they are, side by side, god and devil, mind and matter, king and conspirator, belt and spasm, riding peacefully together in the eye and brain of every man.

So we go—perpetually oscillating between the poles of transcendence and experience, eternity and a Tuesday afternoon, filling the midworld with our progeny.[65]

Why has Emerson structured the universe of "Fate" across these particular "stupendous antagonisms" of Fate and Freedom, Circumstance and Power, Nature and Man, Necessity and Intellect? The whole thrust of his career was not to resist but to accept the modern world, to celebrate technology as a natural evolution, even as he elevated "Nature" into a transcendent principle of value untouchable by human will. The two seem deeply contradictory yet seamlessly articulated. Here I have found Bruno Latour's recent analysis of modernism helpful, for it suggests that the structuring of Emerson's career was not accidental but representative; Emerson becomes, via Latour's analysis, America's representative modern man. To borrow Latour's terminology: modernism has succeeded so well because it designates two sets of cooperating but entirely different practices that must remain distinct if they are to remain effective: "purification," which separates nature and culture into two utterly distinct zones, nonhuman and human, mechanism and passion, objective and subjective, science and poetry; and "translation," which by mixing nature and culture creates the hybrids that constitute our lived environment.

Modernism segregates these two practices of translation and purification, because in a world that demands purity, hybrids are forbidden; therefore, the necessary mediation must be invisible, and the hybrids that fill our universe must be swiftly resolved back into the pure categories of natural or human. The mediators themselves aspire to be transparent (as in Emerson's poet, for example, or his vision of himself as a "transparent eye-ball"), for "they merely transport, convey, transfer the power of the only two beings that are real, Nature and Society"; acquiring, in the familiar Baconian formulation, power through obedience. They become transparent that the Law may shine through them; or, if they must cast the shadow of the material, their axis shall be aligned with the axis of things, and their shadow point to the light that is the Source.[66]

In this analysis, Emerson emerges as a premier modernist because he was instrumental in fashioning and popularizing modernism in the United States. Emerson's dialectical or oscillating movement across the poles of his thought—Man and Nature, Freedom and Fate, Power and Necessity—the movement which produces his essays, is exciting because it *is* productive, and that is the very point of modernism. The more objects are purified into the polar antagonisms of Nature and Society, the more the equatorial center can proliferate in such hybrids as gardens and steam engines: gardens which conduce the products of nature to cooperate with the designs of fine art; steam engines which conduce the elements of nature to cooperate with the designs of machine art. But each must then be whisked back into its pure category, Nature or Society: *ergo,* we haunt ourselves with modernism's dark, recurring drama of the pure and timeless garden of nature, violated repeatedly by the foul engine of human progress. "Balances," as Emerson said, "are kept." Or, quoting again the racial theorist Robert Knox, "Nature respects race, and not hybrids."

Yet every move toward the poles evacuates the center, and so the unbearable emptiness—that "dumb abyss"—must be repopulated with new acts of translation and mediation. The trajectory of "Fate" builds to this insight: "The whole world is the flux of matter over the wires of thought to the poles or points where it would build." As Latour declares, by now our whole world is crammed so full of hybrids built by modernism, hybrids we have no idea any longer how to purify—rainbow coalitions, managed wilderness, ozone holes, global warming, emergent viruses, cloned animals—that we are forced to acknowledge that there simply *is* no pure nature or pure society left anywhere. We have been that successful at Emersonian acts of taking the world into the self and projecting the self into the world. Only Latour's point is that, despite the claims of modernism, there never *was* any pure nature or pure society; all along we have created and lived among hybrids of human will and natural possibility. "We have never," he proclaims, "been modern." If this is the geography of Emerson's universe, then perhaps, as archaeologists turn over the shards of the ideals that bound it, he can be honored as Modernism's own Representative Man, whose transcribed experience renders us visible to ourselves.[67]

And so "Fate" ends, like *Nature,* with blueprints for a construction project. Emerson continues to show how to build the modern world. Yet within this similarity lies a significant difference: in 1836, Emerson had exhorted his readers to "build, therefore, your own world," in confidence that matter would smoothly "unfold" the "pure idea in your mind." In "Fate," the world we are to build is more obviously a hybrid world, a mixture of human will with technical possibility. "Let us build altars to the Blessed Unity," Emerson concludes—the Unity or organic whole "which holds nature and souls in perfect solution, and compels every atom to serve an universal end." But this is not enough, this determinist organicism; and so he continues: "Let us build altars to the Beautiful Necessity, which secures that all is made of one piece . . . to the Necessity which rudely or softly educates him to the perception that there are no contingencies, that Law rules throughout existence, a Law which is not intelligent but intelligence . . . it dissolves persons; it vivifies nature; yet solicits the pure in heart to draw on all its omnipotence." David Van Leer worries that "it is perhaps too exclusively with this beauty of necessity that the essay ends," yet in insisting on "the pictorial dimension of nature—the necessity of beauty as well as the beauty of necessity—Emerson outlines a proto-pragmatic theory of truth that permits both general stability and local freedom." The stable system can be endlessly redescribed, just as any point in the cosmos can round out a horizon or the arch of a rainbow. Beauty lures us in; the law solicits us to turn beauty into necessity. From organic wholeness Emerson has turned, one last time, to gnomic law.[68]

Emerson uses the organic metaphor to bind the universe into just that—a *universe*, a harmonious whole in constant flux, but contained by the controlling force of the gnomic, centering Law. The gnomic object—such as the ecstatic unfurling of the chambered nautilus, its beauty tightly determined by the logarithmic spiral—is a sign of lawfulness throughout nature, a lawfulness that is at once fully natural, unconstructed, and spontaneous, and fully artful, a beautifully wrought, formally perfect object. Such gnomic objects as shells and trees—objects that embody and thereby make self-evident the law of their formation—might be laid on the altar of the "Beautiful Necessity" that ends Emerson's essay; yet the point of the essay is not just to build or adorn altars to natural law, but to *seize* that law, to turn humanity into "lawgivers" who "speak for Nature." Law becomes the copula joining and correlating the purified zones of nature and society, activating each through the other: thus man, that "stupendous antagonism," by marrying the purified opposites of nature and society becomes *himself* a beautiful necessity, the gnomon or necessary continuation of nature. Emerson's concept of law is therefore itself a hybrid joining human intelligence with material reality through the agency of transcendent law, intelligence that is also Intelligence; a hybrid, then, of transcendence and immanence, "natural" and "moral" laws conjoined in a formation that not only commands our obedience, but through which our obedience allows us to command nature and society. Emerson, too, becomes a hybrid, a poet-scientist who would legislate for humanity.

To build altars to the Beautiful Necessity is to sacralize law as transcendent principle even as it turns "transcendent" principle into a very earthly force, one that can assemble bricks and mortar, men and women into real and current institutions, the homes, schools, universities, churches, and offices that housed Emerson, his friends, and his associates as they collectively forged a culture of truth, a culture that turned law into a principle of personal and national growth. To that culture we now turn.

CHAPTER SIX

..

The Solar Eye of Science

WHERE THERE IS LIGHT THERE WILL BE EYES

As Emerson's career drew to a close, his thoughts turned again to science, and to the kind of education in science he wished it were not too late to begin. In 1849 he had addressed another note to Plato: "When our Republic, O Plato! shall begin, the education shall not end with the youth, but shall be as vigorously continued in maturity. We have in nowise exhausted the books. Astronomy invites, and Geology & Geometry & Chemistry. . . . Let us keep the fathers up to as high a point of aim as we do the children. Then you will have a state." Nearly twenty years later, in 1867, he daydreamed about the kind of education he might gain from his scientific friends: "If I were rich, I should get the education I have always wished by persuading Agassiz to let me carry him to Canada; & Dr. Gray to go to examine the trans-Mississippi Flora; & [Jeffries] Wyman should find me necessary to his excavations . . . & I can easily see how to find the gift for each master that would domesticate me with him for a time." One of Emerson's last journal entries, made a few weeks after the traumatic burning of his house in July 1872, extended his longing for world enough and time: "I thought today, in these rare seaside woods, that if absolute leisure were offered me, I should run to the College or the Scientific school which offered best lectures on Geology, chemistry, Minerals, Botany, & seek to make the alphabets of those sciences clear to me. How could leisure or labor be better employed. 'Tis never late to learn them, and every secret opened goes to authorize our aesthetics." Sadly, it was, for Emerson, too late. His memory was already growing clouded, and his power to organize and present the masses of material he had accumulated over many decades was declining. Guided by his daughter Ellen, Emerson would give just a few more honorary lectures in the mid-1870s before withdrawing into the serene privacy of his final years.[1]

Why, in his last productive years, did Emerson's thoughts turn so often to natural science? As the 1872 journal entry indicates, he was interested not in stockpiling the latest scientific information but in enlarging the scope and power of his own aesthetics, grounded as they were in the mind's relation to nature. Science offered the peculiar power of educating the mind, or what Emerson called "the Intellect." Emerson's favorite analogy for this process traces back to Plato's theory of vision in the *Timaeus*. He had found the same idea updated in Goethe's *Theory of Color:* "From among the lesser ancillary organs of the animals, light has called forth one organ to become its like, and thus the eye is formed by the light and for the light so that the inner light may emerge to meet the outer light." Or, as Goethe explained in another of his science writings, self-knowledge comes not from introspection but from engagement with the outer world: "The human being knows himself only insofar as he knows the world; he perceives the world only in himself, and himself only in the world. Every new object, clearly seen, opens up a new organ of perception in us." The light of the eye is latent only until active stimulus brings it out; so the act of attention actually reconfigures the mind, creating "eyes" where none were before.[2]

"Where there is light," Chambers declared, "there will be eyes." Or, as his opponent William Whewell also declared, "As the eye was made for the light, so light must have been made . . . for the eye." The cocreative relationship between eye and light became a figure for the deep affinity of mind and world, the point of articulation between the two created universes. As Emerson mused, "Does not the eye of the human embryo predict the light?" In the same way, every human faculty predicts the universe that will call it out, and reciprocally, nature takes care to plant "an eye wherever a new ray of light can fall." Or the converse: "The fish in the cave is blind; such is the eternal relation between power & use." In Stallo's evolutionary narrative, the metaphor broadened to embrace the creation of life itself: mineral solutions stand "inert and shapeless, until the magic lines of light" polarize individual molecules into organizing activity. "Similarly the warm breath of the Universe impregnates the terrestrial womb with vegetable forms, and the solar eye, as it were, *looks* them into distinct, individual existence. The same with animals." The principle, Stallo asserted, was obvious: life appears only *"when the Whole energizes in a part, or when the Particular enters into relation with the Universal."* Where there is light, then, there *will* be eyes, "looked" into existence by the great "solar eye" of the whole.[3]

This principle—the "solar eye" of science—became the foundation for Emerson's late work. As Kant maintained, mind cannot think by itself, any more than light can be seen without its falling on an object. So the mind thinks through bodies; thought "coins itself indifferently into house or inhabitant, into planet, or man, or fish, or oak, or grain of sand. All are reconvertible into it. . . . Every thing by being comes to see & to know. Work is eyes, & the artist informs himself in *efforming* matter." As attention reconfigures the mind to create new organs of perception, so does "work" create "eyes." The artist, shaping or "*efforming*" matter, shapes himself.[4]

Work *is* eyes, as Emerson had long ago declared to the American Scholar; and the work of science is eyes to the world. In Emerson's late science lectures, this intimate cocreative capacity of mind and nature was embodied in another image: thoughts seem to flow past, carried like objects floating on a river; "Only I have a suspicion, that, as geologists say, 'Every river makes its own valley,' so does this mystic stream. It makes its valley, makes its banks, and makes, perhaps, the observer too." The idea Emerson was developing found support in metaphors drawn from geology, physics, and biological evolution, authorizing for Emerson the truth of his aesthetic insight: "I had rather have a good symbol of my thought, or a good analogy, than the suffrage of Kant or Plato. If you agree with me, or if Locke, or Montesquieu, or Spinoza agree, I may still be wrong: but if the elm tree thinks the same thing,—if running water, if burning coal, if crystals, acids, & alkalis, say what I say, it must be true." The mind that thinks in things thinks the truth: Emerson's scientific universe was made not of atoms but of metaphors.[5]

Emerson was willing to go further still. Analogy, he argued—defined by Plato as "identity of ratio"—was *the* "method" of science, for it resolved "all the variety of structure and element" into a single ratio or design. Through it, all the sciences "illuminate each other"—indeed, through it, all *thought* coheres as one. "All thought analogizes," wrote Emerson. "The symbolism of nature exists, because the mental series exactly tallies with the material series. And who enunciates a law of nature, enunciates a law of the mind." Emerson was confident he had the authority of all science for his analogical method, because the unity of science ensured that one could supplement gaps in one science by analogy with another. In fact Emerson had witnessed this himself, as he reported, when Agassiz, stumped by a problem in embryology, had turned to the mathematician and astronomer Benjamin Peirce with a question about the development of planets, "and presently he got the analogic hint he wanted." Just as attention generates new organs of perception, so does every discovery make "new instruments at the mathematical and philosophical shop," reconstructing the observatory and letting in "new light" on old observations; "Then the eyes of analogy bring it to the students of other sciences." Thus far in modern science, botany, chemistry, geology, and anatomy had all undergone reform, "but our metaphysics still awaits its author. A high analogic mind, a mind which with one *aperçu* penetrates many successive crafts, and strings them as beads on its thread of light, will charm us with disclosing mental structure, as the naturalist with his architectures." What we need, then, is "Wisdom with his solar eye." Since man is "only a sort of compend of the globe with its centrifugence and centripetence, with its chemistry, with its polarity, with its undulation," we will acquire a key to the sublimities of human consciousness "by the solar microscope of *Analogy*. 'Tis the key that opens the Universe."[6]

Emerson's "solar microscope" was the eye of science itself, "looking" the world into existence: "The light by which we see in this world, comes out from the soul of the observer," Emerson wrote. In Plato's original version, the light of the eye allows us to see, because "then like falls upon like, and they coelesce, and one body is formed by natural affinity." Or, as Emerson reiterated, "All knowledge is assimilation to the object of knowledge."

Emerson envisioned science not as a passive observation or accumulation of facts but as an active making of the world to fit our needs and desires: "Thanks to the *hand* of science, the earth fits man. . . . It does not fit him as a farm, or a cave, or a house fits him,—into which he accommodates himself, but as his body fits his mind." Science as *making* forged an alliance between science and poetics, through the Greek root *poiein,* to make or create; however, this act of creation was no longer an active power shaping a passive universe, as in the Neoplatonist Ralph Cudworth's organic theory of creation, but as the essay "Fate" made clear, the *co*creation of opposed powers. What Emerson said of the Imagination and the poet applied equally to Reason and the scientist: "The act of Imagination is, the sharing of the real circulations of the Universe," and the poet "beholds the central identity, and sees an ocean of power roll and stream this way and that, through million channels, and, following it, can detect essential resemblances in things never before named together." What Emerson was describing is the essential creative act, the discovery—or making—of likeness where none was seen before. As he had wished at the start of his career, Emerson by career's end had positioned himself at the source or fountainhead of knowledge, the trunk of the tree of truth from which all subdivisions branch, the point at which science and poetry meet and are one—the core identity or "ratio" at the root of the world, from which all the rest in all its plenitude unfurls.[7]

Truth, in this system, is not a thing found but a thing made, as the river makes its bank or light makes the eye. Both emerge together as twin halves of the process of "*natura naturans,*" or nature becoming. As Carlyle had said, truth "never *is,* alway *is a-being.*" John Burroughs, a close reader of Emerson, picked up this idea and made it central to what he called his "Gospel of Nature":

> Things are not designed; things are begotten. It is as if the final plan of a man's house, after he had begun to build it, should be determined by the winds and the rains and the shape of the ground upon which it stands. The eye is begotten by those vibrations in the ether called light, the ear by those vibrations in the air called sound, the sense of smell by those emanations called odors. . . . Our reason is developed and disciplined by observing the order of Nature; and yet human rationality is of another order from the rationality of Nature. Man learns from Nature how to master and control her. He turns her currents into new channels; he spurs her in directions of his own.[8]

This notion—that things are "begotten" rather than designed, in a process that can be controlled by the disciplined rationality of the mind—is at the heart of what I have been calling "the Culture of Truth": truth is not found but is made, "begotten"—or "created," in a figure tracing back to the Latin root *creare,* "to cause to grow." Truth is "cultured," grown, in a process of "cultivation" that brings all elements of the environment into harmony. Science is intrinsic to this process, for science, by opening nature to our grasp, opens us to ourselves. The solar eye of science "looks" the individual into existence as well, in a vegetative analogy that ties together sun, air, water, and earth, human purpose and natural fecundity, the social whole and every constituent individual. Science, by nature productive and progressive, becomes the key to civilization: "The science of power is forced to remember the power of Science," claims Emerson; "Civilization mounts and climbs." Civilization—a social organization that generates ever more complex wholes composed of ever more "individuated" particles—"looks" each of us into being by virtue of our own existence as participants, our own "obedience . . . to the currents of the world." The metaphor is fundamentally organic, and science, far from being some extraterrestrial intruder, is its sun, its solar center, its beating heart.[9]

EMERSON AND THE CULTURE OF TRUTH

As early as 1838, Emerson declared that he wanted to write "the Natural History of Reason." In "The Method of Nature," his 1841 lecture, he set out to trace the "natural history of the soul." By the late 1840s, Emerson was working out a "new Metaphysics," what he came to call "A Natural History of the Intellect," based on his notion of analogic truth. This ambitious project was conceived during his second visit to England, when, as David Robinson says, he set about "translating the paradigm of the scientific study of nature into an inquiry into the processes of mind and spirit." The first attempt was given in London in 1848, as the lecture series "Mind and Manners of the Nineteenth Century." Emerson developed his ideas through several more lecture series, culminating in "Natural History of the Intellect," a lengthy and demanding philosophy course offered to Harvard students and associates in 1870 and again in 1871. In the normal progress of Emerson's compositional practice, the lectures would have been revised into a book, the capstone of Emerson's lifelong interest in science. It was, as James Elliot Cabot realized while acting as Emerson's literary executor, the "chief task of his life."[10]

Yet already by 1870 Emerson found himself unable to master his own materials, and soon he was forced to turn them over to Cabot and his own son Edward, who were, as Ronald Bosco has shown, unequal to the task. Not that they weren't capable and hardworking editors; rather, Bosco suggests, both were constitutionally unsympathetic to Emerson's approach. The philosophically inclined Cabot, "a strict logician," was ill equipped to appreciate Emerson's wide-ranging spontaneity, and Edward, trained as a medical doctor, was too much the scientific positivist. Indeed, although Emerson appropriated the language of science, his proposed "Natural History" was analogic, not analytic, in approach, hence "more a poetic than a philosophical or scientific construction." As a result, the manuscripts of

Emerson's capstone lectures were dispersed as Cabot and Edward Emerson edited and rearranged them to form several of Emerson's late essays. All that survive from the original lectures are manuscript fragments, notes from several auditors, and two published versions titled "Natural History of Intellect" that bear only a tenuous relation to Emerson's original text.[11]

Although the uncompleted "Natural History of the Intellect" failed to materialize as the keystone of Emerson's last body of work, the structure Emerson was building is nevertheless evident throughout his late essays and lectures. He was, as Bosco observes, much closer to completion than he realized, for the entirety of his work carried forward his central ideas. Moreover, Emerson's later lectures, on which he drew for his Harvard course, did survive, and they provide a trove of riches for any study of Emerson and science, not to mention proof of Emerson's continuing intellectual vitality. In them, Emerson defended his analogic method with an astute observation: anyone attempting to write out the laws of the mind was hampered by the "stupendous peculiarity" that they must be "at once observers and observed, so that it is difficult to hold them fast as objects of examination, or hinder them from turning the professor out of his chair." Given that every man tends "to conquer the whole world to himself," Emerson offered to take up "a modester stand—under protest, however—and . . . content myself with reporting my emphatic experiences." In other words, given that truth must be experienced, *and* that no trustworthy consensus yet existed upon which Emerson could draw for experience beyond his own, Emerson stood firmly by his own sense of truthfulness. As he said, our own experience is spotty, not systematic, and there is "affectation" in assuming that one can chart or model the interior universe by oneself; "But so long as each sticks to his private experiences, he may be interesting and irrefutable."[12]

Thus far Emerson was operating as the steadfast Baconian, suspicious of Okenesque leaps to some higher metaphysical system. He cautioned that "the science of the Intellect . . . intoxicates all who approach it," evoking once again the danger faced by all scientists too much in love with their object to separate themselves from it intellectually. Given the impossibility of separating mind from itself, in studying mind analogy was the only proper scientific method. Emerson understood that the greatest office of natural science was to explain man to himself, and he praised Agassiz for only *pretending* to study turtle eggs: "I can see very well what he is driving at, he means men and women." Just as Agassiz studied mind by way of his turtles, so must one study the mind in the "chamber lined with mirrors" that is nature.[13]

Thus the modest claim that he would offer but a "Farmer's Almanac of mental moods" had a serious purpose: Emerson had a long history of "farming" his mind. As early as "The American Scholar" he was framing the idea of "Culture" as a revolutionary power that would awaken men to "quit the false good and leap to the true." Culture's material metaphor of agriculture, used to link ideas and things, appears in "Circles": ideas "cause the present order of things as a tree bears its apples. A new degree of culture would instantly revolutionize the entire system of human pursuits." A good crop of apples demands the careful culture of each single tree, elevating the individual against the mass "as a drop of water balances the sea; and under this view the problem of culture assumes wonderful interest." Namely, culture allows the mind to possess its powers "as languages to the critic, telescope to the astronomer." Indeed, in the essay "Culture," culture and power become virtually synonymous: "Man's culture can spare nothing, wants all the material. He is to convert all impediments into instruments, all enemies into power . . . until at least culture shall absorb the chaos and gehenna. He will convert the Furies into Muses, and the hells into benefit." The source of this tremendous cultural power is ultimately moral: "The foundation of culture, as of character, is at last the moral sentiment. This is the fountain of power." Science may sweep away the old creeds and necessitate a new faith, "yet it does not surprise the moral sentiment. That was older, and awaited expectant these larger insights." The moral law, then, culminates in science, and the "identity of ratio or design" that relates all the sciences grounds them equally in moral law.[14]

In his late lecture "The Natural Method of Mental Philosophy" (1858), Emerson's review of moral law sweeps together the scientific metaphors that structured his entire career. The first quality of moral law is "Centrality, which we commonly call gravity, and which holds the Universe together. . . . To this central essence answers Truth, in the intellectual world, Truth, whose center is everywhere and its circumference nowhere . . . Truth, the soundness and health of things against which no blow can be struck." After gravity, or Truth, comes "*Polarity*. . . . the principle of *difference, of generation,* of change," which shows itself "in circulation, in undulation," in chemical affinities and in the organic realm, in sex. Emerson continues: "See how the organism of mind corresponds to that of the body. There is the same hunger for food,—we call it curiosity; there is the same swiftness of seizure,—we call it perception. The same assimilation of food to the eater, we call it culture." As food to the eater, so is the world to the knower: the intellect eats the world like a ripe apple, converting the fruit of the earth to the fruit of culture.[15]

Given this physical basis in gravity (or truth) and polarity (or generation), the "culture of truth" that grows out of moral/physical law is at root an agricultural metaphor with a political object. Whewell pondered "by what plans and rules of culture . . . those rich harvests" of scientific knowledge "have been produced which fill our garners." Sir Humphry Davy revealed the political object when he showed that social improvement flowed from the "proprietors of land" down to "the laboring classes" who work it, concluding that the "agriculturalist" and the "patriot" shared the same goal. Men "love their country better, because they have seen it improved by their own talents and industry," and therefore identify their interests with those of their governing institutions. More briskly, Emerson declared: "Intellect; 'Tis a finer vegetation. It has, like that, germination, maturation, crossing, blight, and parasite." Emerson took much pleasure, in several of his late lectures, in playing with this agricultural metaphor: "Man seems a higher plant," he mused, "suggesting that the planter among his vines is in the presence of his ancestors; or, shall I say, that the orchardist is a pear raised to the highest power?"[16]

The notion of man as a higher plant gave Emerson's popular lecture "Country Life" its point and humor, as when the aging speaker suggested that we "draw a moral from the fact that 'tis the old trees that have all the beauty and grandeur." The metaphor could also critique *im*proper culture: "the education of the garden is like the education of the college, or the bound apprentice; its aim is to produce, not sap, but plums or quinces; not the health of the tree, but an overgrown pericarp." Better than such artificial forcing is the education of *nature,* for (echoing Bacon) "it commands, and is not commanded," and so must be trusted: "Nature is forever over education—our famous orchardist once more. [Jean-Baptiste] Van Mons of Belgium, after all his experiments at crossing and refining his fruit, arrived at last at the most complete trust in the native power." Emerson—who spent years cultivating his own pear trees, and claimed some local fame as an orchardist himself—concluded that "the thinker is as much prepared in the general plan of things, as the cultivator of the ground. . . . Nature, provided for the communication of thought by the planting with it in the receiving mind a fury to impart it." In sum, as he claimed in the essay "Power," "A cultivated man, wise to know and bold to perform, is the end to which nature works, and the education of the will is the flowering and result of all this geology and astronomy." Nature "plants" within man both thought and the "fury to impart" it, bearing forth the "cultivated" man as her ultimate crop.[17]

Emerson noticeably placed weight on the third person singular: a cultivated *man,* not a cultivated *society.* The starting point for the culture of truth was the single individual, for, as Emerson asserted, individuality was indeed "the basis" of culture. There would be no good crop without good, strong plants: "Your power in the world must be by and through your individualism. Opinions are organic." The recognition of this "organic" or inborn power of the single individual had become one of the defining features of the nineteenth century: "The modern mind believes that the nation exists for the individual, for the guardianship and education of every man. . . . In the eye of the philosopher, the individual is the World." However, not every human being was, in this worldmaking sense, a true "individual." In an extension of Quetelet's insights, Emerson remarked that "so often amid myriads of invalids, fops, dunces, & all kinds of damaged individuals, one sound healthy brain will be turned out, in symmetry & relation to the system of the world;—eyes that can see, ears that can hear, soul that can feel, mind that can receive the resultant truth." Given this troubling preponderance of fops, dunces, and "damaged individuals," it became all the more crucial to cultivate and protect those whole and healthy few, and to find means for them to associate.[18]

When Emerson looked toward the cultivated "man," did he mean to exclude women? Yes and no. Late in 1837 he acknowledged that when "conversing with a lady," his insistence on "this self sufficiency of man" seemed "a bitterness & unnecessary wound." When she objected that it was "very true but very mournful," Emerson did not yield. "But to women my paths are shut up," he responded, and the women he had known of "genius & cultivation" had also "something tragic in their lot." Their power was diminished, though perhaps (given his logic of power through submission) a little subversive: "A woman's strength is not masculine but is the unresistable might of weakness," he noted in his journal.[19]

Two decades later, his words to the Woman's Rights Convention in Boston in 1855 impressed this might of weakness upon his audience: "the omnipotence of Eve is in humility." The strength of women lay in their sentiment, their sympathy with their husbands, the glory of their husbands and children. Their "organic office in the world" was to educate the young and to be the civilizers of mankind. "What is civilization? I call it the power of good women." That is, in Coleridge's contest between culture—"the showing the harmony of the unshorn landscape"—and civilization,—"the trimming & turfing of gardens"—women's role was in the garden, trimming and turfing, polishing and embellishing. "They finish society, manners, language. Form and ceremony are their realm." Their genius delighted "in decorating life with manners, with properties, with order and grace." As for equal rights, property, and the vote, Emerson was happy to grant them, although he suspected that "the best women do not wish these things" and would find them, if granted, "irksome and distasteful." His support for women's rights did, as Len Gougeon establishes, put him in the vanguard for his time, for Emerson acceded to their demand for economic and political power—the power, that is, to "civilize." However, women's weaknesses and dependencies still excluded them from true *cultural* power. As Phyllis Cole has written, "No external change affected women's 'organic office in the world,' and that office was mediatory."[20]

The nineteenth century that so valued the cultivated male individual also took care, Emerson remarked, to provide the means for such cultivation. In "Manners," he wondered, "What fact more conspicuous in modern history, than the creation of the gentleman?" Such men were a creation of nature, and "in the moving crowd of good society, the men of valor and reality are known, and rise to their natural place," forming "a self-constituted aristocracy, or fraternity of the best." The gentleman was important, for he alone had the necessary independence to be a carrier of *truth:* "The gentleman is a man of truth, lord of his own actions, and expressing that lordship in his behavior, not in any manner dependent and servile either on persons, or opinions, or possessions." Truth, then, did not bubble up from below, but trickled down from the self-constituted higher society above: "Thus we all depend at last on the few heads or the one head that is nearest to the stars, nearest to the fountain of all science, & knowledge runs steadily down from class to class down to the lowest people, from the highest, as water does." As he summed up in 1863, "The doctrine of culture of knowledge is that it comes from the gods into one head, or a few heads, then down to the nearest receivers, & slowly thence to the multitude." Stephen Whicher observes that "Emerson's ethics of culture is inherently patrician," for it is "an ethics for the superior man, the well-born soul," "a gospel for gentlemen." As Whicher notes, Emerson never repudiated his earlier faith in "transcendental democracy, the potential greatness in every man," but his interest now was more in how to find and nourish, in the Queteletan mass of damaged individuals, that saving remnant.[21]

That remnant was, though, more likely to be found in the higher classes, for such individuals were freed from the limiting "dependence" on "possessions." Not in Thoreau's sense, detachment from material desires; rather, because they were far more likely to have possessions, to be those landed proprietors at the top of Davy's social hierarchy. Stallo had asserted that the destiny of mankind was "to spiritualize the Material," and of the individual, to transform "material media into a means and history of his activity." Such media were then "his *property;* they bear the impress of his life, and hence they are his *own,*" the "written document of his action . . . the recorded language of his organic energy." Possessions, then, were not passive accumulations weighing down the soul but essential components of a fully built world.[22]

Not long after reading Stallo, a similar thought came home to Emerson: "It occurred yesterday more strongly than I can now state it, that we must have an intellectual property in all property, & in all action, or they are naught. I must have children, I must have events, I must have a social state & history,—or my thinking & speaking will have no body & background. But having these, I must also have them not, (so to speak), or carry them as contingent and merely apparent possessions to give them any real value." Emerson was always ready to condemn property for its own sake, for then it became merely "brutish"—as had happened, for example, in England. Then (echoing Thoreau) it was not the man who had got the farm, but the other way round: "If a man own land, the land owns him." Yet clearly, ownership did draw one into fuller connection with the world, gave one "body & background." The trick was to "spiritualize" one's material media, as Stallo had said—or, in Emerson's terms, acquire "*intellectual property*" in them. As he reiterated, "Until we have intellectual property in anything, we have no right property in it. Only he has a thing who can use it."[23]

Such ownership demanded knowledge, and so could be obtained only by education: "Who would be my best benefactor," Emerson asked rhetorically, "—he who gave me a loadstone? or he who showed me that it would turn to the north?" Who more truly possessed the land: the Indian who merely roamed across it, or the chemist who drew support for fifty men out of one acre? Indeed, *intellectual* property as acquired by education was necessary to secure the safety of *all* property. As Emerson said in 1852, the two-foot stone wall that guarded his "fine pears & melons, all summer long, from droves of hungry boys & poor men & women" would hardly protect him were they to question his right and pluck the fruit, nor would "the cumbrous machinery of the law" be much practical help. If every passenger helped him- or herself, "my house would not be worth living in, nor my fields worth planting. It is the education of these people into the ideas & laws of property, & their loyalty, that makes these stones in the low wall so virtuous." Emerson might use the stones as a language to record his own organic energy on the land, but only if *others* read and respected that language would Emerson's crops be safe.[24]

However, as Stallo hints, property that hid forever behind stone walls would have no real value. Emerson wrote that the truly rich man should "animate" his possessions by his "quality and energy." This showed another way to construct a scale of being: "The oyster has few wants and is a poor creature. The Mammalia with their manifold wants are rich men," and men, who want "everything," are the richest of all: "They are made of hooks and eyes and put the universe under contribution. Man is rich as he is much-related." Arnold Guyot, in *The Earth and Man,* had built a global economy on the wants of rich men supplied by the labor of the poor, creating a system of exchange whose circulation of wealth animated the planet. The idea was fundamentally Baconian: a frontispiece to *The Advancement of Learning* shows, below the two globes of the visible and intellectual world, a ship sailing an open sea over the legend "Multi pertransifunt et angebitur scientia," Latin for the biblical injunction "Many shall go to and fro and knowledge shall be increased." As Emerson remarked, one of the hallmarks of the progress of science was that it brought "the remote near. The kelp which grew neglected on the roaring sea beach of the Orkneys, now comes to the shops," and so with seal, otter, ermine, the strombus and the pearl, and the green eggs laid by the birds of Labrador: all "must come to the Long Wharf also."[25]

In the essay "Wealth," Emerson developed this idea of intellectual property and exchange at length. Wealth lies in "applications of mind to nature," and man is "tempted out by his appetites and fancies to the conquest of this and that piece of nature, until he find his well-being in the use of his planet, and of more planets than his own." True wealth is not a toy for show and pleasure; rather, "men of sense esteem wealth to be the assimilation of nature to themselves, the converting of the sap and juices of the planet to the incarnation and nutriment of their design." Wealth, then, is the agent of *knowledge,* also defined as the assimilation of nature to self.[26]

Thus knowledge necessarily becomes the distinctive property of the rich, who "take up something more of the world into man's life." Yet Emerson's trickle-down moral economy would not exclude the poor from knowledge. What the single citizen cannot afford, the community should provide for the benefit of all. Emerson has in mind the tools of science: a telescope, "electrical and chemical apparatus," encyclopedias and other expensive reference books, "pictures also of birds, beasts, fishes, shells, trees, flowers, whose names he desires to know." For, Emerson continues, "If properties of this kind were owned by states, towns, and lyceums, they would draw the bonds of neighborhood closer. A town would exist to an intellectual purpose." Wealth, that is, as the ally of *truth,* not only binds the community together but is fundamentally "moral," which means that material and moral value enhance each other. The dollar literally increases in value with the increase in "virtue of the world," such that "a dollar in a university, is worth more than a dollar in a jail." In sum, as all things "ascend," so must economy, to a higher aim: "Thus it is a maxim, that money is another kind of blood . . . the estate of a man is only a larger kind of body," and "bodily vigor" climbs ever higher to become "mental and moral vigor." The body of the world is animated by the circulation of money, the hot blood that mounts from animal sensation to intellectual thought and finally to spiritual creation.[27]

As early as the 1840s Emerson began applying this line of reasoning to what he came to call a "True Aristocracy," one in which "every member commits himself, imparts without reserve the last results of intellect, because he is to receive an equivalent in virtue, in genius, in talent from each other member. None shall join us but on that condition. No idler, no mocker, no counterfeiter, no critic, no frivolous person whatever, can remain in this company." Since Emerson's aristocrats combine material and moral wealth, they become "nobles because they know something originally of the world. If the sun were extinguished & the solar system deranged they could begin to replace it." Naturally enough, given his substantial if not extravagant means, Emerson began to imagine himself part of such elevated company, and he became deeply attracted to the idea of gentlemanly clubs and associations. In 1842 he commented, "I like a meeting of gentlemen; for they also bring each one a certain cumulative result. From every company they have visited, from every business they have transacted, they have brought away some thing which they wear as a certain complexion or permanent coat and their manners are a certificate a trophy of their culture." The power of the "culture" that such a collective of accomplished gentlemen would create together is suggested in *English Traits*. England's practical power is built on "English truth and credit," for "truth is the foundation of the social state."[28]

If truth is the foundation of society, falseness is its undoing. As Emerson said in "Prudence," "Every violation of truth is not only a sort of suicide in the liar, but is a stab at the health of human society." Emerson was once again applying his old principle of compensation, with its automatic paybacks; so, for instance, events will lay on even the most profitable lies "a destructive tax," "whilst frankness invites frankness. . . . Trust men, and they will be true to you." This sounds like a contradiction of his famous injunction in "Self-Reliance," "Trust thyself," yet it is actually an extension of the same logic. Emerson indicated this in the "Divinity School Address" when he wrote, "By trusting your own soul, you shall gain a greater confidence in other men." That is, the self-trusting soul shuns society not to be alone, but to recruit a *higher* society. Other self-trusting souls will find their way to you as the community of self-trusting individuals converges on the truth.[29]

The real difficulty lay in the troubling nature of the individual. He must be trusted, but ultimately, given human limitations, he is not trustworthy: as Emerson had written in his youth, "The individual is always mistaken." The sequence can be traced in his 1848 lecture "The Tendencies and Duties of Men of Thought." To begin with, Emerson asserts that men work best when they work alone, "and always work in society with great loss of power." Unfortunately, if one tries to interest such loners in one's own thought, they will only return to their "private theatre," in which the all-absorbing play is performed "before himself *solus*." Nevertheless, each man should stay the course by following his excess, his partiality, his individualism, for power lies in concentration, in "a certain narrowness. . . . The horse goes better with blinders." While those blinders may limit us, they also protect us. "Truth is our only armour in all passages of life and death," Emerson adds, drawing on the words wrung out of him by Waldo's death: "I will speak the truth in my heart, or think the truth against what is called God."[30]

Then comes the turn: such narrowness or concentration on the truth in one's heart provides an "asylum"; however, that old crack in nature guarantees that our asylum will also confine us. "I confess that everything connected with our personality fails," Emerson admits. "Nature never spares the individual. We are always baulked of a complete success." Thus the final step is deeply social, for private limitation and failure can be redressed only by a turning outward, to "the order of things": "The sky, the sea, the plants, the rocks, astronomy, chemistry, keep their word. Morals and the genius of humanity will also. In short, the whole moral of modern science is the transference of that trust which is felt in Nature's admired arrangements, to the sphere of freedom and of rational life." It takes, then, the full roundness of society to balance the inevitable limitations of the single self, despite the threat that society poses to the individual. Wisdom, thought Emerson in his late essay "Clubs," "is like electricity. There is no permanently wise man, but men capable of wisdom, who, being put into certain company, or other favorable conditions, become wise for a short time, as glasses rubbed acquire electric power for a while." Truth in the heart will be truth at all only if it draws on the universal currents that animate and electrify the whole. Such truth is less individual than social, as Emerson observed in "Fate": "The strongest idea incarnates itself in majorities and nations, in the healthiest and strongest."[31]

Emerson did worry about the pressure this social truth placed on the individual. On the one hand, culture requires the collective. The "certain culture & state of thought" of society exists "only in their congregation: detach them, & there is no gentleman, no lady in the group." On the other hand, association with others could fatally compromise the single gentleman. "The uniform terms of admission to the advantages of civilized society, are,—You shall have all as a member, nothing as a man." The solution was, as usual, intensification of the same logic: *more* aloneness. The *best* society was composed of the union of the "absolutely isolated," who, in going alone, will "go up & down doing the works of a true *member*, and, to the astonishment of all, the whole work will be done with concert, though no man spoke; government will be adamantine without any governor. . . . then would be the culmination of science, useful art, fine art, & culmination on culmination."[32]

Here, once again, England stood as the warning. In this highly technologized society, division of labor had turned the nation into "a tent of caterpillars," whereas "the best political economy is care and culture of men; for, in these crises, all are ruined except such as are proper individuals, capable of thought, and of new choice and the application of their talent to new labor." Proper individuals, *cultivated* individuals, would be armed by truth, in a process that would simultaneously individuate and connect them. But England was not the only example of this failure. As Phyllis Cole shows, Emerson saw in England the modern tendency that would soon characterize America as well. Already in 1845 Emerson worried that proslavery southerners, lacking "culture," had consigned their free-thinking individuality to the state: "It is the inevitable effect of culture it cannot be otherwise to dissolve the animal ties of brute strength, to insulate, to make a country of men not one strong officer but a thousand strong men ten thousand. In all S.C. there is but one opinion but one man Mr. Calhoun. Its citizens are but little Calhouns. In Massachusetts there are many opinions many men." By contrast with the little Calhouns, a "truly" cultivated man would offer shelter to those less free, those prevented by circumstances from becoming gentlemen: "With the truly cultivated man—the maiden, the orphan, the poor man, & the hunted slave feel safe." Furthermore, a gathering of such men would do more than just shelter the unfortunate. They would use their moral leadership to guide public policy: "I need only hint the value of the club for bringing masters in their several arts to compare and expand their views . . . so that their united opinion shall have its just influence on public questions of education and politics." Significantly, Emerson's first example was the British Association for the Advancement of Science, throwing light on Agassiz's motives in founding the elite Academy of Sciences during the Civil War.[33]

The linking of science, societies of scientists, and public policy persisted throughout Emerson's work, from the sermons in which he recommended to his congregation that "cultivation of the intellect" would, by spreading the love of natural science, supplant the base pursuit of wealth with more spiritual pleasures; to the early lectures in which he advised that the moral of science was the "transference of that trust" felt in "nature's admired arrangements" to the "social & moral order," replacing politics with education; to the late lectures, wherein Emerson proposed that science alone could ground a just social and moral order. In 1858, Emerson opened "The Natural Method of Mental Philosophy" by quoting the German Heinrich Steffens: "The view of nature generally prevailing, at any determined time, in a nation, is the foundation of their whole Science; and its influence spreads over every department of life. It has an important influence on all social order, on morals, nay even on religion." Where science led, true humanity would follow. Good men, "scholars and idealists," were set down, "as in a barbarous age," to calm and guide insanity, check self-interest, and force good laws on bad governments. This was indeed what Emerson saw happening around him. Speaking to the Phi Beta Kappa Society in 1867, he confidently declared that when he looked out at "the sound material of which the cultivated class here is made up," those "most distinguished by genius and culture . . . I cannot distrust this great knighthood of virtue, or doubt that the interests of science, of letters, of politics and humanity, are safe. I think their hands are strong enough to hold up the Republic."[34]

When Emerson looked around him, what he saw was material organized through the method of science into practical power: truth materialized into things, institutions, buildings, networks, societies. "Every truth tends to become a power," he told his audiences in 1848. "Every idea from the moment of its emergence, begins to gather material forces, and, after a little while makes itself known in the spheres of politics and commerce. It works first on thoughts, then on things; it makes feet, and afterward shoes; first, hands; then, gloves; makes the men, and so the age and its *materiel* soon after." Emerson had learned exactly how spirit materialized into things by observing the slow and steady growth of trees. "Whence came all these tools, inventions, books, laws, parties, kingdoms? Out of the invisible world, through a few brains. Nineteen-twentieths of their substance do trees draw from the air." The pitch pine planted in a sandbank soon makes a grove and covers the sand with soil; "Not less are the arts and institutions of men created out of thought. The powers that make the capitalist are metaphysical: the force of method and the force of Will makes trade, and builds towns." The lesson was "Trust entirely the thought. Lean upon it; it will bear up thee and thine, and society, and systems, like a scrap of down." Emerson's alleged "idealism" was as keen-edged as the cleaver dividing intellect from object, as strong as the pitch pines building sand into fertile soil, as rock-hard as the low stone fence drawing the law around his tender pears and melons, themselves the realization of his will, a wall dividing law from chaos that was held in place by nothing more nor less than the iron filaments of trust.[35]

The culture of truth was Emerson's own redaction of his life experience in which solitude was the catchword but clubs were the reality, from the Transcendental Club of the late 1830s, to the Town and Country Club of the late 1840s, Concord's Social Circle, and the famous Saturday Club, which he himself helped found in 1854. As Louis Menand writes, "private literary and philosophical societies were one of the venues in which intellectual work got done in the United States in the years before the emergence of the university." One member of the Saturday Club, Dr. Oliver Wendell Holmes, recalled that Emerson had joined the Saturday Club "in reality before it existed as an empirical fact, and when it was only a platonic idea. The Club seems to have shaped itself around him as a nucleus of crystallization," starting with two or three friends meeting Emerson occasionally for dinner in Boston. Holmes recalled that the club brought together "many distinguished persons," from the poet Henry Wadsworth Longfellow at one end of the table—"florid, quiet, benignant, soft-voiced"—to the scientist Agassiz at the other—"robust, sanguine, animated, full of talk, boy-like in his laughter"—and in between, in addition to Emerson and Holmes himself, the likes of Nathaniel Hawthorne; James Russell Lowell, the great arbiter of literary culture; Henry James Sr.; Benjamin Peirce; Jeffries Wyman, the Harvard anatomist and physiologist; Judge Ebenezer Hoar; John S. Dwight, Boston's leading music critic; Senator Charles Sumner; Charles Eliot Norton, America's first art history professor; and Emerson's literary executor, the writer and critic James Elliot Cabot. The Saturday Club was a fraternity of the leading intellectuals of Massachusetts, many of them Boston's first citizens, at a time when Boston was the hub of America's literary and scientific universe.[36]

The Saturday Club did far more than gather around a good supper. Both Len Gougeon and Albert von Frank show that it immediately became "the focus of pro-Union activity and sentiment," in Gougeon's words; von Frank notes that discussion at the club's first meeting "turned largely on the political situation" and especially on a review of the case of Anthony Burns. It grew from there: "The club, which quickly expanded to include a significant portion of Boston's intelligentsia, was the nucleus of a coalescing antislavery culture, out of which developed, within three years, its official organ, the *Atlantic Monthly*"—and which contributed crucially to the founding of the Republican party and the election, then reelection, of President Lincoln. The Saturday Club, then, was modeled on the Emersonian ideal: the congress of educated gentlemen, each bringing his own knowledge and commitment to the benefit of all, using their collective influence to guide and direct both public policy (as the Saturday Club did with abolitionism) and intellectual culture (as the Transcendental Club did with *The Dial*, and the Saturday Club, with spectacular success, with the *Atlantic Monthly*). Emerson's ideal became as solid as the intelligentsia of Boston could make it. He leaned on it, hard, and watched it bear up the developing culture not just of New England but of the republic itself—and more, a global culture of humanity: "But the triumph of culture is to overpower nationality by importing the flower of each country's genius into the humanity of a gentleman." The culture of truth transcended nationality to create a class in common among all civilized nations. Emerson qualified his criticism of England (where the popular press was rife with anti-Union sentiment) by leaving

> out of question the truly cultivated class. They exist in England, as in France, in Germany, in America, in Italy, and they are like Christians, or like poets, or chemists, existing for each other across all possible nationalities, strangers to their people & brothers to you. I lay them out of question. They are sane men as far removed as we from the bluster & mendacity of the London Times, & the shop-tone of Liverpool. They, like us, with to be exactly informed, & to speak & act for the public good & not for party.

On one side, the local bluster of commerce and the media; on the other, the universality of truth.[37]

The members of Emerson's transcendent culture were, notably, "brothers" to such as himself but "strangers to their people." The culture of truth was deliberately, pointedly, exclusive. Early on Emerson worried over whether he was living in a fool's paradise. In 1838 he observed, shrewdly, that "I please myself with the thought that my accidental freedom by means of a permanent income is nowise essential to my habits." Would he not, "rich or poor," do the same things? His answer, as Robinson notes, is evasive. "If I did not think so, I should never dare to urge the doctrines of human Culture on young men. The farmer, the laborer, has the extreme satisfaction of seeing that the same livelihood he earns, is within the reach of every man. The lawyer, the author, the singer, has not." Farmers and laborers might *aspire* to truth, but only those in the highest rank, such as himself, had the leisure and the freedom to *attain* it: "Sincerity is the luxury allowed, like diadems and authority, only to the highest rank, *that* being permitted to speak truth, as having none above it to court or conform unto."[38]

Late in life Emerson defended to himself the Saturday Club's strict admission policies: "Is the Club exclusive? 'Tis made close to give value to your election." In public he defended the value of "social barriers," since every "highly organized person" knows the best society can be spoiled by the intrusion of "bad companions"; "there are people who cannot be culti- vated. . . . Bolt these out." Dr. Holmes also registered some awareness that the nonelect might resent their exclusion, although he, too, found no great cause for concern: "Some outsiders furnished still another name for this much-entitled Club. They called it 'The Mutual Admiration Society,' and sometimes laughed a little, as though the designation were a trifle deroga- tory. Yet the brethren within the pale were nowise disturbed by this witti- cism." More seriously, Emerson's certainty in the transcendent virtue of ex- clusive culture allowed him to dismiss the many it did not embrace. "Certainly I go for culture, & not for multitudes," he reminded himself; "I suppose that a cultivated laborer is worth many untaught laborers," a sci- entific engineer worth "many thousands," and an Archimedes or Napoleon "worth for labor a thousand thousands"; given a "wise & genial soul" one could "well dispense with populations of paddies." Worse, as he exclaimed in 1852, "The calamity is the masses," not worth the preserving. Two years later he added, "Shall we judge the country by the majority or by the mi- nority? Certainly, by the minority. The mass are animal, in state of pupi- lage, & nearer the chimpanzee. We are used as brute atoms, until we think. Then we instantly use self-control, & control others."[39]

Yet the contempt expressed here for those shut out by the exclusive cul- ture of truth is belied by its key foundational principle: education. The masses may be still "in state of pupilage," but the responsibility of intellec- tual leaders like Emerson was not just to control those chimpanzee "others" through thought, but to lead *them* to think as well—at least, insofar as pos- sible. Here, too, science showed the way—and showed the impassable limi- tations of that way. Agassiz had celebrated science for the people, whether in his charismatic appearances before lecture audiences, his many books and articles (particularly those published in the *Atlantic Monthly,* which be- came virtually a house organ for Agassiz's popular essays), or his scheme for educating high school teachers, women as well as men. Yet Agassiz also worked to professionalize American science, which meant instituting de- manding graduate education programs designed to exclude all but the most talented and dedicated of men, and Agassiz recognized that few could fol- low him across the threshold from lecture room to scientific laboratory.

Emerson throughout his career urged the man of science to bring his learning down to the people, for the benefit of both parties, and he praised the science of his day for doing exactly that. In his journal in 1836 he mused, "Does it not seem that the tendency of Science is now from hard fig- ures & marrowless particulars—dead analysis back to synthesis, that now Ideology mixes therewith, that the Education of the people forces the savant to show the people something of his lore which they can comprehend, &

that he looks for what humanity there remains in his science, & calls to mind by finding it valued, much that he had forgotten." The notion had been the heart of his 1834 lecture "The Humanity of Science," in which Emerson put his hopes before the very people he had in mind; and nearly everything he published used concepts from science to ground the understanding and fire the imagination of his reading audience.[40]

One of his last journal entries restated his goal for science: having just learned that stone arrowheads like those found in Concord were found as well in Italy and Africa, pushing back the age of man from six thousand to a hundred thousand years, Emerson marveled that the "new facts of science" were valuable not just in themselves but for the "electric shock" they gave the mind: "each new law of nature speaks to a related fact in our thought," and however skillful the scientist may be, "& how much soever he may push & multiply his researches, he is a superficial trifler in the presence of the student who sees the strict analogy of the experiment to the laws of thought & of morals." The people *needed* science, and science needed people like Emerson himself to electrify the arcane research of the laboratory and send it abroad on the currents of popular thought.[41]

Emerson lived in the golden age of science, as he declared any number of times. "Ours is the *Zymosis* of Science," he exulted late in life; "the heavens open, & the earth, & every element, & disclose their secrets. . . . What beams of light have shown upon men now first in this Century! The Genius, Nature, is ever putting conundrums to us, & the savans, as in the girls' game of 'Twenty questions,' are every month solving them successively by skilful exhaustive method." Looking back on the progress achieved during his own lifetime, Emerson judged that "the splendors of this age outshine all other recorded ages." Above all were those "five miracles," the steamboat, the railroad, the electric telegraph, the spectroscope applied to astronomy, and the photograph; then there were all those others, from cheap postage and the sewing machine to power looms and presses, anesthesia, Oersted's discovery of the unity of electricity and magnetism, Agassiz's of the correspondence of stratified organic remains "to the ascending scale of structure in animal life," and finally the emerging science of weather prediction. It was indeed a zymosis, a yeasty fermentation of ideas.[42]

Emerson celebrated the same theme in public. His popular lecture "The Spirit of the Times" detailed the revolution wrought by the "Age of Science," which had replaced the barren and repulsive fact-gathering of English and French science with "modern science," which studied "under the light of ideas" and "with all its tongues, teaches unity," hunting out "endless analogies." In his vision, science was hardly a remote affair confined to laboratories and specialists. It was a popular movement: as democracy had "conquered the State," so had "humanity rushed into Science," first in the lower forms of phrenology and mesmerism, then—guided by such revolutionary leaders as Goethe, Schelling, Oken, Owen, and Agassiz—into all nature. Since 1836, science had indeed realized the ambition Emerson voiced in the opening of *Nature:* it had *found* its "theory of nature":

> This science aimed at thoroughness to give a theory to elements, gases, earths, liquids, crystals, plants, animals, men; to show inevitable were all the steps; that a vast plan was successively realised; and that the last types were in view from the beginning, and were approached in all the steps; that each circle of facts, as, chemistry, astronomy, botany, repeated in a new plane the same law; that the same advancement which natural history showed on successive planes, civil history also showed; that event was born of event, that one school of opinion generated another; that the wars of history were inevitable; that victory always fell where it was due; no party conquered that ought not to conquer; for ideas were the real combatants in the field, and the truth was always advancing, and always victorious.

As Emerson had prophesied, so he concluded: "The idea of the age, as we say, was return to nature; conquest of nature; conversion of nature into an instrument; perfect obedience to nature, and thereby perfect command of nature; perfect representation of the human mind in nature." Intrinsic to this "idea" was the "gracious lesson taught by science to this Century," that the history of nature was "*melioration*," the incessant advance from lower to higher, matter conspiring with spirit. Evolution, then, was to organic life what culture was to humanity: "Melioration in nature answers to Culture in man." As a young man, Emerson had vowed to listen to the truth at the bottom of the heart, and now he generalized from himself to humanity: when man "listens to the impulses of the heart, in him, and sees the progressive refinement in nature, out of him, he cannot resist the belief that every new race and moral quality impresses its own purpose on the atoms of matter, refines and converts them into ministers of human knowledge and virtue; that cultivated bodies, instructed blood, ennobled brains" will yet obtain "a new command" of nature, converting it to ever more "magnificent instrumentalities . . . to run on errands still more magnificent." If man was the necessary continuation of nature on a new, a spiritual plane, "culmination on culmination," culture was his instrument, culture that converted the natural process of growth and fruition into the American narrative of endless progress.[43]

The Puritan "errand into the wilderness" had been, by now, thoroughly Baconianized. Emerson deployed a reclaimed wilderness to run new errands into the future, that peculiarly American territory. From coast to coast, North America was a continental laboratory of science, and every citizen who worked according to the principles of the culture of truth became a kind of Emersonian scientist, translating matter into mind, resource into achievement. Emerson saw in this scientific culture not the decline of religion, but the basis for a religious renewal. "You say, there is no religion now," he joked in his late essay "Worship": " 'Tis like saying in rainy weather, there is no sun." One must know where and how to look: "Our recent culture has been in natural science," Emerson continued. The knowledge gained by science has been the key unlocking human history and moral truth, showing that human freedom and individuality are balanced by the prefiguration of "the primordial atoms" to "moral issues," so that the very atoms "are in search of justice, and ultimate right is done."[44]

In 1848, heady with the evolutionary ideas of Richard Chambers, the pleasures of Hooker's Kew Gardens, and the lectures of Robert Owen, Emerson declared, "Yes, there will be a new church founded on moral science, at first cold & naked, a babe in a manger again, a Copernican theory, the algebra & mathematics of Ethical law, the church of men to come without shawms or psaltery or sackbut, but it will have heaven & earth for its beams & rafters, all geology & physiology, botany, chemistry, astronomy for its symbol & illustration, and it will fast enough gather beauty, music, picture, poetry." As he added in "Tendencies and Duties," "The religion which is to guide and satisfy the present and coming ages, whatever else it be, must be intellectual. The scientific mind must have a faith which is science." Clearly, Emerson saw, his age was troubled because it was still in transition, a kind of "temporary anarchy" between the religion of seventy years before which was "a belt to the mind, giving it a concentration & force," and the coming abandonment "to the all sufficiency & beatitude of Ethics." Help was on the way, however, for popular science was steadily advancing the religious mind: "The perceptions which metaphysical & natural science cast upon the religious traditions, are every day forcing people in conversation to take new & advanced positions. We have been building on the ice, & lo! the ice has floated. And the man is reconciled to his losses, when he sees the grandeur of his gains."[45]

Emerson placed tremendous responsibility on the shoulders of science, which meant that he could be intensely critical when scientists did not live up to their mission. This was especially true during the 1850s, the heyday of Thoreau's own deep engagement with the practice and purpose of natural science. Emerson's journal shows that he took frequent educational walks with his naturalist friend, and the two were often in conversation about both the nature of science and Thoreau's purposes in investing so much of his own time and intellectual powers in a detailed field study of the natural history of Middlesex County. Some of Emerson's critiques of science could have been ghostwritten by his friend Thoreau:

> The science is false by not being poetical. It assumes to explain a reptile or mollusk, & isolates it; which is hunting for life in graveyards: reptile or mollusk only exists in system, in relation. The metaphysics, the poet only, sees it as inevitable step in the path of the Creator.

> For this reason, the scientific men of Cambridge & of London appear as amateurs. There is an affectation in their talking & knowing about a fish or a bird. An Indian or a hunter, or a poet & walker, has the sweetest right to his knowledge of these.

The "poet and walker" at Emerson's side had himself been put on the defensive by the professional scientists of Cambridge, and it must have been sweet to think of them as amateurs who knew less than the locals about Concord's fish and birds. Too often, Emerson added, such pretenders were interested only in the *utility* of their objects: "is there no right wishing to know what is," he asked, "without reaping a rent or commission? Now their natural history is profane. They do not know the bird, the fish, the tree, they describe. The ambition that 'hurries them after truth, takes away the power to attain it.' "[46]

For example, the great Lyell, to Emerson's astonishment, "did not know by sight . . . the shells he has described in his Geology"—even as it was Lyell's fundamental principle that enabled the kind of science Emerson valued, science that aimed "to transport our consciousness of cause & effect into those remote & by us uninhabited members, & see that they all proceed from 'causes now in operation' from one mind, & that Ours." As Emerson added in a passage that he reworked several times, "I do not wish to know that my shell is a strombus or my moth a Vanessa, but I wish to unite the shell & the moth to my being." Presumably, the very shells Lyell could not name, Emerson would not have wished to name, yet he did expect the scientist to be responsible both for the details that Emerson himself might want to ignore and for the general principles that related such details to the broader truths Emerson valued. "All science must be penetrated by poetry," he added in his final rewrite of this passage: the scientist was prone to mistake facts as ends in themselves and scientific truth as "a finality," whereas the poet would recall that no fact is of value in itself, but only as it was related to all else, and hence convertible "into every other fact & system, & so indicative of First Cause."[47]

What Emerson meant by "relatedness" or "poetry" here is not a kind of ecological interconnectedness that links facts as material specifics in a web of interactions, but facts as dematerialized representations of a single, endlessly reconvertible Idea. Such a line of reasoning might label Emerson's approach to science as "idealist," but what such a label obscures is both the tendency of much science in Emerson's day to treat objects as representations of ideas, and the sheer earthy physicality with which Emerson addressed science, rendering it not mystical but as practical as a ship's compass. Emerson understood science as the art of *leaning* on ideas, making them bear the weight of the world: "Mother wit animates mountains of facts by turning them to human use,—milking the cow, suspending the loadstone, pouring human will & human wit through things till the world is a second self." Science is "a house held up by magnetism,—draw out the magnet, & the house falls & buries the inhabitant."[48]

Magnetism, or more generally polarity, was both a physical force and a spiritual resource. As gravity, the physical analogue of truth, upheld the world, so polarity animated it with the energies of circulation and exchange. The two formed the physical basis of Emerson's "culture of truth," with organic analogues in body and blood or earth and water, and spiritual analogues in matter and spirit. Analogy stacked the universe into an ordered system, mirrors upon mirrors, and the culture of truth showed how the developing human being participated in—became part of—the systemic whole, the whole that "looked" each individual into being, from the undifferentiated masses to distinctive singleness. The great whole became a way to conceptualize truth, metaphorized as a "solar eye" that placed mind at the center of a new Copernican revolution.

Imagined as the light- and life-giving whole, truth could be simultaneously transcendent above all particular things and events, and immanent in each waterdrop or animalcule. This simultaneity led to the most telling contradiction within the culture of truth: truth as transcendent had to be available to any and all, equally, through a universality of experience, Emerson's notorious mystical leap into transparent and all-seeing vision. Yet grounding truth in nature as apprehended by modern, Western science forced concepts of truth into the domain of the expert, for no one person could hope to apprehend all of natural truth through personal experience. "Truth" had to be taken on "trust," generating a secondary economy based on the moral values of sincerity, veracity, social independence, and an increasingly intensive process of education.

The hierarchy thus created promised access to all comers, but it reinforced a strict discipline on its aspirants while excluding those necessarily compromised by circumstances beyond their control. At the top of the hierarchy would be the "savants," "scholars," and "priests," who could constitute a Coleridgean clerisy, there to protect, guide, and enlighten the otherwise calamitous masses. At its best, Emerson's model republic did just that: witness the cultivated circle of New England abolitionists who in the 1850s successfully fought the ignorance of politicians and the public with potent appeals to "higher law."[49]

As for nature, there, too, the culture of truth featured a peculiar doubleness. In its transcendent mode, nature was "higher," beyond human contact and certainly beyond the ability of human beings to construct or alter it. As Emerson established in *Nature*, nature "refers to essences unchanged by man." Yet the whole direction of *Nature*, and of Emerson's work right to the end, was to reimagine nature as the instrument of human will, such that ultimately all nature would be nothing more nor less than our living "double," our second self reconstructed in the image of our desire. Science, as the key that released natural power for human use, also participated in this doubleness. Allied with transcendent nature, science was elevated into a new religion, mystical and untouchable. Allied with the global project to reconstruct nature, science was creative, the technical means by which we literally *do* "build our own world"—in short, *poetry*. Or, as put memorably by one of today's most visible scientists, "Science is poetry that is true." The power of science marries poetry and religion, beauty and truth, converting natural fate into human freedom.[50]

Yet both attributions—science as religion, and science as poetry—mask the extent to which science is neither. For all Emerson said, what he showed was science as a civil agreement constituted by a self-selected society, which cooperated in a shared quest to master nature by learning obedience to nature's laws. Emerson proposed a modern form of interdisciplinarity grounded in a transcendent moral agency—what intellectuals around him typically called "God"; yet that agency was in fact composed of human beings, members of a human community, whose shared belief in nature as transcendent truth converted facts hammered out by generations of hard scientific work into truths that energized and constrained individual members with all the power of law—law not as observed regularities of nature, the minimalist definition recommended by scientists like John Herschel, but law as divine legislation, the meaning that Herschel's caution attempted, with limited success, to defend against. To recover those laws from the veil of appearances demanded a form of social organization that approached Emerson's vision of an ideal human community: a working whole within which each member knew his role, and knew that in fulfilling it he was contributing to a greater cause, as fingers join to make a hand.

Such a community of trust was built first of all on *self*-trust, not the solipsism that today sells consumer goods (although one might indeed draw a line from Emersonian ideas of property to a consumerism that asserts one is what one buys), but the necessary self-discipline by which one connects to that whole greater than the self, a whole achieved only by the actions and beliefs of all those who cooperate by also cultivating self-truth. At stake here is the creation of a moral economy that determines both what truth is, and who can be trusted to speak the truth. Self-discipline creates such a one: the sober, modest, plain-speaking gentleman; the "self-possessed" individual, Emerson's "beautiful soul," who is beholden to no man hence free to speak the truth as he sees it, whose work will be his bond; whose word, given, bonds the community of those who accept it, on exactly similar grounds. As Emerson exhorted America: "In self-trust, all the virtues are comprehended. Free should the scholar be,—free and brave."[51]

Emerson constructed his argument against the loss of trust in the old institutions of church and state. Emerson himself swept aside the old, "dead" institutions in favor of the new, "dynamic" philosophy he helped to formulate by synthesizing the Newtonian lawfulness of Anglo-American natural theology with the Romanticism adapted from Germany and England. As

Steven Shapin points out, every acceptance of authoritative knowledge also modifies existing usage; that is, trust is a creative as well as a conservative force. "One must be an inventor to read well," Emerson reminded his audience. In Emerson's inventive "reading" of the planted plots of the Paris Botanical Garden in 1833, the old static lawfulness of nature was suddenly animated by a surging life force, showing a passive collection of "facts" to be an active assimilation of matter by truth. Although Emerson took his place with the poets, he did so after vowing to "be a naturalist," positioning himself as the ally of scientists, their moral interpreter and spiritual leader, he who can "kindle science with the fire of the holiest affections" and so send God forth anew into the creation.[52]

Emerson's culture of truth sought to erect in the midst of a disordered world a palace of order, whose precincts enabled a gracious, rational, and civilized life otherwise impossible, and whose boundaries were to expand until the enlightened principles of self-discipline and self-education had embraced everyone capable of cultivation—a kind of global church estate or university campus, networked from nation to nation by means of language. Emerson re-created scientific discourse not as a technical jargon for the few but as a spiritual language for the many: his physical and moral universes did articulate, just as he said they would, exactly to the extent that the reading public read the inked marks on the white page, and so looked them into being as well, as words—as *things*.

In his last great essay, "Poetry and Imagination," Emerson mixes word and thing, poetry and science, like "hydrogen and oxygen to yield a new product, which is not these, but water . . . a new and transcendent whole." Such analogies are the matrix of the universe. With all nature a "trope," it becomes "the use of life to learn metonymy." Such play is not whimsical, but fatal: "There is no choice of words for him who clearly sees the truth." This makes the poet "the lawgiver," who knows "that he did not make his thought,—no, his thought made him, and made the sun and the stars." The cycle is completed by *study* of the sun and stars, which returns us to the thought that made them: "Slowly, by comparing thousands of observations, there dawned on some mind a theory of the sun,—and we found the astronomical fact. But the astronomy is in the mind: the senses affirm that the earth stands still and the sun moves." Astronomy, an act of intellect, of imagination, of mind not the senses, translates thought back into fact, "a fact which perfectly represents it, and is thereby education." Facts, therefore, are made things that "educe" the soul, educate us in the world. This is education in the "truth" that defies all sensual appearances, "namely, that the soul generates matter. And poetry," Emerson reiterates, "if perfected, is the only verity; is the speech of man after the real, and not after the apparent." Or, so to say, poetry is *science* that is true.[53]

Emerson sees that hope must precede science, and science is the "realization" of that hope. By translating thought into fact, poetry/science reforms the philosophy that will rule "its religion, poetry, politics, arts, trades, and whole history" by "the marrying of nature and mind, undoing the old divorce in which poetry had been famished and false, and nature had been suspected and pagan." In the copula of this marriage, as "man's life comes into union with truth" and his thoughts parallel natural law, he will express himself "by natural symbols . . . the ecstatic or poetic speech." Ecstatic science "strings worlds like beads upon his thought"; once it has returned to Earth, ecstatic science separates into two, complementary "powers," embodied once again in the old vegetable metaphor: one power "shoots down as rootlet, and one upward as tree. You must have eyes of science to see in the seed its nodes; you must have the vivacity of the poet to perceive in the thought its futurities." The whole future, Emerson had long before decided, lay in the bottom of the heart; from that prophetic seed of science would unfold the universe of human life—the empire of humanity. In the vast trope of nature, poetry turns things into words; then science turns words into new things. First poetry spiritualizes matter, then science crystallizes spirit around the nucleus of idea into the instruments of power.[54]

Emerson continued reading science until he could no longer read at all, returning to old friends to catch up on their latest work (such as John Herschel's *Familiar Lectures on Scientific Subjects*, 1866) and discovering in new voices a second generation whose thoughts echoed and extended his own in surprising ways: George Perkins Marsh's *Man and Nature*, Ludwig Büchner's *Force and Matter*, the popular essays of the British physicist John Tyndall. Far from seeing his "transcendentalism" pushed aside by the next generation, Emerson lived long enough to see the formation of a scientific culture that built on his own form of naturalism. For Emerson's Transcendentalism helped science become a substitute for institutionalized religion: first, by separating "permanent" spiritual truth from "transient" religious doctrines; then, by establishing a foundation on which the next generation of scientific naturalists could build a secular faith in the creative processes of nature, of which they would be the appointed priests and interpreters. Of this unchurched faith, Emerson was the first minister, for he was one of the earliest—perhaps *the* earliest—to untangle the institutional elements of religion from the two-hundred-year-old marriage of religion and science, thereby inventing a new natural theology. Whereas the old pointed to the ingenious design of nature's mechanisms as evidence of a transcendent designing intelligence that governed all, Emerson's new natural theology pointed to dynamic processes of nature as evidence of an inflowing and overarching spirit or intelligence, a force which precipitated itself as matter according to laws that governed simultaneously the physical and moral worlds.

In other words, when Emerson broke from the church he not only turned to science, but he leveraged science as a new form of dissent that, unlike the terrible skeptical science of the French Revolutionaries, was in strict obedience to the laws of nature, laws that corresponded to those governing society. This put all the authority of the universe on Emerson's side: the law of gravity, the polar powers of galvanism and electricity, the inexorable dynamic balance of nature, and the power of organic growth, which battened on death and defeat. Transcendence was located not above the Earth but everywhere around us, in the human power to remake nature in the image of our desire.

This Emerson is the prophet and midwife of the twentieth—and even the twenty-first—centuries. It is revealing to learn that John Tyndall, a British physicist and one of the leading Victorian voices for scientific naturalism, was quoted in the *Boston Journal*'s obituary of Emerson: "If any one can be said to have given the impulse to my mind, it is Emerson. Whatever I have done the world owes to him." Tyndall's popular book *Fragments of Science* reprinted an address to students in which he credited his scientific career to reading, as a young man, the works of two men, Carlyle and Emerson: "I never should have become a physical investigator, and hence without them I should not have been here to-day. [Carlyle and Emerson] told me what I ought to do in a way that caused me to do it, and all my consequent intellectual action is to be traced to this purely moral source. To Carlyle and Emerson I ought to add Fichte, the greatest representative of pure idealism. These three unscientific men made me a practical scientific worker." When Tyndall came to evaluate Emerson, he honored him as a man "entirely undaunted by the discoveries of science, past, present or prospective. . . . By Emerson scientific conceptions are continually transmuted into the finer forms and warmer lines of an ideal world." Such praise raises a new question: to what extent does the role of science as arbiter of truth carry forward a permanent Transcendental legacy?[55]

Emerson's science remained deeply optimistic to the end, for at no point could it be defeated; the lawfulness that bound the universe into one whole armed all the future against chaos and the dark. First, the organicism Emerson had adapted from Cudworth, Coleridge, and Goethe had placed each human being securely within the expanding circle of nature, part or parcel of the transcendent whole. Society itself was effectively naturalized, embosomed within the deep rhythms and cycles of nature. Then "gnomicism," belief in the gnomic truth at the core of nature from which all nature's diversity unfurled, placed humanity in an even more hopeful position: as the gnomon, or necessary completing figure to all nature, with the emergence of man marking the exact point where the old order of nature concluded and the new order of science and technology began. Man was not the usurper of God's throne but his inheritor. As he aged, Emerson could even view with placid equanimity the very nightmare that had haunted Mary Shelley and has haunted every generation since: that science would assume the power of God to create life. Emerson's faith turned the horror of Frankenstein into humanity's coming of age:

I do not know that I should feel threatened or insulted if a chemist should take his protoplasm or mix his hydrogen, oxygen & carbon, & make an animalcule incontestably swimming & jumping before my eyes. I should only feel that it indicated that the day had arrived when the human race might be trusted with a new degree of power, & its immense responsibility; for these steps are not solitary or local, but only a hint of an advanced frontier supported by an advancing race behind it.

At the end, Emerson's faith not only embraced the ultimate potentiality of science but also resolved the long-standing hostility between matter and mind, the very division that once undergirded his philosophy and that had troubled his idealism with the fear that it threw stones at his beautiful mother. Materialists such as the German physiologist Karl Vogt had scandalized Emerson's generation by proposing that even mind was material and that thought was nothing more than a product of the brain, exactly as bile was a product of the liver. By 1871 this, too, had ceased to trouble Emerson: "What at first scares the Spiritualist in the experiments of natural Science,—as if thought were only finer chyle, fine to aroma,—now redounds to the credit of matter, which it appears, is impregnated with thought & heaven, & is really of God, & not of the Devil, as he had too hastily believed. All is resolved into Unity again." In his late lecture "The Natural Method of Mental Philosophy," Emerson concluded that "it makes no difference herein whether you call yourself materialist or spiritualist." If matter can sing, and matter can write poetry, and matter can make its own law, then what difference could that old—that antiquated!—division make? "If there be but one substance or reality, and that is body, and it has the quality of creating the sublime astronomy, of converting itself into brain, and geometry, and reason; if it can reason in Newton, and sing in Homer and Shakspeare, and love and serve as saints and angels, then I have no objection to transfer to body all my wonder and allegiance." Poetry and science were as close as light and object, speech and air, wings and the wind that lifted them. Emerson could rest content: for all time would the Earth carry his thought, born under the solar eye of science.[56]

Abbreviations

CompW   *The Complete Works of Ralph Waldo Emerson,* ed. Edward Waldo Emerson, 12 vols. (Boston: Houghton Mifflin, 1903–1904)

CW   *The Collected Works of Ralph Waldo Emerson,* ed. Robert E. Spiller et al., 5 vols. to date (Cambridge: Harvard University Press, 1971–)

CWSTC   *The Collected Works of Samuel Taylor Coleridge,* ed. Kathleen Coburn et al., 14 vols. to date (London: Routledge, 1969–)

E&L   Ralph Waldo Emerson, *Essays and Lectures,* ed. Joel Porte (New York: Library of America, 1983)

EAW   *Emerson's Antislavery Writings,* ed. Len Gougeon and Joel Myerson (New Haven: Yale University Press, 1995)

EL   *The Early Lectures of Ralph Waldo Emerson, 1833–1842,* ed. Stephen E. Whicher, Robert E. Spiller, and Wallace E. Williams, 3 vols. (Cambridge: Harvard University Press, 1959–1971)

J   *The Writings of Henry David Thoreau: Journal,* ed. John C. Broderick et al., 7 vols. to date (Princeton: Princeton University Press, 1981–)

JMN   *The Journals and Miscellaneous Notebooks of Ralph Waldo Emerson,* ed. William H. Gilman et al., 16 vols. (Cambridge: Harvard University Press, 1960–1982)

L   *The Letters of Ralph Waldo Emerson,* ed. Ralph L. Rusk and Eleanor M. Tilton, 10 vols. (New York: Columbia University Press, 1939–1995)

LL   *The Later Lectures of Ralph Waldo Emerson, 1843–1871,* ed. Ronald A. Bosco and Joel Myerson, 2 vols. (Athens: University of Georgia Press, 2001)

LSA   Ralph Waldo Emerson, *Letters and Social Aims,* new and revised ed. (Boston: Houghton Mifflin, 1886)

NHI   Ralph Waldo Emerson, *Natural History of Intellect and Other Papers* (Cambridge: Riverside, 1904)

S *The Complete Sermons of Ralph Waldo Emerson,* ed. Albert J. von Frank et al., 4 vols. (Columbia: University of Missouri Press, 1989–1992)

S&S Ralph Waldo Emerson, *Society and Solitude* (Boston: Houghton Mifflin, 1904)

TN *The Topical Notebooks of Ralph Waldo Emerson,* ed. Ralph H. Orth et al., 3 vols. (Columbia: University of Missouri Press, 1990–1994)

Notes

1. *J* 1:469.

2. Richard Garnett, *Life of Ralph Waldo Emerson* (London: Walter Scott, 1988), 199; Sarah Ann Wider, *The Critical Reception of Emerson: Unsettling All Things* (Rochester, N.Y.: Camden House, 2000), 14; Harry Hayden Clark, "Emerson and Science," *Philological Quarterly* 10.3 (1931): 225, 258–60; Gay Wilson Allen, "A New Look at Emerson and Science," in *Literature and Ideas in America: Essays in Memory of Harry Hayden Clark*, ed. Robert Falk (Athens: Ohio University Press, 1975), 58, 75.

3. David Robinson, "Emerson's Natural Theology and the Paris Naturalists: Toward a Theory of Animated Nature," *Journal of the History of Ideas* 41 (1980): 69; *JMN* 4:406; *EL* 1:317. Others who have focused on the Paris revelation of 1833 are Elizabeth Dant, "Composing the World: Emerson and the Cabinet of Natural History," *Nineteenth-Century Literature* 44 (1989): 18–44; and Lee Rust Brown, *The Emerson Museum: Practical Romanticism and the Pursuit of the Whole* (Cambridge: Harvard University Press, 1997); see also Robert D. Richardson Jr., *Emerson: The Mind on Fire* (Berkeley: University of California Press, 1995), 152–54.

4. Stephen E. Whicher, *Freedom and Fate: An Inner Life of Ralph Waldo Emerson* (Philadelphia: University of Pennsylvania Press, 1953), 124–25; on Emerson as "hardly scientific," see 89–90.

5. For a historical and theoretical overview of this common intellectual culture, see George Levine, "One Culture: Science and Literature," in *One Culture: Essays in Science and Literature*, ed. Levine (Madison: University of Wisconsin Press, 1987), 3–32. The expression "the two cultures" is derived from C.P. Snow's essay of that title, given first as a lecture in 1959. See C.P. Snow, *The Two Cultures and a Second Look* (Cambridge: Cambridge University Press, 1964).

CHAPTER 1. THE SPHINX AT THE CROSSROADS

1. *CW* 1:34; Leonard N. Neufeldt, "The Science of Power: Emerson's Views on Science and Technology in America," *Journal of the History of Ideas* 38 (1977): 333. I argue that Emerson's project from its outset was Baconian in scope and conception, meaning that he drew not just his vocabulary but his organizing ideas from science. Here I am in fundamental agreement with Lee Rust Brown; see *The Emerson Museum: Practical Romanticism and the Pursuit of the Whole* (Cambridge: Harvard University Press, 1997), 142. Although it did not appear until this book

was virtually complete, the reader interested in Emerson and science will find that William Rossi's essay repays close attention; see "Emerson, Nature, and Natural Science," in *A Historical Guide to Ralph Waldo Emerson,* ed. Joel Myerson (New York: Oxford University Press, 2000), 101–50.

2. CW 1:8.

3. CW 1:21.

4. Joel Porte, *Emerson and Thoreau: Transcendentalists in Conflict* (Middletown, Conn.: Wesleyan University Press, 1966), 68–73. In *The Friend,* Coleridge notes the astonishing success of a ship commander who disciplined his men not by fear but by the "invisible" and "irresistible" power of law. Strength may be met with strength and rage with defiance, but since the power of law "is the same with that of my own permanent Self," it becomes "the true necessity, which compels man into the social state, now and always, by a still-beginning, never ceasing force of mutual cohesion" far stronger than the material chains of a slave ship; *CWSTC* 4.1:169–71.

5. *JMN* 2:49; David Robinson, *Apostle of Culture: Emerson as Preacher and Lecturer* (Philadelphia: University of Pennsylvania Press, 1982), 51–53.

6. George Barrel Cheever, "Coleridge," *North American Review* 40.87 (1935): 334, 310–11; Coleridge, *Biographia Literaria,* in *CWSTC* 7.1:191–92; *JMN* 5:71. Cheever was a Congregationalist minister and prolific author and editor whose career ranged from Maine to New Jersey.

7. *JMN* 5:71; Cheever, "Coleridge," 327–28.

8. *JMN* 1:331, 2:136.

9. Dugald Stewart, *Dissertation: Exhibiting the Progress of Metaphysical, Ethical, and Political Philosophy since the Revival of Letters in Europe,* in *Collected Works of Dugald Stewart,* ed. William Hamilton (Edinburgh: Thomas Constable, 1854), 1:526–27; idem, *Elements of the Philosophy of the Human Mind,* in *Collected Works,* 3:338; Samuel Taylor Coleridge, *On the Constitution of Church and State,* in *CWSTC* 10:69–70; Sir Humphry Davy, *Elements of Agricultural Chemistry,* in *The Collected Works of Sir Humphry Davy,* ed. John Davy (London: Smith, Elder, 1839–1840), 8:87–88; Anne Marie Louis Necker de Staël, *Germany* (1813; reprint, Boston: Houghton, Osgood, 1879), 2:138; Thomas Carlyle, "Signs of the Times," in *Works of Thomas Carlyle,* vol. 27 (London: Chapman and Hall, 1905), 58.

10. De Staël, *Germany,* 2:122, 122–23. The comparison of Kant to Copernicus is de Staël's; Kant himself nowhere describes his work this way. See I. Bernard Cohen, *Revolution in Science* (Cambridge: Harvard University Press, 1985): 237–53.

11. Andrew Cunningham and Nicholas Jardine, "Introduction," in *Romanticism and the Sciences* (Cambridge: Cambridge University Press, 1990), 1–2. For Emerson's account of Kant's so-called Copernican turn, or "Transcendental Idealism," see "The Transcendentalist," in *CW* 1:206–7. David Van Leer chides Emerson for calling Kant's categories "intuitions of the mind," but praises the correctness of Emerson's summary; see *Emerson's Epistemology: The Argument of the Essays* (Cambridge: Cambridge University Press, 1986), 55–56.

12. De Staël, *Germany,* 2:217–18. Both de Staël and Emerson were indebted to an earlier source in Bacon; see *The Advancement of Learning,* in *Works of Francis Bacon,* ed. James Spedding, Robert Leslie Ellis, and Douglas Denon Heath (New

York: Hurd and Houghton, 1869–72), 6:221–22. In *The New Organon* Bacon established that natural philosophy was no enemy to religion but "rightly given to religion as her most faithful handmaid, since the one displays the will of God, the other his power"; *Works*, 8:126. See John F. W. Herschel, *Preliminary Discourse on the Study of Natural Philosophy* (1830; facsimile ed., Chicago: University of Chicago Press, 1987), 221–361.

13. Quoted in L.S. Jacyna, "Romantic Thought and the Origins of Cell Theory," in *Romanticism and the Sciences,* ed. Andrew Cunningham and Nicholas Jardine (Cambridge: Cambridge University Press, 1990), 166; Philip F. Rehbock, "Transcendental Anatomy," ibid., 144–45; Justus von Liebig, *Animal Chemistry, or Organic Chemistry in Its Application to Physiology and Pathology* (1842; reprint, New York: Johnson Reprint, 1964), ix.

14. Joseph Henry, *A Scientist in American Life: Essays and Lectures of Joseph Henry* (Washington, D.C.: Smithsonian, 1980), 46; Walter D. Wetzels, "Johann Wilhelm Ritter: Romantic Physics in Germany," in Cunningham and Jardine, *Romanticism and the Sciences,* 199; Tony Rothman, "Irreversible Differences," *The Sciences* 37.4 (1997): 26.

15. David Knight, "Romanticism and the Sciences," in Cunningham and Jardine, *Romanticism and the Sciences,* 22.

16. De Staël, *Germany,* 2:122–23, 216, 226.

17. De Staël, *Germany,* 2:226; CW 1:39, 40.

18. CW 1:31, 3:12, 1:38, 17, 41, 44.

19. CW 1:43, 44. The editors of Emerson's *Early Lectures* suggest that Emerson's interest in science faded; *EL* 1:3. David Robinson suggests that it remained constant if static; "Fields of Investigation: Emerson and Natural History," in *American Literature and Science,* ed. Robert Scholnick (Lexington: University Press of Kentucky, 1992), 94–109. Both Leonard Neufeldt in *The House of Emerson* (Lincoln: University of Nebraska Press, 1982) and Robert D. Richardson Jr. in *Emerson: The Mind on Fire* (Berkeley: University of California Press, 1995) suggest that it grew and changed, a view with which I heartily agree. Thoreau's interest in and involvement with science has been largely documented, especially in recent years; see Walls, *Seeing New Worlds: Henry David Thoreau and Nineteenth-Century Natural Science* (Madison: University of Wisconsin Press, 1995). For Sarah Ripley, see Joan Goodwin, "Sarah Alden Ripley, Another Concord Botanist," *Concord Saunterer,* n.s., 1.1 (1993): 76–86. The distinction between science as practice and science as discourse is explored by Ian Hacking, *Representing and Intervening: Introductory Topics in the Philosophy of Natural Science* (Cambridge: Cambridge University Press, 1983).

20. Neufeldt, "Science of Power," 330, 344; *House of Emerson,* 99; "Science of Power," 333, 335–36; *CompW* 7:161. The "marriage" metaphor is offered by Jaroslav Pelikan in his article "Natural History Married to Human History: Ralph Waldo Emerson and the 'Two Cultures,' " in *The Rights of Memory: Essays on History, Science, and American Culture,* ed. Taylor Littleton (University: University of Alabama Press, 1986), 35–75.

21. Neufeldt, "Science of Power," 344; Richardson, *Emerson: Mind on Fire,* 489 (see also 449–50); NHI 427, 4–5. Michael Lopez is particularly sensitive to "the recurring, agonistic pattern beneath [Emerson's] prose," which in his view pulls Emerson toward Nietzsche's "search after power" and away from the "monumental mu-

seum piece" of traditional scholarship. See *Emerson and Power: Creative Antagonism in the Nineteenth Century* (De Kalb: Northern Illinois University Press, 1996), 10, 9. For a particularly innovative and exciting reading of Emerson and science from this perspective—a reading which emphasizes Emerson's indebtedness to alchemy and the physical sciences—see Eric Wilson, *Emerson's Sublime Science* (New York: St. Martin's, 1999).

22. Raymond Williams, *Keywords: A Vocabulary of Culture and Society* (1976; New York: Oxford University Press, 1983), 278, 279. The overlap between the two uses, "rational knowledge" generally and the study of nature specifically, emphasized the naturalization of all knowledge as grounded in God's created world, "nature," which, as Raymond Williams notes, was altered in the eighteenth and nineteenth centuries "from an absolute to a constitutional monarch" operating through natural laws or "Reason."

23. Herschel, *Preliminary Discourse,* 16; *EL* 2:39. Pease suggests that American Renaissance writers attempted "to overcome a division of cultural realms" by establishing "an American public sphere" to enact political decisions—making their work "implicitly political because both artists and politicians shared a common task" of trying to shape the "public will"; see *Visionary Compacts: American Renaissance Writings in Cultural Context* (Madison: University of Wisconsin Press, 1987), ix–x, 45. Simon Schaffer shows that Romantic scientists who, like Oersted in 1806, declared that "the laws of 'material' and 'spiritual' nature were identical" were helping to establish a cultural agreement about the just ordering of the natural and social universe; see "Genius in Romantic Natural Philosophy," in Cunningham and Jardine, *Romanticism and the Sciences,* 94. Similarly, James L. Drummond, writing for children, admonished that "knowledge has been truly said, by Bacon, to be power; with equal, if not greater truth, it may be asserted, that when pursued with a reference to the God of all knowledge, it is virtue"; see *Letters to a Young Naturalist on the Study of Nature and Natural Theology* (London: Longman, Rees, Orme, Brown, and Green, 1831), 321. As Schaffer notes, "Such professional appeals made cultural politics an integral part of the nineteenth-century transformation of natural philosophy into natural sciences" (94).

24. Herschel, *Preliminary Discourse,* 18, 36–37. "*Nature,*" as Pease comments, "exists as a thought experiment Emerson intended to be used in a specific way: to dissolve a world of disconnected agents and dissociated actions back into their source in the nation's principles, which are nature's laws"; *Visionary Compacts,* 217.

25. Steven Shapin, *A Social History of Truth: Civility and Science in Seventeenth-Century England* (Chicago: University of Chicago Press, 1994), 19, xxv.

26. Ibid., 6. For more on the metamorphosis of "sincerity" or "truth" into competent social performance in nineteenth-century middle-class America, see Karen Halttunen, *Confidence Men and Painted Women: A Study of Middle-Class Culture in America, 1830–1870* (New Haven: Yale University Press, 1982). Rather than revealing the truth, such performances instead revealed who could be trusted.

27. *JMN* 5:410; Carlyle, "Characteristics," in *Works,* 28:38. Emerson quotes Carlyle's passage in *JMN* 4:18. For the concept of self-culture, see Robinson, *Apostle of Culture* and *Emerson and the Conduct of Life: Pragmatism and Ethical Purpose in the Later Work* (Cambridge: Cambridge University Press, 1993), esp. 8–12; on culture, see Naomi Quinn, "The Cultural Basis of Metaphor," in *Beyond Metaphor,* ed. James W. Fernandez (Stanford: Stanford University Press, 1991), 57;

Raymond Williams, "Culture and Civilization," in *Encyclopedia of Philosophy*, ed. Paul Edwards (New York: Macmillan, 1967), 2:274; Jonathan Bate, *The Song of the Earth* (Cambridge: Harvard University Press, 2000), 4–5. My argument here intersects with Eduardo Cadava's when he maintains that Emerson's writings "resituate" the "value of truth" within "a more powerful, more stratified context, but a context which cannot be saturated"; see *Emerson and the Climates of History* (Stanford: Stanford University Press, 1997), 65–70.

28. Robinson, *Apostle of Culture*, 25–29; Sampson Reed, *Observations on the Growth of the Mind*, in *Emerson the Essayist*, ed. Kenneth Walter Cameron (Hartford: Transcendental Books, 1972), 2:19; Joseph-Marie de Gérando, *Self-Education; or the Means and Art of Moral Progress*, trans. Elizabeth Peabody (Boston: Carter and Hendee, 1830), 3; *S* 1:280. As Richardson says of Emerson's late essay "Culture," "Culture for Emerson still means cultivation, or, most simply, education"; *Emerson: Mind on Fire*, 493.

29. *S* 3:202; *CW* 1:63; *S* 3:202.

30. *CWSTC* 10:42–43; Thomas De Quincey, "The Poetry of Pope," in *Collected Writings of Thomas De Quincey* (London: A. C. Black, 1897), 11:54–56.

31. *JMN* 5:410–11.

32. *JMN* 1:192; William Whewell, *Selected Writings on the History of Science*, ed. Yehuda Elkana (Chicago: University of Chicago Press, 1984), 132–33; *CW* 1:19. Jonathan Levin makes a similar observation when he writes, "Husbandry integrates with the world . . . and cultivates both self and the world through labor that reforms nature"; see *The Poetics of Transition: Emerson, Pragmatism, and American Literary Modernism* (Durham: Duke University Press, 1999), 68.

33. Shapin, *Social History*, 27; Herschel, *Preliminary Discourse*, 18.

34. *S* 3:255; *JMN* 5:89.

35. Stewart, *Elements*, in *Works*, 2:209.

36. *S* 3:118.

37. *S* 3:127.

38. Shapin, *Social History*, 36. The "colonization" metaphor is taken from Mary Douglas, *Implicit Meanings: Essays in Anthropology* (London: Routledge, 1975), 28.

39. *JMN* 4:306. In this sense, I am agreeing with Foucault when he asserts that " 'Truth' is linked in a circular relation with systems of power which produce and sustain it, and to effects of power which it induces and which extends it." What Foucault calls a " 'regime' of truth" is a succinct description of the reciprocal process described by, say, Sampson Reed in *Observations on the Growth of the Mind*. As Foucault remarks, the real problem is not one of detaching truth from power (should we wish to make "truth" powerless and "power" wholly arbitrary?) but "of detaching the power of truth from the forms of hegemony, social, economic, and cultural, within which it operates at the present time"; see "Truth and Power," in *The Foucault Reader*, ed. Paul Rabinow (New York: Pantheon, 1984), 74–75.

40. *S* 1:199; *CW* 1:17.

41. *CWSTC* 10:47–48.

42. Mark Johnson, *The Body in the Mind: The Bodily Basis of Meaning, Imagination, and Reason* (Chicago: University of Chicago Press, 1987); *CW* 1:31; James Elliot Cabot, "On the Relation of Art to Nature, Part I," *Atlantic Monthly* 13.76 (1864): 199.

43. George Lakoff and Mark Johnson, *Metaphors We Live By* (Chicago: University of Chicago Press, 1980), 3, 193, 197, 193. The watershed book for revisionist theories of metaphor is Max Black's *Models and Metaphors: Studies in Language and Philosophy* (Ithaca: Cornell University Press, 1962). Black claims that metaphors work by *interaction*. To say, for example, "Man is a wolf" evokes a system of related commonplaces about wolves, "suppresses some details, emphasizes others—in short, *organizes* our view of man"—and, Black reminds us, organizes our view of wolves as well, by humanizing them. "Interactive" metaphors cannot be decoded into a literal meaning; rather, they use a "principle subject" to foster insight into a "subsidiary subject" that demands "simultaneous awareness of both subjects but [is] not reducible to any comparison between the two"; 30–46.

44. Quinn, "Cultural Basis," 60. On mixed metaphors, see Dale Pesmen, "Reasonable and Unreasonable Worlds: Some Expectations of Coherence in Culture Implied by the Prohibition of Mixed Metaphor," in Fernandez, *Beyond Metaphor*, 213–43.

45. *S* 3:37.

46. Jonathan Bishop, *Emerson on the Soul* (Cambridge: Cambridge University Press, 1964), 52; James J. Bono, "Science, Discourse, and Literature: The Role/Rule of Metaphor in Science," in *Literature and Science: Theory and Practice*, ed. Stuart Peterfreund (Boston: Northeastern University Press, 1990), 73, 76. Bishop's observation predates Thomas Kuhn's classic argument that metaphors constitute the scientific theories they express. According to Kuhn, "the same interactive, similarity-creating process which Black has isolated in the functioning of metaphor is vital also to the function of models in science." Metaphors call forth "a network of similarities which help to determine the way in which language"—including scientific language—"attaches to the world." Metaphors are thus "substantive or cognitive"; Kuhn, "Metaphor in Science," in *The Road Since Structure* (Chicago: University of Chicago Press, 2000), 203–4.

47. Martin Eger, "Hermeneutics and the New Epic of Science," in *The Literature of Science*, ed. Murdo William McRae (Athens: University of Georgia Press, 1993), 203, 204. On Emerson's Newtonian optics, see Carl M. Lindner, "Newtonianism in Emerson's *Nature*," *ESQ* 20 (1974): 260–69.

48. *JMN* 5:353; *S* 3:35; *JMN* 2:224.

49. Pease, *Visionary Compacts*, 226, 231; *JMN* 2:5–10.

50. *CW* 1:10, 22; Lee Rust Brown, "Emersonian Transparency," *Raritan* 9.3 (1990): 127.

51. *JMN* 3:58; Van Leer, *Emerson's Epistemology*, 22; Reed, *Observations*, 16.

52. Immanuel Kant, *The Philosophy of Kant: Immanuel Kant's Moral and Political Writings*, ed. Carl J. Friedrich (New York: Random House, 1949), 29; *CW* 1:127.

53. *CW* 1:45; Bacon, *New Organon*, in *Works*, 8:99.

54. *CW* 1:44–45.

55. Bacon, "Sphinx; or Science," in *De Sapientia Veterum*, in *Works*, 13:161. For an illuminating discussion of "The Sphinx" as a "threshold poem," see Saundra Morris, "The Threshold Poem, Emerson, and 'The Sphinx,' " *American Literature* 69.3 (1997): 547–70.

56. Kenneth Walter Cameron, *Emerson the Essayist*, 2 vols. (Hartford: Transcendental Books, 1972), 2:417; *Poems*, in *CompW* 9:412. According to David

Porter, " 'The Sphinx' was the essential statement and act of all the other poems that accompanied it"; *Emerson and Literary Change* (Cambridge: Harvard University Press, 1978), 80–81. After attempting his own interpretation of "The Sphinx," Thoreau concluded, "You may find this as enigmatical as the Sphinx's riddle—Indeed I doubt if she could solve it herself"; *J* 1:286.

57. Bacon, *Works*, 13:160–62.

CHAPTER 2. CONVERTING THE WORLD

1. *CW* 1:52, 53.

2. *JMN* 5:347.

3. Emerson preserved Warren's words in his journal: *JMN* 5:376; *CW* 1:53–54.

4. Francis Bacon, *The Advancement of Learning*, in *The Works of Francis Bacon*, ed. James Spedding, Robert Leslie Ellis, and Douglas Denon Heath, 15 vols. (New York: Hurd and Houghton, 1869–72), 6:207, 209, quoted in *JMN* 3:360; Bacon, *Advancement of Learning*, in *Works*, 6:210–11; *JMN* 3:360. On Emerson and Bacon, see Vivian C. Hopkins, "Emerson and Bacon," *American Literature* 29 (1958): 408–30. As Hopkins states and the *JMN* context makes clear, it was Emerson's reading in Joseph-Marie de Gérando in 1830 that led him back to Bacon's "First Philosophy" (414).

5. Bacon, "The Great Instauration," in *Works*, 8:40–41, 26; *CW* 1:56.

6. Bacon, *The New Organon*, in *Works*, 8:63–64; "Great Instauration," in *Works*, 8:24.

7. Bacon, "Great Instauration," in *Works*, 8:25; *New Organon*, in *Works*, 8:62; "Great Instauration," in *Works*, 8:34.

8. *EL* 1:333, 187, 335.

9. Bacon, "Great Instauration," in *Works*, 8:34, 44; *New Organon*, in *Works*, 104–5; *Advancement of Learning*, in *Works*, 6:231.

10. Bacon, *New Organon*, in *Works*, 8:106, 140; *Advancement of Learning*, in *Works*, 6:236.

11. Bacon, *New Organon*, in *Works*, 8:89. See also *Advancement of Learning*, in *Works* 8:396: "For the cripple in the right way (as the saying is) outstrips the runner in the wrong." Or, as Emerson often repeated, "A cripple in the right way will beat a racer in the wrong"; see *JMN* 3:124, 4:254; *EL* 1:290; *CW* 1:22.

12. Bacon, "Great Instauration," in *Works*, 8:36, *New Organon*, in *Works* 8:156–57.

13. Bacon, *Advancement of Learning*, in *Works*, 6:166.

14. Bacon, *New Organon*, in *Works*, 8:162–63, 53; *CW* 1:55. The American Puritan John Winthrop used a parallel argument when putting down the antinomianism of Anne Hutchinson in 1645: liberty "is maintained and exercised in a way of subjection to authority," just as a woman who chooses a husband is subject to him, "yet in a way of liberty, not of bondage," and a Christian is subject to the "easy and sweet" authority of Christ. See *Winthrop's Journal, "History of New England,"* *1630–1649*, ed. James Kendall Hosmer, 2 vols. (New York: Scribner's, 1908), 2:238–39.

15. "Ode, Inscribed to W.H. Channing," in *Poems*, in *CompW* 9:76–79; *E&L* 958, 972. As Robert M. Young has asked, "When will we look deeply enough to see that science embodies values in theories, therapies, and things, and that that embod-

iment must always be looked after, outwardly and inwardly?" See *Darwin's Metaphor: Nature's Place in Victorian Culture* (Cambridge: Cambridge University Press, 1985), 247.

16. Steve Fuller, "Disciplinary Boundaries and the Rhetoric of the Social Sciences," *Poetics Today* 12.2 (1991): 308; *JMN* 5:168. Realism in this sense asserts that the categories we apply to the world are "really" there, whereas nominalism counters that our human categories are necessarily arbitrary and limited, "in name only."

17. Bacon, "Great Instauration," in *Works,* 8:35–36, and see also *Advancement of Learning,* in *Works,* 6:92; "Great Instauration," in *Works,* 8:17; *New Organon,* in *Works,* 8:350; *CW* 1:45. For more on the imperial remaking of Eden, see Carolyn Merchant, "Reinventing Eden: Western Culture as a Recovery Narrative," in *Uncommon Ground: Toward Reinventing Nature,* ed. William Cronon (New York: Norton, 1995), 132–45.

18. *CW* 1:54–55; Bacon, *Advancement of Learning,* in *Works,* 6:165. Emerson also found the metaphor of seal and print used by the Cambridge Neoplatonist Ralph Cudworth, who observed that there can be only *one* "original mind," of which all others partake, "being, as it were, stamped with the impression or signature of one and the same seal"; *The True Intellectual System of the Universe,* 4 vols. (London, 1820), 3:415.

19. *CW* 1:55, 56, 67; *EL* 1:335.

20. *CW* 1:59; *EL* 1:187; *CW* 1:59.

21. *S* 2:202, 234–35.

22. Bacon, *New Organon,* in *Works,* 8: 131; *CW* 1:58.

23. *CW* 1: 60, 61, 63.

24. *CW* 1: 67, 68, 70.

25. *CW* 1: 54; John Playfair, *Dissertation Second: Exhibiting a General View of the Progress of Mathematical and Physical Science, since the Revival of Letters in Europe,* reprinted in *Dissertations on the Progress of Knowledge* (New York: Arno, 1975), 93.

26. Playfair, *Dissertation,* 93–94; Joyce Appleby, Lynn Hunt, and Margaret Jacob, *Telling the Truth about History* (New York: Norton, 1994), 178–79.

27. Gary R. Deason, "Reformation Theology and the Mechanistic Conception of Nature," in *God and Nature,* ed. David C. Lindberg and Ronald L. Numbers (Berkeley: University of California Press, 1986), 185–86.

28. Cudworth, *True Intellectual System,* 1:51–58, 3:292; Margaret Jacob, "Christianity and the Newtonian Worldview," in Lindberg and Numbers, *God and Nature,* 240.

29. Perry Miller, *The New England Mind: The Seventeenth Century* (Cambridge: Harvard University Press, 1939), 224, 226; Cudworth, *True Intellectual System,* 1:384, 337.

30. Cudworth, *True Intellectual System,* 3:292. For God's two books, see Ernst Robert Curtius, *European Literature and the Latin Middle Ages,* trans. Willard R. Trask (Princeton: Princeton University Press, 1953), 319–26; James Bono, *The Word of God and the Languages of Man: Interpreting Nature in Early Modern Science and Medicine,* vol. 1: *Ficino to Descartes* (Madison: University of Wisconsin Press, 1995), 5, 6.

31. Edward Grant, "Science and Theology in the Middle Ages," in Lindberg and

Numbers, *God and Nature*, 50; Bacon, *New Organon*, in *Works*, 8:126; *Advancement of Learning*, in *Works*, 6:144.

32. James R. Moore, "Geologists and Interpreters of Genesis in the Nineteenth Century," in Lindberg and Numbers, *God and Nature*, 323; *CW* 1:59.

33. *CW* 1:84, 89, 80, 86. Emerson's position would place him on the side of professional science even as the compromise was beginning to break down, resulting in the "two truths" solution, which finally split religion and science into two very different realms. See Moore, "Geologists and Interpreters," 335–36, 340, 344.

34. *CW* 1:85, 88, 90, 93.

35. James Ferguson, *Ferguson's Astronomy, Explained upon Sir Isaac Newton's Principles*, ed. David Brewster, 2 vols. (Philadelphia, 1817), 1:1–2; Humphry Davy, *Elements of Chemical Philosophy*, in *The Collected Works of Sir Humphry Davy*, ed. John Davy, 9 vols. (London: Smith, Elder, 1839–40), 4:1–2; John F.W. Herschel, *A Preliminary Discourse on the Study of Natural Philosophy* (1830; facsimile ed., Chicago: University of Chicago Press, 1987), 4–5; Charles Lyell, "Review of *Transactions of the Geological Society of London*," *Quarterly Review* 34.58 (1826): 538. Emerson read the 1817 edition of *Ferguson's Astronomy* in 1821 and recalled this passage in 1832, but without recalling its source; see *JMN* 4:25. Lyell's anonymously authored article on geology is the one that so impressed Emerson when he read it in 1826; see *JMN* 3:51. For an extremely useful introduction to natural theology, see John Hedley Brooke, *Science and Religion: Some Historical Perspectives* (Cambridge: Cambridge University Press, 1991), 192–225.

36. William Paley, *Natural Theology, or Evidences of the Existence and Attributes of the Deity*, in *Works* (London: C. and J. Rivington et al., 1825), 5:2, 8; *S* 1:298–99. Writing of the religious beliefs of the Harvard botanist Asa Gray, A. Hunter Dupree states that even in the 1850s, "Paley was a symbol for a live system of ideas widely diffused, strongly felt, thoroughly in tune with the traditions of science, and thoroughly accepted by evangelical Christianity"; *Asa Gray: American Botanist, Friend of Darwin* (1959; reprint, Baltimore: Johns Hopkins University Press, 1988), 137–38.

37. Coleridge, *Aids to Reflection*, in *CWSTC* 9:405.

38. Bacon, *Advancement of Learning*, in *Works*, 6:93; *CW* 1:21; Anne Marie Louise Necker de Staël, *Germany*, 2 vols. (1813; reprint, Boston: Houghton, Osgood, 1879), 2:217–18; Emerson first refers to the Swedenborg quotation in *JMN* 4:33.

39. *EL* 2:25.

40. *CW* 1:54; *JMN* 5:233; *CW* 1:17.

41. Robert Spiller, "Introduction," in *CW* 1:xiv.

42. Dugald Stewart, *Elements of the Philosophy of the Human Mind*, in *The Collected Works of Dugald Stewart*, ed. William Hamilton (Edinburgh: Thomas Constable, 1854) 3:296; see also 1:289.

43. Bacon, *Advancement of Learning*, in *Works*, 6:276; *New Organon*, in *Works* 8:76–90.

44. Bacon, *Advancement of Learning*, in *Works*, 6:202, 203; *CW* 1:31.

45. *CW* 1:56–57.

46. Cudworth, *True Intellectual System*, 1:317, 322, 332–34.

47. *JMN* 6:213–14 and Cudworth, *True Intellectual System*, 1:335; Cudworth, *True Intellectual System*, 1:337; *CW* 1:31, 33.

48. Stewart, *Dissertation: Exhibiting the Progress of Metaphysical, Ethnical, and Political Philosophy since the Revival of Letters in Europe,* in *Works,* 1:86 n. 2 (quoting Cudworth), 134–35, 398–400. Stewart quotes Cudworth approvingly: "the power of knowing or understanding" is "an active exertion from the mind itself"; *Works,* 1:407.

49. De Staël, *Germany,* 2:122–23, 159, 162, 203.

50. Stewart, *Dissertation,* in *Works,* 1:75–76; Sampson Reed, *Observations on the Growth of the Mind* (1821), in *Emerson the Essayist,* ed. Kenneth Cameron, 2 vols. (Hartford: Transcendental Books, 1972), 2:13.

51. William Whewell, *Selected Writings on the History of Science,* ed. Yehuda Elkana (Chicago: University of Chicago Press, 1984), 249; Theodore Parker's review reprints several of Whewell's aphorisms, including Aphorism III, in *The Dial,* 2.4 (1842): 529–30. William B.O. Peabody, "Study of Natural History," *North American Review* 41 (October 1835), 423.

52. Herschel, *Preliminary Discourse,* 14–15. The Shakespeare quotation is from *As You Like It,* II.i.16–17.

53. Barbara Packer, *Emerson's Fall: A New Interpretation of the Major Essays* (New York: Continuum, 1982), 112.

54. *EL* 1:327, 2:27. Or, as Whewell put it, while Bacon "was not the first to tell men that they must collect their knowledge from observation, he had no rival in his peculiar office of teaching them *how* science must thus be gathered from experience"; *Selected Writings,* 224.

55. *EL* 1:330–31; Bacon, *Advancement of Learning,* in *Works,* 6:221–22; *New Organon,* in *Works,* 8:71–72.

56. Bacon, *New Organon,* in *Works,* 8:137; also *Advancement of Learning,* in *Works,* 6:215; *New Organon,* in *Works,* 8:138. For more on Bacon in nineteenth-century Anglo-America, see Susan Faye Cannon, *Science in Culture: The Early Victorian Period* (New York: Dawson and Natural History Publications, 1978); Richard Yeo, "An Idol of the Market-Place: Baconianism in Nineteenth-Century Britain," *History of Science* 23 (1985): 251–98; Jonathan Smith, *Fact and Feeling: Baconian Science and the Nineteenth-Century Literary Imagination* (Madison: University of Wisconsin Press, 1994); Walls, *Seeing New Worlds: Henry David Thoreau and Nineteenth-Century Natural Science* (Madison: University of Wisconsin Press, 1995).

57. Immanuel Kant, *The Philosophy of Kant: Immanuel Kant's Moral and Political Writings,* ed. Carl J. Friedrich (New York: Random House, 1949), 41; Whewell, *Selected Writings,* 231. See Playfair, *Dissertation,* 55; Stewart, *Works,* 1:63–64; Herschel, *Preliminary Discourse,* 114; Whewell, *Selected Writings,* 226.

58. Whewell, *Selected Writings,* 228; Stewart, *Works,* 2:417; Herschel, *Preliminary Discourse,* 114, also 198–201.

59. James Secord, *Victorian Sensation: The Extraordinary Publication, Reception, and Secret Authorship of Vestiges of the Natural History of Creation* (Chicago: University of Chicago Press, 2000), 407; William Whewell, "Modern Science—Inductive Philosophy," *Quarterly Review* 45 (July 1831): 398, 379–80. For an instance of Herschel's influence on Emerson, when Herschel "earnestly" recommended William Charles Wells's *Essay on Dew* "as one of the most beautiful specimens we can call to mind of inductive experimental enquiry lying within a moderate compass" (*Preliminary Discourse,* 163), Emerson made sure to seek it out

and read it, in 1834. As for Whewell's two major works, *History of the Inductive Sciences* (1837) and *The Philosophy of the Inductive Sciences* (1840), there is no record that Emerson encountered either except at second hand.

60. Coleridge, *The Friend*, in CWSTC 4.1:482–92. Herein lies a controversy, in which idealists tried to rehabilitate and claim Bacon for Platonism, while nominalists (who maintained that names were too arbitrary to correspond to real essences) flatly rejected Platonic ideals.

61. Packer, *Emerson's Fall*, 44–45; Samuel Taylor Coleridge, *Aids to Reflection* (Port Washington, N.Y.: Kennikat, 1971), 211; idem, *The Friend*, in *CWSTC* 4.1:157–58, 499. Coleridge's "Method" is fully explained in "Essays on the Principle of Method" in *The Friend*, which Emerson read in 1829.

62. *CWSTC* 4.1:464, 470; Packer, *Emerson's Fall*, 44.

63. *CWSTC* 4.1:471, 497–98.

64. *CW* 1:31, 34, 33–34; *EL* 2:33; *CW* 3:12, 13, 1:22.

65. *CW* 1:108; *JMN* 5:114.

66. *JMN* 5:145; *CW* 1:40; *EL* 2:36; *JMN* 5:221; *EL* 2:37, 1:79.

67. Stewart, *Works*, 1:161; *CW* 1:106.

68. For the specifics of Emerson's reading—texts, editions, and dates, when known—see Kenneth Walter Cameron, *Ralph Waldo Emerson's Reading* (Raleigh, N.C.: Thistle, 1941); and Walter Harding, *Emerson's Library* (Charlottesville: University Press of Virginia, 1967).

69. *JMN* 5:146; David Knight, *The Age of Science* (New York: Basil Blackwell, 1986), 133–34.

70. William Whewell, Review of Mary Somerville, *On the Connexion of the Physical Sciences*, *Quarterly Review* 51 (March 1834): 59; Richard Yeo, *Defining Science: William Whewell, Natural Knowledge, and Public Debate in Early Victorian Britain* (Cambridge: Cambridge University Press, 1993), 5. The geologist Adam Sedgwick was so alarmed by Whewell's neologism that he wrote in the margins of a volume in which Whewell stressed the need for such a word, "better die of this want than bestialize our tongue by such barbarisms"; Secord, *Victorian Sensation*, 404–5. For a full history of the word *scientist*, see Sydney Ross, "*Scientist*: Story of a Word," *Annals of Science* 18.2 (1962): 65–85.

71. Richard Owen, in defending the purposeful design of nature, goes so far as to suggest that those who do not believe in it suffer from "some, perhaps congenital, defect of mind" analogous to color blindness or tone deafness. See *Palaeontology; or, a Systematic Summary of Extinct Animals and Their Geological Relations* (Edinburgh: Adam and Charles Black, 1860), 314.

72. For Emerson's reading of Roget's treatise, see William Rossi, "Emerson, Nature, and the Natural Sciences," in *A Historical Guide to Ralph Waldo Emerson*, ed. Joel Myerson (New York: Oxford University Press, 2000), 125–26; indeed, this essay is the best account of Emerson, natural theology, and science. Rossi makes the point that natural theology remained integral to both science and religious belief even after the "triumph of scientific naturalism led by Darwin." The shift from Paleyan functionalism to explanations of natural phenomena in terms of natural law was not the "bugle call to the final battle between the forces of science and religion," but accompanied the professionalization of science and sustained scientific inquiry both epistemologically and rhetorically (105–7).

73. Brewster's biography of Newton was republished in a greatly expanded edition

in 1855 (*Memoirs of the Life, Writings, and Discoveries of Sir Isaac Newton,* 2 vols. [Edinburgh, 1855]); Emerson read both editions and owned a copy of the earlier one.

74. Immanuel Kant, *Metaphysical Foundations of Natural Science,* trans. James Ellington (Indianapolis: Bobbs-Merrill, 1970), 3; CW 3:103–4; Kant, *Metaphysical Foundations,* 4.

75. Kant, *Metaphysical Foundations,* 4; Georges Cuvier, *A Discourse on the Revolutions of the Surface of the Globe, and the Changes Thereby Produced in the Animal Kingdom* (Philadelphia: Carey and Lea, 1831), 3.

76. Candolle recommended Carl Linnaeus's *Lachesis Lapponica, or a Tour in Lapland;* see *Elements of the Philosophy of Plants* (Edinburgh: William Blackwood, 1821), 166. Emerson, perhaps following his suggestion, looked into this work several times from 1839 to 1858.

77. Jonathan Bishop, *Emerson on the Soul* (Cambridge: Cambridge University Press, 1964), 53. For Emerson and Goethe, see Gustaaf Van Cromphout, *Emerson's Modernity and the Example of Goethe* (Columbia: University of Columbia Press, 1990). Robert D. Richardson Jr. dates American Transcendentalism to the moment Emerson read Hedge's 1823 article on Coleridge and the Germans, in 1834; *Emerson: The Mind on Fire* (Berkeley: University of California Press, 1995), 166. The meetings of the loose group who came to be known as Transcendentalists were held when Hedge, who lived in Bangor, Maine, was in town; hence the group was known to insiders as "the Hedge Club."

78. Johann Gottfried Herder, *Outlines of a Philosophy of the History of Man,* trans. T. Churchill (1830; reprint, Chicago: University of Chicago Press, 1987), 112, 177–78.

79. Stewart, *Works,* 2:72–80.

80. *JMN* 1:253–54, 253 n. 6.

CHAPTER 3. GNOMIC SCIENCE

1. *JMN* 2:248, 72, 252. For an illuminating account of young Emerson's life, see Mary Kupiec Cayton, *Emerson's Emergence: Self and Society in the Transformation of New England, 1800–1845* (Chapel Hill: University of North Carolina Press, 1989); and Evelyn Barish, *Emerson: The Roots of Prophecy* (Princeton: Princeton University Press, 1989).

2. Plato, *Timaeus,* trans. Benjamin Jowett, in *The Collected Dialogues,* ed. Edith Hamilton and Huntington Cairns (Princeton: Princeton University Press, 1961), 1173.

3. *JMN* 1:192, 260, 2:219.

4. *JMN* 1:336, 61, 189, 194, 2:71.

5. *JMN* 3:282, 2:128, 136, 167, 3:74, 4:41–42. In 1840 Emerson recorded a dream in which this metaphor was most suggestively literalized: "I dreamed that I floated at will in the great Ether, and I saw this world floating also not far off, but diminished to the size of an apple. Then an angel took it in his hand & brought it to me and said 'This must thou eat.' And I ate the world" (*JMN* 7:525).

6. *JMN* 3:367 (Emerson's source was Joseph-Marie de Gérando); 2:55–56, 65.

7. *JMN* 2:145, 50–51, 340–41, 52; S 1:58, 156. Although Emerson did not connect them, his temporary blindness was probably a symptom of his tuberculosis. See Barish, *Emerson,* 177–84.

8. *JMN* 2:141.

9. *JMN* 2:141.

10. *JMN* 2:120, 130.

11. *JMN* 2:205.

12. *JMN* 2:209, 1:298, 2:231, 238–42.

13. *JMN* 2:167, 168.

14. *JMN* 2:264–65, 278–79, 318. Barbara Packer notes that this forecast "shows considerable prescience"; see *Emerson's Fall: A New Interpretation of the Major Essays* (New York: Continuum, 1982), 32.

15. *JMN* 2:291.

16. *JMN* 2:109, 277, 275.

17. *JMN* 2:222.

18. *S* 2:139.

19. *JMN* 3:45–49.

20. *JMN* 3:51, 61.

21. *JMN* 3:69, 78.

22. *JMN* 3:70.

23. *JMN* 3:93, 4:132, 3:98–99.

24. *JMN* 3:107, 130, 135, 139, 256.

25. *JMN* 3:148–49, 152.

26. *JMN* 3:166–67, 191, 168.

27. *JMN* 3:171–72.

28. *JMN* 3:199, 178, 185; see also *JMN* 3:55.

29. *JMN* 3:168.

30. *L* 1:256; *JMN* 3:149; Gay Wilson Allen, *Waldo Emerson: A Biography* (New York: Viking, 1981), 120–24, 140–44, 165–71; Robert D. Richardson Jr., *Emerson: The Mind on Fire* (Berkeley: University of California Press, 1995), 84–88, 91–92, 108–11.

31. *JMN* 3:193, 194, 199, 209, 216–17, 220, 214 (see also *JMN* 3:248, 4:24); *CWSTC* 10:40; Dugald Stewart, *Dissertation: Exhibiting the Progress of Metaphysical, Ethical, and Political Philosophy since the Revival of Letters in Europe,* in *The Collected Works of Dugald Stewart,* ed. William Hamilton, vol. 1 (Edinburgh: Thomas Constable, 1854), 524.

32. *CW* 1:53; *JMN* 3:218, 224, 230–31, 226.

33. *JMN* 3:235–38.

34. *JMN* 3:239, 250, 253, 268, 270, 2:200, 3:280, 4:96; and *CW* 1:22. Wesley T. Mott (taking issue with the editors of *Early Lectures,* vol. 1) makes the important point that while Emerson's closest study of science was coincident with his resignation from the church, the study of science in itself does not fully account for his "shift" from "a theological to a secular basis for his moral philosophy." As Mott writes, "The issue . . . was rather the personal appropriation of the concept of revelation, which for Emerson predated and verified what was of value in science"; see *"Strains of Eloquence": Emerson and His Sermons* (University Park: Pennsylvania State University Press, 1989), 68.

35. Immanuel Kant, *Philosophy of Kant: Immanuel Kant's Moral and Political Writings,* ed. Carl J. Friedrich (New York: Random House, 1949), 134–36.

36. *JMN* 3:318, 325, 326. There is no evidence that Emerson read Kant's essay, but Kant's ideas about scholarly freedom had enormous influence on Fichte, whom Emerson did read. Kant's statement remains the classic defense of academic freedom.

37. *JMN* 4:7, 24, 25–26; Richardson, *Emerson: Mind on Fire,* 3.

38. *JMN* 3:150, 4:26, 27, 30.

39. *S* 4:305; *JMN* 4:40, 45.

40. *JMN* 4:48, 274.

41. See *JMN* 4:197, 204; Harry Hayden Clark, "Emerson and Science," *Philological Quarterly* 10 (July 1931): 251–52.

42. *JMN* 4:405, 406, 200; Lee Rust Brown, *The Emerson Museum: Practical Romanticism and the Pursuit of the Whole* (Cambridge: Harvard University Press, 1997), 86; Jonathan Bishop, *Emerson on the Soul* (Cambridge: Cambridge University Press, 1964), 54; *EL* 1:10.

43. On Goethe's ur-plant, see Johann Bernhard Stallo, *General Principles of the Philosophy of Nature* (Boston: Crosby and Nichols, 1848), 19. Brown, *Emerson Museum,* 19–20.

44. *JMN* 4:47, 48, 52, 58, 6:197; 4:87.

45. *JMN* 4:278–79.

46. *JMN* 4:30. Goethe's expression came to Emerson through Carlyle; see *JMN* 4:87 n. 185.

47. *EL* 2:355, 356; *CW* 1:93. This idea that a turn to the "heart" is a turn outward is, of course, repeated in the famous line from "Self-Reliance": "To believe your own thought, to believe that what is true for you in your private heart, is true for all men,—that is genius"; *CW* 2:27.

48. *CW* 2:173–75.

49. *CW* 2:161.

50. *EL* 2: 355. The concept of the "modern" here is derived from Bruno Latour, *We Have Never Been Modern,* trans. Catherine Porter (Cambridge: Harvard University Press, 1993).

51. *EL* 1:7, Bishop, *Emerson on the Soul,* 54.

52. *EL* 1:7, 10, 7, 6, 7–10.

53. *J* 1:465.

54. *EL* 1:11–13.

55. *EL* 1:6, 15.

56. *EL* 1:20, 23.

57. *EL* 1:23, 24.

58. *EL* 1:24, 25–26.

59. *EL* 1:26; Packer, *Emerson's Fall,* 46.

60. *EL* 1:29.

61. *EL* 1:29–31, 34.

62. *EL* 1:34, 33–34, 36.

63. *EL* 1:41–3, 46–48, 49.

64. *EL* 1:53, 52.

65. *EL* 1:51.

66. *EL* 1:55, 59.

67. *EL* 1:63, 66.

68. *EL* 1:67, 68.

69. *EL* 1:66; *CW* 3:164, 4:103; *EL* 1:68. In "Intellect" this becomes: "One soul is a counterpoise of all souls, as a capillary column of water is a balance for the sea"; *CW* 2:203.

70. *CW* 2:59.

71. *EL* 1:70–71; *CW* 1:53; *EL* 1:72.

72. *Comp W* 9:4.

73. *EL* 1:76.

74. *EL* 1:80, 78, 79.

75. *EL* 1:83.

76. *CW* 1:8. When in 1849 Emerson drew up a list of "Bigendians" vs. "Littleendians," he paired himself with Goethe, and Thoreau with Napoleon (*JMN* 11:173).

77. *CW* 1:1, 8. Compare Ralph Cudworth, *True Intellectual System*, 4 vols. (London: 1820), 1:337: Nature "is a living stamp or signature of the Divine wisdom; which, though it act exactly according to its archetype, yet it doth not at all comprehend nor understand the reason of what itself doth." Emerson may have derived some of his reassuring tone here from Charles Lyell, who, after listing several pages of British species extirpated by human beings, nevertheless concluded that "the aggregate force exerted by man is truly insignificant"; *Principles of Geology*, 3 vols. (1830–1833; facsimile ed., Chicago: University of Chicago Press, 1990–1992), 2:207.

78. *CW* 1:10, 22.

79. *CW* 1:22–23; Kenneth Burke, "I Eye, Aye—Emerson's Early Essay on 'Nature': Thoughts on the Machinery of Transcendence," in *Emerson's Nature: Origin, Growth, Meaning*, ed. Merton M. Sealts Jr. and Alfred R. Ferguson (Carbondale: Southern Illinois University Press, 1979), 161; *CW* 1:37, 35–36; David Van Leer, *Emerson's Epistemology: The Argument of the Essays* (Cambridge: Cambridge University Press, 1986), 53, 30; *CW* 1:38.

80. Van Leer, *Emerson's Epistemology*, 53, 30; *CW* 1:38. Van Leer praises Emerson's "advanced, even modern, understanding of truth" here: "a true theory *is* 'its own evidence' and does 'explain all phenomena.' For, as Emerson carefully demonstrates, phenomena exist only inside a theoretical framework, never outside one," and so "the measure of truth becomes not external correspondence but internal coherence. Perhaps Emerson lacked the control to go no further. But the wonder is that he had the intelligence and discipline to go so far" (58).

81. *CW* 1:8.

82. *CW* 1:11, 15–16, 17.

83. *CW* 1:17, 19, 21, 23.

84. *CW* 1:23, 25–26, 27. Emerson's first use of Coleridge's distinction between Reason and Understanding was in June 1834; see *JMN* 4:297–98, 299. Herschel compares nature to a horse that will throw the rider who thwarts or contests it; John F. W. Herschel, *A Preliminary Discourse on the Study of Natural Philosophy* (1830; facsimile ed., Chicago: University of Chicago Press, 1987) 66; Sampson Reed offers the image of nature as the animal on which the "King of Zion rode into Jerusalem; at once free and subject to the will of the rider"; *Observations on the Growth of the Mind*, in *Emerson the Essayist*, ed. Kenneth Cameron, 2 vols. (Hartford: Transcendental Books, 1972) 2:22.

85. *CW* 1:31.

86. *CW* 1:34–35, 36.

87. Van Leer, *Emerson's Epistemology*, 47; *CW* 1:37, 38.

88. *CW* 1:38–39.

89. Van Leer, *Emerson's Epistemology*, 65; *CW* 1:39, 40.

90. *CW* 1:43, 44, 45.

91. *EL* 1:26; *JMN* 15:406; *CW* 1:23.

92. For the religious background to Emerson's ecstatic states, see Alan H. Hodder, " 'After a High Negative Way': Emerson's 'Self-Reliance' and the Rhetoric of Conversion,' " *Harvard Theological Review* 84 (1991): 423–46.

93. *CW* 1:120.

94. David Jacobson, *Emerson's Pragmatic Vision: The Dance of the Eye* (University Park: Pennsylvania State University Press, 1993), 97, 2; Jonathan Levin, *The Poetics of Transition: Emerson, Pragmatism, and American Literary Modernism* (Durham: Duke University Press, 1999), 30, 37, 42; David Robinson, *Emerson and the Conduct of Life: Pragmatism and Ethical Purpose in the Later Work* (Cambridge: Cambridge University Press, 1993), 4–5; Jacobson, 98; Robinson, 7. As David Van Leer notes, Emerson's optimism was "not temperamental, but an emotional discipline in the face of real pain"; *Emerson's Epistemology*, 8.

95. *CW* 1:124, 126, 125, 127, 120–21, 122, 128, 130. For a longer and more theoretical version of the argument that follows, see Laura Dassow Walls, "The Anatomy of Truth: Emerson's Poetic Science," *Configurations* 5 (1997): 425–61. David Robinson suggests that this address grew directly out of the "unfinished business" of *Nature* and that this passage is, therefore, a self-indictment: "The new book says, 'I will give you the key to nature,' and we expect to go like a thunderbolt to the center. But the thunder is a surface phenomenon, makes a skin-deep cut, and so does the sage" (*CW* 1:123). Robinson, "*The Method of Nature* and Emerson's Period of Crisis," in *Emerson Centenary Essays,* ed. Joel Myerson (Carbondale: Southern Illinois University Press, 1982), 83.

96. *EL* 1:382.

97. *CW* 1:131, 9.

98. *CW* 1:131–32, 120, 122.

99. *CW* 1:131; *JMN* 5:135. As discussed earlier, the marriage metaphor can be traced to Bacon.

100. *CW* 1:132; Francis Bacon, "Great Instauration," in *The Works of Francis Bacon,* ed. James Spedding, Robert Leslie Ellis, and Douglas Denon Heath, 15 vols. (New York: Hurd and Houghton, 1872), 8:53.

101. *CW* 2:193; Georges Cuvier, *A Discourse on the Revolutions of the Surface of the Globe, and the Changes Thereby Produced in the Animal Kingdom* (Philadelphia: Carey and Lea, 1831), 59, 61, quoted in *TN* 1:31.

102. James J. Bono, "Science, Discourse, and Literature: The Role/Rule of Metaphor in Science," in *Literature and Science: Theory and Practice,* ed. Stuart Peterfreund (Boston: Northeastern University Press, 1990), 59–89; *CW* 1:21. Important works that do ask this question are Richard P. Adams, "Emerson and the Organic Metaphor," *PMLA* 69 (March 1954): 117–30; and F.O. Matthiessen, *American Renaissance: Art and Expression in the Age of Emerson and Whitman* (Oxford: Oxford University Press, 1941). For a critique of the traditional organicist approach, see Michael Lopez, *Emerson and Power: Creative Antagonism in the Nineteenth Century* (De Kalb: Northern Illinois University Press, 1996), 175–85.

103. The saying "Truth is one, error many" originated with Hume and was widely cited, seldom with attribution. See, for example, Stewart, *Dissertation,* 247, 515; Dale Pesmen, "Reasonable and Unreasonable Worlds: Some Expectations of Coherence in Culture Implied by the Prohibition of Mixed Metaphor," in *Beyond*

Metaphor, ed. James W. Fernandez (Stanford: Stanford University Press, 1991), 216, 213, 218.

104. Robert Koch, "The Case of Latour," *Configurations* 3.3 (1995), 337; CW 1:123, 120; Koch, 338 (alluding to Bruno Latour).

105. As detailed earlier, Coleridge follows a tradition established by the Cambridge Neoplatonists of the seventeenth century, particularly Ralph Cudworth, who established the organic metaphor in everything but name. I also discuss this process in Walls, *Seeing New Worlds: Henry David Thoreau and Nineteenth-Century Natural Science* (Madison: University of Wisconsin Press, 1995), 70–76.

106. CW 1:8.

107. Hero of Alexandria quoted in Robert Lawlor, *Sacred Geometry* (New York: Crossroad, 1982), 65.

108. CW 2:201, 1:27. Compare Herschel: "As truth is single, and consistent with itself, a principle may be as completely and as plainly elucidated by the most familiar and simple fact, as by the most imposing and uncommon phenomenon"; *Preliminary Discourse,* 14. In different form this becomes: "A leaf is a compend of Nature, and Nature a colossal leaf. An animal is a compend of the World, and the World is an enlargement of an animal. There is more family likeness than individuality. Hence Goethe's striving to find the Arch-plant"; *JMN* 5:137–38.

109. Roland Barthes, *Mythologies,* trans. Annette Lavers (New York: Farrar, Straus, and Giroux, 1972), 69; CW 1:34; György Doczi, *The Power of Limits: Proportional Harmonies in Nature, Art and Architecture* (Boulder: Shambhala, 1981), 49–50. Eric Wilson's discussion suggests deep roots for the magic of the double helix: alchemists "often symbolized nature's 'unrollings' and 'rollings,' its *solve et coagula,* with a double spiral or with the caduceus, two dragons or serpents winding in contrary directions around a staff or tree. The double spiral or the caduceus is a symbol of the synthesis of opposites, be they solution or dissolution, centrifugal or centripetal force, positive and negative charges"; *Emerson's Sublime Science* (New York: St. Martin's, 1999), 56.

110. CW 2:194; Herschel, *Preliminary Discourse,* 358, 360–61; CW 1:27, 3:105.

111. Herschel, *Preliminary Discourse,* 37; CW 2:195. See Trevor H. Levere, "Coleridge and the Sciences," in *Romanticism and the Sciences,* ed. Andrew Cunningham and Nicholas Jardine (Cambridge: Cambridge University Press, 1990), 298.

112. Quoted in Lawlor, *Sacred Geometry,* 65; Oliver Wendell Holmes, *The Poetical Works of Oliver Wendell Holmes* (Boston: Houghton Mifflin, 1891), 2:107–8; CW 1:40. For more on gnomons, see D'Arcy Thompson, *On Growth and Form* (Cambridge: Cambridge University Press, 1961), 181–87. The dimensions of the Golden Rectangle are such that the ratio of the smaller to the larger is the same as that of the larger to the sum of the two (roughly three to five).

113. *JMN* 7:324–25; CW 1:43; Albert E. Waugh, *Sundials: Their Theory and Construction* (New York: Dover, 1973), 4–29; "Gnomon," in *Encyclopaedia Britannica,* 11th ed. (New York, 1910), 12:152. According to George Willis Cooke, Alcott proposed the name at a meeting of the Transcendental Club on September 18, 1839, at which Margaret Fuller gave her views about the goals of their new journal. "Alcott spoke of his diary, to which he had given the title of 'The Dial,' and proposed that this should be the name of the new periodical"; see Cooke, *Historical*

and Biographical Introduction to Accompany "The Dial," 2 vols. (New York: Russell and Russell, 1961), 1:61. Compare Thoreau in August 1841: "The landscape contains a thousand dials which indicate the natural divisions of time—the shadows of a thousand styles point to the hour"; *J* 1:321.

114. *CW* 2:179, 186, 1:21. The editors of Emerson's *Collected Works* note that Emerson found this quotation from Swedenborg in Samuel Sandels's article "Emanuel Swedenborg," *New Jerusalem Magazine* 5 (July 1832): 437. See *CW* 1:249; "Gnomon," 152; *CW* 2:180.

115. *CW* 1:187. See also Lee Rust Brown, who makes similar observations in *The Emerson Museum,* 66, 127. See also Elizabeth A. Dant, "Composing the World: Emerson and the Cabinet of Natural History," *Nineteenth-Century Literature* 44 (1989): 18–44.

116. The title of Lawlor's book on the subject, *Sacred Geometry;* Theodore Andrea Cook, *The Curves of Life* (1914; reprint, New York: Dover, 1979), 424. For a nineteenth-century version of this, see Stallo, *General Principles,* 224–25. James Watson, *The Double Helix: A Personal Account of the Discovery of the Structure of DNA* (New York: Norton, 1980), 120.

117. Goethe's ur-plant, then, can be seen as an attempt to unite both the organic, cyclical side and the gnomic, timeless side of the metaphor of organicism into one figure, the ur-plant, which would enfold all possible plants into one single organism, all links of the chain into one primal link. Here is the point where my argument intersects most closely with David Robinson's, when he suggests that "The Method of Nature" opposes the atemporal quality of ecstasy "to nature's growth or progress through time," with its associated process "of growth or culture of the soul." The two are synthesized when the mind in its own growth grasps the "cause of nature as it grows": an ecstatic experience that is both necessary and dangerously fragile and unwilled (*"The Method of Nature,"* 88). I am suggesting that the synthesis is somewhat sturdier, for though paradoxical, it is nevertheless modeled by science's growing insight into nature's principle of growth.

118. Susan Stewart, *On Longing: Narratives of the Miniature, the Gigantic, the Souvenir, the Collection* (Durham: Duke University Press, 1993), 55, 53.

119. Ibid., 172; "Fate," *E&L* 958, 967.

120. *CW* 2:72, 73, 1:28–29. On the compensations of calamity, see also Richard Grusin, *Transcendental Hermeneutics: Institutional Authority and the Higher Criticism of the Bible* (Durham: Duke University Press, 1991), 44–46. As Michael Lopez says, critics have increasingly attended to the costs of such a philosophy, which uses loss as a catalyst to power; *Emerson and Power,* 58.

121. *CW* 2:193.

122. *CW* 1:53, 59. Mary Poovey writes that by the early nineteenth century the image of the "social body" was used to refer both to the poor as an isolated population and to English society as an "organic whole." The term *"social body* therefore promised full membership in a whole (and held out the image *of* that whole) to a part identified as needing both discipline and care"; *Making a Social Body: British Cultural Formation, 1830–1864* (Chicago: University of Chicago Press, 1995), 7–8.

123. Christopher Newfield, *The Emerson Effect: Individualism and Submission in America* (Chicago: Chicago University Press, 1996), 4; Howard Horwitz, *By the Law of Nature: Form and Value in Nineteenth-Century America* (New York: Oxford University Press, 1991), 173. For a related argument, see Cyrus R. Patell,

"Emersonian Strategies: Negative Liberty, Self-Reliance, and Democratic Individuality," *Nineteenth-Century Literature* 48 (March 1994): 440-79.

124. My argument is worded to evoke Latour's *We Have Never Been Modern*.

125. *CW* 1:123.

126. *CW* 2:201, 179, 185.

127. *CW* 2:194, 193-94.

128. *EL* 2:37-38.

129. *EL* 2:39-40; *CW* 2:218. For the notion that this moment marks a historical change in the concept of "environment," I am indebted to Lee Sterrenburg, particularly to "A Narrative Overview: The Making of the Concept of the Global 'Environment' in Literature and Science," manuscript. I have developed this idea further in Walls, "Believing in Nature: Wilderness and Wildness in Thoreauvian Science," in *Thoreau's Sense of Place: Essays in Environmental Criticism*, ed. Richard Schneider (Iowa City: University of Iowa Press, 2000), 15-27.

CHAPTER 4. GLOBAL POLARITY AND THE SINGLE LIFE

1. *EL* 1:74; *JMN* 3:93, 99. For an extensive discussion of polarity in Emerson and in other American Romantics, see Eric Wilson, *Romantic Turbulence: Chaos, Ecology, and American Space* (New York: St. Martin's, 2000). James Elliot Cabot fell under the spell of polarity, too. According to Nancy Craig Simmons, "The attempt to perceive reality as a reconciled dualism became central to Cabot's thought"; see "Man without a Shadow: The Life and Work of James Elliot Cabot, Emerson's Biographer and Literary Executor" (Ph.D. diss., Princeton University, 1980). In his *Massachusetts Quarterly Review* article on J. B. Stallo, Cabot declared that polarity was the "great idea which the active thought of this epoch, in every domain of life, is pledged to substantiate"; "Man without a Shadow," 88.

2. *JMN* 5:30, 304.

3. *JMN* 4:284-85.

4. *JMN* 5:337; *E&L* 953; *CW* 2:115.

5. Immanuel Kant, *Metaphysical Foundations of Natural Science*, trans. James Ellington (Indianapolis: Bobbs-Merrill, 1970), 91, 78, 92.

6. Ibid., 13-14.

7. Friedrich Wilhelm Joseph von Schelling, *Ideas for a Philosophy of Nature*, 2d ed., trans. Errol E. Harris and Peter Heath (Cambridge: Cambridge University Press, 1988), x, 177, ix-x.

8. Johann Bernhard Stallo, *General Principles of the Philosophy of Nature* (Boston: Crosby and Nichols, 1848), 48.

9. Raimonda Modiano, *Coleridge and the Concept of Nature* (Tallahassee: Florida State University Press, 1985), 154; *CWSTC* 7.1:252, 255, 256; *EL* 1:26.

10. *CWSTC* 7.1:257-58, 261-62. Even today, defenders of objectivity occasionally use Coleridge's argument, offering smashed plates (and other bodies) as demonstrations not of the fragility of crockery, but of the reality of law. Richard Dawkins's saying was widely circulated during the recent science wars: "Show me a cultural relativist at 30,000 feet and I'll show you a hypocrite"; *River Out of Eden* (New York: Basic Books, 1995), 31-32.

11. George Lakoff and Mark Johnson, *Metaphors We Live By* (Chicago: University of Chicago Press, 1980), 192-93; *CWSTC* 7.1:273; *CW* 2:201. Such com-

monplaces have a long history, but their most notable authority is Alfred North Whitehead, who declared in *Science and the Modern World* (1925; reprint, New York: Macmillan, 1953) that the nineteenth century was characterized by a "Romantic Reaction" against science: "The literature of the nineteenth century, especially its English poetic literature, is a witness to the discord between the aesthetic intuitions of mankind and the mechanism of science" (87). For Coleridge and the concept of objectivity, see Peter Galison, "Objectivity Is Romantic," in Jerome Friedman et al., *The Humanities and the Sciences,* Occasional Paper No. 47 (New York; American Council of Learned Societies, 1999); and Lorraine Daston, "Baconian Facts, Academic Civility, and the Prehistory of Objectivity," *Annals of Scholarship* 8.3/4 (1991): 337–63.

12. *CWSTC* 7.1:282, 285, 286.

13. Samuel Taylor Coleridge, "On the Definitions of Life Hitherto Received. Hints towards a More Comprehensive Theory" [hereafter "The Theory of Life"], in *Miscellanies, Aesthetic and Literary, to Which Is Added The Theory of Life,* ed. T. Ashe (London: George Bell and Sons, 1885), 369–70. For a particularly useful explication of Coleridge's science, see Craig W. Miller, "Coleridge's Concept of Nature," *Journal of the History of Ideas* 25 (1964): 77–96. As Timothy Corrigan shows, *Biographia Literaria* also documented Coleridge's growing concern with science: "that the *Biographia Literaria* employs much of the scientific language used in the 'Theory of Life' and implicitly derives many of its critical models from the scientific models sketched in that work is, however, extremely significant. . . . Probably no other alteration in the language of literary criticism has affected the practice of criticism more"; "*Biographia Literaria* and the Language of Science," *Journal of the History of Ideas* 41.3 (1980): 399–400.

14. Coleridge, "Theory of Life," 383, 391, 386–87, 391.

15. *JMN* 6:38; *CW* 2:181. See also Trevor Levere's helpful discussion of Coleridge's "compass" in "Coleridge and the Sciences," in *Romanticism and the Sciences,* ed. Andrew Cunningham and Nicholas Jardine (Cambridge: Cambridge University Press, 1990), 299–306.

16. Coleridge, "Theory of Life," 423.

17. Ibid., 423–24.

18. Robert Chambers, *Vestiges of the Natural History of Creation and Other Evolutionary Writings,* ed. James A. Secord (Chicago: Chicago University Press, 1994), 166–68.

19. *EL* 1:51, 63.

20. Arnold Guyot, *The Earth and Man: Lectures on Comparative Physical Geography, in Its Relation to the History of Mankind* (Boston: Gould, Kendall, and Lincoln, 1849), 16, 3.

21. Ibid., 11, 16, 72, 74.

22. Ibid., 184, 197, 207, 208–10, 212–13, 215.

23. Ibid., 237–39, 297.

24. Ibid., 299–300, 306–8, 309. For a fascinating—and troubling—treatment of Guyot and Thoreau, see Richard Schneider, " 'Climate Does Thus React on Man': Wildness and Geographic Determinism in Thoreau's 'Walking,' " in *Thoreau's Sense of Place: Essays in Environmental Writing,* ed. Richard Schneider (Iowa City: University of Iowa Press, 2000), 44–60.

25. Kant, *Metaphysical Foundations,* 4.

26. Alexander von Humboldt, *Cosmos: A Sketch of the Physical Description of the Universe,* trans. Elizabeth C. Otté, 5 vols. (New York: Harper and Brothers, 1850–1870), 3:14–15.

27. CW 1:8; CWSTC 7.1:268.

28. Humboldt, *Cosmos,* 1:49.

29. Coleridge, "Theory of Life," 403.

30. Humboldt, *Cosmos,* 1:49. This is the point where my line of argument connects most closely with Bruno Latour's in *We Have Never Been Modern,* trans. Catherine Porter (Cambridge: Harvard University Press, 1993).

31. Guyot, *Earth and Man,* 305; William Apess, "An Indian's Looking-Glass for the White Man," in *On Our Own Ground: The Complete Writings of William Apess, A Pequot,* ed. Barry O'Connell (Amherst: University of Massachusetts Press, 1992), 158, 156, 160.

32. Apess, "Indian's Looking Glass," 160–61.

33. Barbara Packer, *Emerson's Fall: A New Interpretation of the Major Essays* (New York: Continuum, 1982), 40; CW 1:68.

34. JMN 7:399, CW 3:39.

35. JMN 5:475, 7:363, 8:82; CW 4:85; JMN 8:194. As Eric Wilson says, "the abyss . . . *plays;* that is, manifests itself as necessity as well as chance, follows rules as well as breaks them. The cosmic game proceeds by way of bipolar processes, which are required to keep everything moving along"; *Romantic Turbulence,* 12–13.

36. JMN 5:481, 7:200.

37. JMN 8:137, 10.

38. JMN 5:482; CW 2:11, 3, 6, 3–4, 18–19.

39. CW 2:22–23.

40. CW 2:29, 33.

41. CW 2:37; JMN 8:327–28. The latter passage found its way into "Politics"; see CW 3:123–24.

42. Stanley Cavell, "An Emerson Mood," in *The Senses of Walden,* rev. ed. (San Francisco: North Point, 1981), 154–55, 159; CW 2:50.

43. CW 1:10; JMN 2:340–41, 345–46; S 1:61; 3:123–24. See also S 1:78, 2:198.

44. CW 2:57.

45. CW 2:58, 59, 60.

46. CW 2:62–63, 64.

47. CW 2:72, 73.

48. JMN 8:199–200, 205.

49. JMN 8:232.

50. CW 3:32, 29.

51. CW 2:183, also JMN 7:494; CW 2:184–85, 188.

52. CW 3:47, 40.

53. CW 3:27; David Van Leer, *Emerson's Epistemology: The Argument of the Essays* (Cambridge: Cambridge University Press, 1986), 152; CW 3:30–31, 2:193–94.

54. Van Leer, *Emerson's Epistemology,* 155; Packer, *Emerson's Fall,* 170.

55. Stanley Cavell calls this "the epistemology, or say the logic, of moods"; "Thinking of Emerson," *New Literary History* 11.1 (1979): 168. See also Packer: "The real difficulty in arriving at an epistemology of moods is that moods are likely to dictate beforehand the shape of one's epistemology"; *Emerson's Fall,* 161.

56. CW 3:31, 32.

57. CW 3:29, 44, 29.

58. David Robinson, *Emerson and the Conduct of Life: Pragmatism and Ethical Purpose in the Later Work* (Cambridge: Cambridge University Press, 1993), 61; CW 3:34–35, 36–37.

59. CW 3:37, 46; Robinson, *Emerson and the Conduct of Life,* 69, 63; CW 3:35.

60. JMN 8:179, 183, 168; CW 3:34, 28, 40.

61. CW 3:38–39.

62. CW 3:40–41.

63. CW 3:43, 45, 46.

64. CW 3:49.

65. CW 3:100, 103, 104; Schelling, *Ideas for a Philosophy,* 50; CW 3:104.

66. CW 3:104–5, 106, 107.

67. CW 3:110, 111, 112.

68. CW 3:141, 136, 135. The debate between nominalism and realism was acquiring new relevancy during the 1840s and 1850s as the question of the origin of species heated up, for that question turned on whether species were "real" or only "nominal" categories. The dominant opinion was that they were quite real, permanent, and immutable, an idea that Darwin worked hard, in *On the Origin of Species,* to shatter.

69. Van Leer, *Emerson's Epistemology,* 190; CW 3:139.

70. CW 3:140, 142, 143–44.

71. JMN 3:93, 98–99; L 1:221; JMN 6:72; Eric Wilson, *Emerson's Sublime Science* (New York: St. Martin's, 1999), 78, 88–89; S 2:20–21, 46.

72. CW 1:43; John F. W. Herschel, *A Preliminary Discourse on the Study of Natural Philosophy* (1830; facsimile ed., Chicago: University of Chicago Press, 1987), 253; Packer, *Emerson's Fall,* 76; JMN 5:233. For Barbara Packer's alternative reading, see *Emerson's Fall,* 72–75. Packer does not mention Herschel, attributing the theory instead to David Brewster in his *Life of Newton* (1831), which Emerson read the following year, in 1832. As Herschel acknowledged, the polar theory of light was actually neither Brewster's nor his own, but Biot's, developed to explain Newton's "fits of easy transmission and reflection"; *Preliminary Discourse,* 262.

73. JMN 5:304; CW 2:37, 84, 3:135 (see also JMN 9:67); JMN 9:226.

74. JMN 7:476–77, 16:172.

75. JMN 16:6, 286.

76. CW 3:57.

77. CW 3:124, 58.

78. CW 4:3, 16.

79. CW 4:6, 7–8.

80. CW 4:8, 104.

81. CW 4:43, 79.

82. CW 4:29–30, 31–32, 66–67; Robinson, *Emerson and the Conduct of Life,* 100; CW 4:75.

83. CW 4:88, 110, 129.

84. CW 4:152–53, 157, 159.

85. CW 4:163, 164, 165–66.

CHAPTER 5. TRUTH AGAINST THE WORLD

1. *JMN* 9:123–24.

2. *JMN* 9:124–25, 126.

3. William Rossi has observed that "in his antislavery writings Emerson also links 'fate' with the doctrine of black racial inferiority and thus implicitly with racial science"; "Emerson, Nature, and the Natural Sciences," in *A Historical Guide to Ralph Waldo Emerson,* ed. Joel Myerson (New York: Oxford University Press, 2000), 149, n. 86.

4. *EAW* 32.

5. Rossi, "Emerson, Nature, and the Natural Sciences," 120. For an excellent, still informative treatment of Emerson and evolution, see Joseph Warren Beach, "Emerson and Evolution," *University of Toronto Quarterly* 3 (1934): 474–97.

6. *EL* 2:24. The original journal passage is somewhat more negative: here Emerson calls Lamarck's system an "imperfect result" of this rage for unity; *JMN* 5:220.

7. Charles Lyell, *Principles of Geology,* 3 vols. (1830–1833; facsimile ed., Chicago: University of Chicago Press, 1990–1992), 2:12, 21; for Lyell on Lamarck, see 2:3–35.

8. See the extensive treatment of this question in Adrian Desmond, *The Politics of Evolution: Morphology, Medicine, and Reform in Radical London* (Chicago: University of Chicago Press, 1989).

9. Rossi, "Emerson, Nature, and the Natural Sciences," 127; *JMN* 9:233; *LSA* 13; Gay Wilson Allen, *Waldo Emerson: A Biography* (New York: Viking, 1981), 575. See also the discussion in Philip L. Nicoloff, *Emerson on Race and History: An Examination of "English Traits"* (New York: Columbia University Press, 1961), 282, n. 37. On the reception of *Vestiges,* see James A. Secord, *Victorian Sensation: The Extraordinary Publication, Reception, and Secret Authorship of "Vestiges of the Natural History of Creation"* (Chicago: University of Chicago Press, 2000). The book's authorship quickly became an open secret among British gentleman-scientists, but was not widely known until the 1850s. Chambers himself never admitted authorship. Emerson met Chambers in Edinburgh in 1848 and identified him in a letter to Lidian as the author of *Vestiges* (*L* 4:19; also *JMN* 13:113). Emerson's lecture tour had been arranged by Chambers's publisher, Alexander Ireland, one of the few in on the secret, but since Ireland had taken an oath of secrecy, Emerson was probably going on popular hunch. Emerson's last coherent entry on science shows how badly his memory was slipping, as well as how long the book held his imagination: " 'Vestiges of Creation,' who is the author?" (*JMN* 16:309).

10. *JMN* 10:353; Robert Chambers, *Vestiges of the Natural History of Creation and Other Evolutionary Writings,* ed. James A. Secord (Chicago: University of Chicago Press, 1994), 197, 198, 199, 218, 219, 222.

11. Chambers, *Vestiges,* 231; *JMN* 13:250.

12. Nicoloff, *Emerson on Race,* 114, 117; *JMN* 11:211.

13. *JMN* 11:158; Johann Bernhard Stallo, *General Principles of the Philosophy of Nature* (Boston: Crosby and Nichols, 1848), 234–35. After publishing *General Principles,* Stallo returned to Cincinnati, where he became a lawyer in 1849, then a judge, active in science education and liberal politics; in 1885 President Cleveland appointed him minister to Italy, where he spent the rest of his life.

14. Stallo, *General Principles,* 291, and *JMN* 11:200; *JMN* 11:157, also 11:154; *JMN* 11:416; Stallo, *General Principles,* 16, 304, and *JMN* 11:200, 16:298.

15. Louis Menand, *The Metaphysical Club* (New York: Farrar, Straus and Giroux, 2001), 158. For more on Agassiz, see Edward Lurie, *Louis Agassiz: A Life in Science* (1960; reprint, Baltimore: Johns Hopkins University Press, 1988); and Mary P. Winsor, *Reading the Shape of Nature: Comparative Zoology at the Agassiz Museum* (Chicago: University of Chicago Press, 1991).

16. Louis Agassiz, *Methods of Study in Natural History* (1863; reprint, Boston: Houghton Mifflin, 1896), 102.

17. *JMN* 10:525, 527, 14:128. On one particular lecture of Owen's that Emerson attended, see Secord, *Victorian Sensation,* 424–25. In a witty passage written in 1857, after Agassiz published the first volume of his monumental but rather dry *Contributions to the Natural History of the United States of America*—which would feature one volume on turtle classification and another on their embryology—Emerson suggested in his journal that "the turtles in Cambridge, on the publication of this book of Agassiz, should hold an indignation meeting, & migrate from the Charles River, with Chelydra serpentina marching at the head, and 'Death to Agassiz!' inscribed on their shields"; *JMN* 14:122–23.

18. *JMN* 11:153–54, 13:113, 11:199. The recapitulation discovery was often attributed to Agassiz rather than to Oken or to Chambers; *TN* 1:32; *JMN* 14:123, 16:266; Stallo, *General Principles,* 303.

19. *JMN* 9:270, 16:160–61, 14:164. See also Emerson's lecture "Humboldt," in *CompW* 11:455–59.

20. *LSA* 13; *JMN* 16:298; *CW* 1:7; *Poems,* in *CompW* 9:281.

21. *JMN* 11:397–98.

22. *CW* 5:24, 25, 26–27, 29. I am uncertain where Emerson finds that Humboldt distinguishes three races. In his discussion of race in *Cosmos* Humboldt uses an argument similar to Emerson's, that the inability of science to agree on the number or distinguishing characteristics of races shows the lack of "any general or well-established principle in the division of these groups," suggesting that race does not exist as a scientific category; Alexander von Humboldt, *Cosmos: A Sketch of the Physical Description of the Universe,* trans. Elizabeth C. Otté, 5 vols. (1850–1870; facsimile ed., vols. 1–2, Baltimore: Johns Hopkins University Press, 1997), 1:356.

23. Nicoloff, *Emerson and Race,* 3, 29 (paraphrasing Henry A. Pochmann); David Robinson, *Emerson and the Conduct of Life: Pragmatism and Ethical Purpose in the Later Work* (Cambridge: Cambridge University Press, 1993), 112; *CW* 5:165. The scientists Nicoloff lists are Charles Lyell, Michael Faraday, William Buckland, Richard Owen, Robert Brown, William Spence, Edward Forbes, George Combe, and Sir William Hamilton (29–30). See also Nicoloff's "Historical Introduction," in *English Traits, CW* 5:xiii–liii. Emerson also mentions meeting the geologists Sir Henry Thomas de la Beche and Adam Sedgwick, and dining with the mathematician, astronomer, and science writer Mary Somerville (*L* 4:49, 51, 94). He also became friends with Samuel Brown, the promising young Edinburgh doctor and chemist who died tragically at the age of thirty-nine.

24. Nicoloff, *Emerson and Race,* 30, 46 (quoting Emerson), 48–49; *CW* 5:172.

25. *JMN* 9:125–26; *EAW* 32.

26. *JMN* 7:393, 11:385.

27. Humboldt, *Cosmos,* 1:358; James Cowles Prichard, *The Natural History of Man,* 2 vols., 4th ed. (London: H. Bailliere, 1855), 2:714; Robert Knox, *Races of Man: A Fragment* (Philadelphia: Lea and Blanchard, 1850), 13, 7, 20, 119. Accord-

ing to Nicoloff, "Among the thinkers of the age who employed racial designations in a significant manner, Emerson knew of Thierry, Guizot, Sir Walter Scott, Blumenbach, Buffon, Bulwer-Lytton, Carlyle, Herder, Hegel, Fichte, Friedrich von Schlegel, Humboldt, John Motley, Gall and Spurzheim, John Kemble, Taine, Goethe, Marx, Michelet, Robert Knox, Agassiz, Friedrich Müller, Pickering, Niebuhr, Ranke, and Prichard—to provide a suggestive list only"; *Emerson and Race*, 95–96.

28. *E&L* 950.

29. Chambers, *Vestiges*, 277–78, 307.

30. Ibid., 267, 276; *JMN* 9:212. The MacLeay or Quinary System had been put forward by William Sharp MacLeay in his *Horae Entomologicae, or Essays on Annulose Animals* (1819, 1821), and popularized by his disciple William Swainson (*JMN* 9:211, n. 87). In his attempts to make his book more acceptable to scientists, Chambers deemphasized the MacLeay system in his third edition (the one Emerson read), and in the fifth edition of 1846 removed all reference to it; *Vestiges*, app. A, 217–18.

31. Stallo, *General Principles*, 325, 315.

32. Arnold Guyot, *The Earth and Man: Lectures on Comparative Physical Geography, in Its Relation to the History of Mankind* (Boston: Gould, Kendall, and Lincoln, 1849) 241, 232, 237, 238, 244.

33. The excerpt from Agassiz's letter is from Stephen Jay Gould, *The Mismeasure of Man* (New York: Norton, 1981), 44–45. The long letter was written in December 1846, when Agassiz was staying in Philadelphia with George Morton. It is reproduced, with this passage removed, in Elizabeth Cary Agassiz, *Louis Agassiz: His Life and Correspondence* (Boston: Houghton Mifflin, 1886), 409–29. For an excellent and succinct account of Agassiz and polygenesis, see Menand, *Metaphysical Club*, 101–16.

34. Louis Agassiz, "The Diversity of Origin of the Human Races," *Christian Examiner* 49 (July 1850): 128, 110, 118, 132, 133–34.

35. Ibid., 110, 112, 136.

36. Ibid., 124, 142, 144.

37. I am indebted in this discussion to William Stanton, *The Leopard's Spots: Scientific Attitudes toward Race in America, 1815–1859* (Chicago: University of Chicago Press, 1960), 104–12 and passim. Stanton notes that Agassiz's articles, though widely endorsed, provoked a moral crisis in the Virginia transcendentalist Moncure Conway. After accepting Agassiz's arguments, Conway realized he'd been willing to declare that Negroes were not men. Conway concluded that Agassiz's "spiritual" equality was hollow, "for where Agassiz 'feared to tread, my crudity rushed in.' " His moral crisis resulted in his conversion to the antislavery movement (111). See also Mary P. Winsor, "Louis Agassiz and the Species Question," *Studies in the History of Biology* 3 (1979): 89–117. For a more recent account of John Bachman and southern racial science, see Lester D. Stephens, *Science, Race, and Religion in the American South: John Bachman and the Charleston Circle of Naturalists, 1815–1895* (Chapel Hill: University of North Carolina Press, 2000).

38. Louis Agassiz, "Sketch of the Natural Provinces of the Animal World and their Relation to the Different Types of Man," in Josiah C. Nott and George R. Gliddon, *Types of Mankind* (Philadelphia: Lippincott, 1854), lxxiv; Nott and Gliddon, *Types of Mankind*, xiv, 415; Stanton, *Leopard's Spots*, 196. According to Mary Winsor, Agassiz's chart represents Caucasian man with a portrait of Cuvier; "Agassiz and the Species Question," 113, n. 6.

39. Agassiz, "Sketch," lxxvi.

40. *JMN* 13:54, 10:357, 13:87; *E&L* 960-61.

41. *JMN* 13:466; *CW* 5:27.

42. Cornel West, *The American Evasion of Philosophy: A Genealogy of Pragmatism* (Madison: University of Wisconsin Press, 1989), 34; Nicoloff, *Emerson and Race,* 245-46; Eduardo Cadava, *Emerson and the Climates of History* (Stanford: Stanford University Press, 1997), 58.

43. Robinson, *Emerson and the Conduct of Life,* 116.

44. Len Gougeon, " 'Fortune of the Republic': Emerson, Lincoln, and Transcendental Warfare," *ESQ* 45 (1999): 278; Albert J. von Frank, *The Trials of Anthony Burns: Freedom and Slavery in Emerson's Boston* (Cambridge: Harvard University Press, 1998), 327; *EAW* 102, 98; *JMN* 8:199-200. The standard work on Emerson and antislavery is Len Gougeon, *Virtue's Hero: Emerson, Antislavery, and Reform* (Athens: University of Georgia Press, 1990).

45. Chambers, *Vestiges,* 376-77.

46. Ibid., 327-29, 331; *Explanations* (bound with *Vestiges*), 24; see also Pierre Simon, Marquis de Laplace, *A Philosophical Essay on Probabilities,* trans. Frederick Wilson Truscott and Frederick Lincoln Emory (1902; reprint, New York: Dover, 1952), 62.

47. Laplace, *Philosophical Essay,* 1, 4, 9. For an insightful treatment of Laplace and Quetelet, see Barbara Packer, "Emerson and the Terrible Tabulations of the French," in *Transient and Permanent: The Transcendentalist Movement and its Contexts,* ed. Charles Capper and Conrad E. Wright (Boston: Northeastern University Press, 1999), 148-67.

48. Laplace, *Philosophical Essay,* 108.

49. See Adolphe Quetelet, *A Treatise on Man and the Development of His Faculties* (1842; reprint, New York: Burt Franklin, 1968), 9; Theodore M. Porter, *The Rise of Statistical Thinking, 1820-1900* (Princeton: Princeton University Press, 1986), 42, 69-70. I am indebted in this discussion to the work of Theodore Porter; see also ibid., 46, 104, 23-27. According to Porter, Quetelet pirated the name for his new science, "physique sociale," from a very annoyed Comte, who was forced to invent a new term, *sociology,* as a replacement (156).

50. Quetelet, *Treatise on Man,* 6, 108, 83, 7.

51. Ibid., 96-98.

52. Ibid., 8, 96, 99.

53. Ibid., 100-101 (as Robert Richardson observes, Emerson didn't quite say this—but he certainly could have; *Emerson: The Mind on Fire* [Berkeley: University of California Press, 1995], 468); Quetelet, *Treatise on Man,* x, 108.

54. *JMN* 11:91; John Herschel, "Quetelet on Probabilities," *Edinburgh Review* 185 (July 1850): 42; *CW* 4:62.

55. *E&L* 950-51.

56. *JMN* 11:210, 218.

57. *JMN* 11:256; *E&L* 945; *JMN* 9:403.

58. Packer, "Emerson and the Terrible Tabulations," 161.

59. *JMN* 13:340: Packer, "Emerson and the Terrible Tabulations," 161; *JMN* 13:440.

60. *E&L* 946.

61. *E&L* 949, 952.

62. *E&L* 953, 954–55.

63. *E&L* 958.

64. *E&L* 960, 962, 961, 962; *CW* 4:91.

65. *CW* 1:128; *E&L* 953.

66. Bruno Latour, *We Have Never Been Modern*, trans. Catherine Porter (Cambridge: Harvard University Press, 1993), 80.

67. *E&L* 961, 950, 965.

68. *CW* 1:45; *E&L* 967–68; David Van Leer, *Emerson's Epistemology: The Argument of the Essays* (Cambridge: Cambridge University Press, 1986), 206.

CHAPTER 6. THE SOLAR EYE OF SCIENCE

1. *JMN* 11:167–68, 16:55, 281. Emerson made his 1872 entry at the seaside summer home of fellow Saturday Club member John Murray Forbes, while recovering from the shock of the fire; Forbes's home on Naushon Island, Massachusetts, was often visited by Emerson and by other club members. See Edward Emerson, *The Early Years of the Saturday Club, 1855–1870* (Boston: Houghton Mifflin, 1918), 230–31.

2. Johann Wolfgang von Goethe, *Scientific Studies*, ed. and trans. Douglas Miller (New York: Suhrkamp, 1988), 164, 39.

3. Robert Chambers, *Vestiges of the Natural History of Creation and Other Evolutionary Writings*, ed. James A. Secord (Chicago: University of Chicago Press, 1994), 163; William Whewell, *Astronomy and General Physics, Considered with Reference to Natural Theology* (Philadelphia: Carey, Lea, and Blanchard, 1836), 74; *JMN* 5:236, also *CW* 2:21, 3:140; *JMN* 10:154; Johann Bernhard Stallo, *General Principles of the Philosophy of Nature* (Boston: Crosby and Nichols, 1848), 14.

4. *JMN* 13:448.

5. *LL* 2:122, also *JMN* 10:307. On Emerson's "homeorrhetic" river imagery, see also Eric Wilson, *Romantic Turbulence: Chaos, Ecology, and American Space* (New York: St. Martin's, 2000), 41–42.

6. *LL* 2:86, 90, 88, 87, 97, 74, 82, 89.

7. *LL* 2:76; Plato, "Timaeus," trans. Benjamin Jowett, in *The Collected Dialogues*, ed. Edith Hamilton and Huntington Cairns (Princeton: Princeton University Press, 1961), 1151–1211, 1173; *LL* 2:89, also *JMN* 7:430; *LL* 1:121–22, 163, 162–63. My definition of creativity is derived from Jacob Bronowski. For an interesting explication of creativity in science, see his *Science and Human Values*, 2d ed. (New York: Harper and Row, 1972): "The discoveries of science, the works of art are explorations—more, are explosions, of a hidden likeness. The discoverer or the artist presents in them two aspects of nature and fuses them into one. This is the act of creation, in which an original thought is born, and it is the same act in original science and original art" (19). This makes discovery in science a fundamentally metaphorical process.

8. John Burroughs, *Time and Change: The Writings of John Burroughs*, vol. 12 (Boston: Houghton Mifflin, 1912), 260–61.

9. *LL* 2:342, 1:163. At the furthest end of this sequence is the full-fledged pragmatism of William James, who famously said in his 1907 lecture "Pragmatism": "Truth *happens* to an idea." See Louis Menand, *The Metaphysical Club* (New York: Farrar, Straus and Giroux, 2001), 353 and passim.

10. *JMN* 5:482; *CW* 1:130, 137; David Robinson, *Emerson and the Conduct of Life: Pragmatism and Ethical Purpose in the Later Work* (Cambridge: Cambridge University Press, 1993), 182; Ronald A. Bosco, " 'His Lectures Were Poetry, His Teaching the Music of the Spheres': Annie Adams Fields and Francis Greenwood Peabody on Emerson's 'Natural History of the Intellect' University Lectures at Harvard in 1870," *Harvard Library Bulletin*, n.s., 8.2 (1997): 8.

11. Bosco, " 'His Lectures Were Poetry,' " 18, 19, 16–17. For a definitive biography of James Elliot Cabot, see Nancy Craig Simmons, "Man without a Shadow: The Life and Work of James Elliot Cabot, Emerson's Biographer and Literary Executor" (Ph.D. diss., Princeton University, 1980). Cabot published his version of "Natural History of Intellect" in volume 12 of the Riverside Edition of *The Works of Ralph Waldo Emerson* (Boston: Houghton Mifflin, 1893), 3–81. Edward Emerson published his expanded version of this essay in *CompW* 12:3–110. Since the published versions of "Natural History of Intellect" do not reflect its original structure, I base my discussion of Emerson's late science lectures instead on the newly published *Later Lectures of Ralph Waldo Emerson*. For an astute analysis of "Natural History of Intellect," see Robinson, *Emerson and the Conduct of Life*, 181–200.

12. *LL* 1:137, 2:69, 70.

13. *LL* 1:137, 2:66, 1:155.

14. *CompW* 12:11 and *JMN* 11:438; *CW* 1:65, 2:184; *LSA* 206; *E&L* 1033–34; *LSA* 216–17.

15. *LL* 2:90, 91, 93.

16. William Whewell, *Selected Writings on the History of Science*, ed. Yehuda Elkana (Chicago: University of Chicago Press, 1984), 133; Sir Humphry Davy, *Elements of Agricultural Chemistry*, in *The Collected Works of Sir Humphry Davy*, ed. John Davy, 9 vols. (London: Smith, Elder, 1839–40), 7:196–97, 8:88; *LL* 2:87, 1:156.

17. *LL* 2:57, 1:158, 181, 148; *E&L* 971.

18. *E&L* 1016; *LL* 2:120, 1:107; *JMN* 13:374–75.

19. *JMN* 5:410, 505.

20. *LL* 2:23, 20–21, 26; Len Gougeon, "Emerson and the Woman Question: The Evolution of His Thought," *New England Quarterly* 71.2 (1998): 582; Phyllis Cole, "Woman Questions: Emerson, Fuller, and New England Reform," in *Transient and Permanent: The Transcendentalist Movement and Its Contexts*, ed. Charles Capper and Conrad E. Wright (Boston: Northeastern University Press, 1999), 434. For a more favorable, less ambivalent view, see Armida Gilbert, "Emerson in the Context of the Woman's Rights Movement," in *A Historical Guide to Ralph Waldo Emerson*, ed. Joel Myerson (Oxford: Oxford University Press, 2000), 211–49. Gougeon reprints the text of a brief address Emerson delivered in 1869, "to celebrate the anniversary of the New England Woman's Suffrage Association." By then Emerson's support for political rights for women was unqualified: "She asks for her property; she asks for her rights, for her vote; she asks for her share in education, for her share in all the institutions of society, for her half of the whole world; and to this she is entitled." He was forthwith elected a vice-president of the association. Yet he still sees their role as "the proper mediators between those who have knowledge and those who want it"; "Emerson and the Woman Question," 589–90.

21. *CW* 3:72–73; *JMN* 13:287, 15:356; Stephen E. Whicher, *Freedom and Fate:*

An Inner Life of Ralph Waldo Emerson (Philadelphia: University of Pennsylvania Press, 1953), 163–64. As Secord observes of Victorian England, "one of the most important changes taking place in the 1840s was that the term 'gentleman' increasingly referred to character as much as birth"; *Victorian Sensation: The Extraordinary Publication, Reception, and Secret Authorship of "Vestiges of the Natural History of Creation"* (Chicago: University of Chicago Press, 2000), 406.

22. Stallo, *General Principles*, 150.

23. *JMN* 11:291; *E&L* 1005; *LL* 2:80, also 2:94 and *E&L* 1029.

24. *LL* 2:80–81; *JMN* 13:100.

25. *LL* 1:170; Daniel 12:4 and *JMN* 6:61, 5:443.

26. *E&L* 989, 991, 993.

27. *E&L* 994, 995–96, 998, 1010.

28. *JMN* 10:153, 13:262, 8:235; *CW* 5:65.

29. *CW* 2:139, 28; 1:90.

30. *LL* 1:185–87.

31. *LL* 1:187; *S&S* 250; *E&L* 948.

32. *JMN* 9:16, 13, 8:252.

33. *CW* 5:94–95; Phyllis Cole, "Emerson, England, and Fate," in *Emerson: Prophecy, Metamorphosis, and Influence*, ed. David Levin (New York: Columbia University Press, 1975), 83–105; *JMN* 9:174, 16:140; *S&S* 248–49.

34. *S* 2:229–30; *EL* 2:39, *JMN* 9:271, 403, and *LL* 1:187, 2:86; *LSA* 221–22.

35. *LL* 1:169, 183–84.

36. Menand, *Metaphysical Club*, 215; *S&S* 413–18; Len Gougeon, " 'Fortune of the Republic': Emerson, Lincoln, and Transcendental Warfare," *ESQ* 45 (1999): 293; see also Edward Emerson, *Early Years*.

37. Gougeon, " 'Fortune of the Republic,' " 294 and passim; Albert J. von Frank, *The Trials of Anthony Burns: Freedom and Slavery in Emerson's Boston* (Cambridge: Harvard University Press, 1998), 330; *JMN* 13:255, 15:393.

38. *JMN* 7:71; Robinson, *Emerson and the Conduct of Life*, 48; *CW* 2:119.

39. *JMN* 16:244; *LSA* 89, 95, also *S&S* 245; *S&S* 425; *JMN* 13:141, 112, 302.

40. *JMN* 5:217–18.

41. *JMN* 16:296.

42. *JMN* 16:132, 242. The discovery of anesthesia was widely credited to Emerson's brother-in-law, Charles T. Jackson, a claim that was hotly contested by Samuel Morton.

43. *LL* 1:119–20, 122–23.

44. *E&L* 1061, 1064.

45. *JMN* 10:306; *LL* 1:189; *JMN* 16:26, 101.

46. *JMN* 13:261, 14:78.

47. *JMN* 8:250, 7:111, 71, and 111, 16:251, 14:360. In his poem "Blight," Emerson mourns that "these young scholars, who invade our hills . . . Love not the flower they pluck, and know it not, / And all their botany is Latin names" (*CompW* 9:140). Emerson's topical "Notebook Naturalist," begun in 1850, shows considerable evidence of a sustained dialogue with Thoreau on matters scientific. See the introduction by Susan Sutton Smith, *TN* 1:7–8.

48. *JMN* 13:412.

49. For a full treatment of this, see von Frank, *Trials*. Elinor S. Shaffer gives a brief history of Kant's notion of the *Klerisei*, translated by Coleridge as "clerisy,"

and makes the point that Fichte's ideas on the subject "were propagated in the United States even more effectively than elsewhere through Emerson"; see "Romantic Philosophy and the Organization of the Disciplines," in *Romanticism and the Sciences,* ed. Andrew Cunningham and Nicholas Jardine (Cambridge: Cambridge University Press, 1990), 40.

50. CW 1:8; Richard Dawkins, "The Selfish Interactor," lecture delivered at the annual meeting of the British Association for the Advancement of Science, Leeds, September 10, 1997.

51. CW 1:63.

52. Steven Shapin, *A Social History of Truth: Civility and Science in Seventeenth-Century England* (Chicago: University of Chicago Press, 1994), 25; CW 1:58; EL 1:10; CW 1:44.

53. LSA 21, 20, 37, 42, 28, 26, 31, 25.

54. LSA 67, 69, 71–72.

55. *Boston Journal,* April 28, 1882; John Tyndall, *Fragments of Science,* 6th ed., 2 vols. (1868–69; reprint, New York: Collier, 1902), 2:104–5; *London World,* April 30, 1882; also quoted in Richard Garnett, *Life of Ralph Waldo Emerson* (1888; reprint, London: Walter Scott, 1988), 199–200. That Tyndall cites Emerson suggests that Emerson's writing provided an important resource to the formation of Victorian scientific naturalism. Joseph Warren Beach notices this, too, concluding that "under cover of Transcendentalism naturalism was enabled to make great advances"—although in Emerson himself, Beach grumbles, "naturalism suffered an 'arrested development' "; "Emerson and Evolution," *University of Toronto Quarterly* 3 (1934): 497. Tyndall heard Emerson lecture on "Napoleon as man of action" at the Halifax Mechanics' Institution on January 5, 1848. According to James Secord, "Although Emerson read carelessly and spoke with a nasal twang, his address was rapturously received," and he became a hero for a generation of young dissenters (*Victorian Sensation,* 338, 354). I was led to the Tyndall passage by Frank M. Turner's essay "Victorian Scientific Naturalism and Thomas Carlyle," in *Contesting Cultural Authority: Essays in Victorian Intellectual Life* (Cambridge: Cambridge University Press, 1993), 101–50. Turner argues that post-1850 Victorian scientific naturalists separated themselves from the radical science communities (which had long been advocating secular and naturalistic ideas) and fought their battle against natural theology by, first, moving science outside a doctrinal framework and, second, moving social authority from the church to the "culture of truth-seeking gentlemen." Emerson was already pioneering both moves in the 1830s.

56. JMN 16:232–33; LL 2:98.

Works Cited

Adams, Richard P. "Emerson and the Organic Metaphor." *PMLA* 69 (March 1954): 117–30.

Agassiz, Elizabeth Cary. *Louis Agassiz: His Life and Correspondence*. 2 vols. Boston: Houghton Mifflin, 1886.

Agassiz, Louis. "Contemplations of God in the Cosmos." *Christian Examiner* 50 (January 1851): 1–17.

———. "The Diversity of Origin of the Human Races." *Christian Examiner* 49 (July 1850): 110–45.

———. "Geographical Distribution of Animals." *Christian Examiner* 48 (March 1850): 181–204.

———. *Methods of Study in Natural History*. 1863; reprint, Boston: Houghton Mifflin, 1896.

———. "Sketch of the Natural Provinces of the Animal World and Their Relation to the Different Types of Man." In Josiah C. Nott and George R. Gliddon, *Types of Mankind*. Philadelphia: Lippincott, 1854, lviii–lxxix.

Allen, Gay Wilson. "A New Look at Emerson and Science." In *Literature and Ideas in America: Essays in Memory of Harry Hayden Clark*. Edited by Robert Falk. Athens: Ohio University Press, 1975, 58–78.

———. *Waldo Emerson: A Biography*. New York: Viking, 1981.

Apess, William. *On Our Own Ground: The Complete Writings of William Apess, A Pequot*. Edited by Barry O'Connell. Amherst: University of Massachusetts Press, 1992.

Appleby, Joyce, Lynn Hunt, and Margaret Jacob. *Telling the Truth about History*. New York: Norton, 1994.

Bacon, Francis. *The Works of Francis Bacon*. Edited by James Spedding, Robert Leslie Ellis, and Douglas Denon Heath. 15 vols. New York: Hurd and Houghton, 1872.

Barish, Evelyn. *Emerson: The Roots of Prophecy*. Princeton: Princeton University Press, 1989.

Barthes, Roland. *Mythologies*. Translated by Annette Lavers. New York: Farrar, Straus and Giroux, 1972.

Bate, Jonathan. *The Song of the Earth*. Cambridge: Harvard University Press, 2000.

Beach, Joseph Warren. "Emerson and Evolution." *University of Toronto Quarterly* 3 (1934): 474–97.

Bell, Charles. *The Hand: Its Mechanism and Vital Endowments, as Evincing Design.* Philadelphia: Carey, Lea, and Blanchard, 1835.

Berger, Michael. *Thoreau's Late Career and "The Dispersion of Seeds."* Rochester, N.Y.: Camden House, 2000.

Bishop, Jonathan. *Emerson on the Soul.* Cambridge: Harvard University Press, 1964.

Black, Max. *Models and Metaphors: Studies in Language and Philosophy.* Ithaca: Cornell University Press, 1962.

Bono, James J. "Science, Discourse, and Literature: The Role/Rule of Metaphor in Science." In *Literature and Science: Theory and Practice.* Edited by Stuart Peterfreund. Boston: Northeastern University Press, 1990, 59–89.

——. *The World of God and the Languages of Man: Interpreting Nature in Early Modern Science and Medicine.* Vol. 1: *Ficino to Descartes.* Madison: University of Wisconsin Press, 1995.

Bosco, Ronald A. " 'His Lectures Were Poetry, His Teaching the Music of the Spheres': Annie Adams Fields and Francis Greenwood Peabody on Emerson's 'Natural History of the Intellect' University Lectures at Harvard in 1870." *Harvard Library Bulletin,* n.s., 8.2 (1997): 1–79.

Brewster, David. *The Life of Sir Isaac Newton.* London: John Murray, 1831.

——. *Memoirs of the Life, Writings, and Discoveries of Sir Isaac Newton.* 2 vols. 1855; reprint, New York: Johnson Reprint, 1965.

Bronowski, Jacob. *Science and Human Values.* 2d ed. New York: Harper and Row, 1972.

Brooke, John Hedley. *Science and Religion: Some Historical Perspectives.* Cambridge: Cambridge University Press, 1991.

Brown, Lee Rust. "Emersonian Transparency." *Raritan* 9.3 (1990): 127–44.

——. *The Emerson Museum: Practical Romanticism and the Pursuit of the Whole.* Cambridge: Harvard University Press, 1997.

Burke, Kenneth. "I, Eye, Aye—Emerson's Early Essay 'Nature': Thoughts on the Machinery of Transcendence." In *Emerson's Nature: Origin, Growth, Meaning.* Edited by Merton M. Sealts Jr. and Alfred R. Ferguson. 2d ed. Carbondale: Southern Illinois University Press, 1979, 150–63.

Burroughs, John. *Time and Change: The Writings of John Burroughs.* Vol. 12. Boston: Houghton Mifflin, 1912.

Cabot, James Elliot. "On the Relation of Art to Nature." Parts I and II. *Atlantic Monthly* 13.76 (February 1864): 183–99; 13.77 (March 1864): 313–29.

Cadava, Eduardo. *Emerson and the Climates of History.* Stanford: Stanford University Press, 1997.

Cameron, Kenneth Walter. *Emerson the Essayist.* 2 vols. Hartford: Transcendental Books, 1972.

——. *Ralph Waldo Emerson's Reading.* Raleigh, N.C.: Thistle, 1941.

Candolle, Alphonse de, and D. Sprengel. *Elements of the Philosophy of Plants.* Edinburgh: William Blackwood, 1821.

Cannon, Susan Faye. *Science in Culture: The Early Victorian Period.* New York: Dawson and Natural History Publications, 1978.

Carlyle, Thomas. *The Works of Thomas Carlyle.* Edited by Henry Duff Traill. 30 vols. London: Chapman and Hall, 1898–1907.

Cavell, Stanley. "An Emerson Mood." In *The Senses of Walden*. Rev. ed. San Francisco: North Point, 1981, 141–60.

———. "Thinking of Emerson." *New Literary History* 11 (autumn 1979): 167–76.

Cayton, Mary Kupiec. *Emerson's Emergence: Self and Society in the Transformation of New England, 1800–1845*. Chapel Hill: University of North Carolina Press, 1989.

Chambers, Robert. *Vestiges of the Natural History of Creation and Other Evolutionary Writings*. Edited by James A. Secord. Chicago: University of Chicago Press, 1994.

Cheever, George Barrel. "Coleridge." *North American Review* 40.87 (April 1835): 299–351.

Clark, Harry Hayden. "Emerson and Science." *Philological Quarterly* 10 (July 1931): 225–60.

Cohen, I. Bernard. *Revolution in Science*. Cambridge: Harvard University Press, 1985.

Cole, Phyllis. "Emerson, England, and Fate." In *Emerson: Prophecy, Metamorphosis, and Influence*. Edited by David Levin. New York: Columbia University Press, 1975, 83–105.

———. "Woman Questions: Emerson, Fuller, and New England Reform." In *Transient and Permanent: The Transcendentalist Movement and Its Contexts*. Edited by Charles Capper and Conrad E. Wright. Boston: Northeastern University Press, 1999, 408–46.

Coleridge, Samuel Taylor. *Aids to Reflection*. 1840; reprint, Port Washington, N.Y.: Kennikat, 1971.

———. *The Collected Works of Samuel Taylor Coleridge*. Edited by Kathleen Coburn et al. 14 vols. to date. London: Routledge, 1969–.

———. "On the Definitions of Life Hitherto Received. Hints towards a More Comprehensive Theory" ["The Theory of Life"]. In *Miscellanies, Aesthetic and Literary, to Which Is added The Theory of Life*. Edited by T. Ashe. London: George Bell and Sons, 1885, 351–430.

Cook, Theodore Andrea. *The Curves of Life*. 1914; reprint, New York: Dover, 1979.

Cooke, George Willis. *An Historical and Biographical Introduction to Accompany The Dial*. 2 vols. New York: Russell and Russell, 1961.

Corrigan, Timothy J. "*Biographia Literaria* and the Language of Science." *Journal of the History of Ideas* 41 (July–September 1980): 399–419.

Cudworth, Ralph. *The True Intellectual System of the Universe*. 4 vols. 1678. London, 1820.

Cunningham, Andrew, and Nicholas Jardine, eds. *Romanticism and the Sciences*. Cambridge: Cambridge University Press, 1990.

Curtius, Ernst Robert. *European Literature and the Latin Middle Ages* (1948). Translated by Willard R. Trask. Princeton: Princeton University Press, 1953.

Cuvier, Georges, Baron. *A Discourse on the Revolutions of the Surface of the Globe, and the Changes Thereby Produced in the Animal Kingdom*. Philadelphia: Carey and Lea, 1831.

Dant, Elizabeth A. "Composing the World: Emerson and the Cabinet of Natural History." *Nineteenth-Century Literature* 44 (1989): 18–44.

Daston, Lorraine. "Baconian Facts, Academic Civility, and the Prehistory of Objectivity." *Annals of Scholarship* 8.[3Q] (1991): 337–63.

Davy, Sir Humphry. *Elements of Agricultural Chemistry* (1813). Parts I and II. In *The Collected Works of Sir Humphry Davy*. Edited by John Davy. 9 vols. London: Smith, Elder, 1839–40.

——. *Elements of Chemical Philosophy* (1812). In *The Collected Works of Sir Humphry Davy*. Edited by John Davy. 9 vols. London: Smith, Elder, 1839–40.

Dawkins, Richard. *River Out of Eden.* New York: Basic Books, 1995.

——. "The Selfish Interactor." Lecture, British Association for the Advancement of Science, Leeds. September 10, 1997.

Deason, Gary R. "Reformation Theology and the Mechanistic Conception of Nature." In *God and Nature*. Edited by David C. Lindberg and Ronald L. Numbers. Berkeley: University of California Press, 1986, 167–91.

De Quincey, Thomas. "The Poetry of Pope." In *Collected Writings of Thomas De Quincey*. Edited by David Masson. London: A. C. Black, 1897, 11:51–95.

Desmond, Adrian. *The Politics of Evolution: Morphology, Medicine, and Reform in Radical London*. Chicago: University of Chicago Press, 1989.

de Staël, Anne Marie Louise Necker, Baronne. *Germany*. 2 vols. 1813; reprint, Boston: Houghton, Osgood, 1879.

Doczi, György. *The Power of Limits: Proportional Harmonies in Nature, Art, and Architecture*. Boulder: Shambhala, 1981.

Douglas, Mary. *Implicit Meanings: Essays in Anthropology*. London: Routledge, 1975.

Drummond, James L. *Letters to a Young Naturalist on the Study of Nature and Natural Theology*. London: Longman, Rees, Orme, Brown, and Green, 1831.

Dupree, A. Hunter. *Asa Gray: American Botanist, Friend of Darwin.* 1959; reprint, Baltimore: Johns Hopkins University Press, 1988.

Eger, Martin. "Hermeneutics and the New Epic of Science." In *The Literature of Science*. Edited by Murdo William McRae. Athens: University of Georgia Press, 1993, 186–209.

Egerton, Frank N., and Laura Dassow Walls. "Rethinking Thoreau and the History of American Ecology." *Concord Saunterer*, n.s., 5 (fall 1997): 5–20.

Emerson, Edward Waldo. *The Early Years of the Saturday Club, 1855–1870.* Boston: Houghton Mifflin, 1918.

Emerson, Ralph Waldo. *The Collected Works of Ralph Waldo Emerson*. Edited by Alfred R. Ferguson et al. 5 vols. to date. Cambridge: Harvard University Press, 1971–.

——. *The Complete Sermons of Ralph Waldo Emerson*. Edited by Albert J. von Frank et al. 4 vols. Columbia: University of Missouri Press, 1989–1992.

——. *The Complete Works of Ralph Waldo Emerson*. Edited by Edward Waldo Emerson. 12 vols. Boston: Houghton Mifflin, 1903–04.

——. *The Early Lectures of Ralph Waldo Emerson, 1833–1842*. Edited by Stephen E. Whicher, Robert E. Spiller, and Wallace E. Williams. 3 vols. Cambridge: Harvard University Press, 1959–1972.

——. *Emerson's Antislavery Writings*. Edited by Len Gougeon and Joel Myerson. New Haven: Yale University Press, 1995.

——. *English Traits*. Edited by Philip Nicoloff, Robert E. Burkholder, and Douglas Emory Wilson. Cambridge: Harvard University Press, 1994.

——. *Essays: First Series*. Edited by Joseph Slater, Alfred R. Ferguson, and Jean Ferguson Carr. Cambridge: Harvard University Press, 1979.

——. *Essays: Second Series*. Edited by Joseph Slater, Alfred R. Ferguson, and Jean Ferguson Carr. Cambridge: Harvard University Press, 1983.

——. *Essays and Lectures*. Edited by Joel Porte. New York: Library of America, 1983.

——. *The Journals and Miscellaneous Notebooks of Ralph Waldo Emerson*. Edited by William Gilman et al. 16 vols. Cambridge: Harvard University Press, 1960–1982.

——. *The Later Lectures of Ralph Waldo Emerson, 1843–1871*. Edited by Ronald A. Bosco and Joel Myerson. 2 vols. Athens: University of Georgia Press, 2001.

——. *Letters and Social Aims*. 1875. Rev. ed., Boston: Houghton Mifflin, 1886.

——. *The Letters of Walph Waldo Emerson*. Edited by Ralph L. Rusk. 6 vols. New York: Columbia University Press, 1939.

——. *Nature, Addresses, and Lectures*. Edited by Robert E. Spiller and Alfred R. Ferguson. Cambridge: Harvard University Press, 1971.

——. *Representative Men*. Edited by Wallace E. Williams and Douglas Emory Wilson. Cambridge: Harvard University Press, 1987.

——. *The Topical Notebooks of Ralph Waldo Emerson*. Edited by Ralph H. Orth et al. 3 vols. Columbia: University of Missouri Press, 1990–1994.

Enfield, William. *Institutes of Natural Philosophy, Theoretical and Experimental*. 1785; reprint, Boston: 1802.

Ferguson, James. *Ferguson's Astronomy, Explained upon Sir Isaac Newton's Principles*. Edited by David Brewster. 2 vols. Philadelphia, 1817.

Foucault, Michel. "Truth and Power." In *The Foucault Reader*. Edited by Paul Rabinow. New York: Pantheon, 1984, 51–75.

Fuller, Steve. "Disciplinary Boundaries and the Rhetoric of the Social Sciences." *Poetics Today* 12.2 (1991): 301–25.

Galison, Peter. "Objectivity Is Romantic." In Jerome Friedman et al., *The Humanities and the Sciences*, Occasional Paper No. 47. New York: American Council of Learned Societies, 1999.

Garnett, Richard. *Life of Ralph Waldo Emerson*. 1888; reprint, London: Walter Scott, 1988.

Gérando, Joseph-Marie de. *Self-Education, or the Means and Art of Moral Progress*. 1824. Translated by Elizabeth Peabody. Boston: Carter and Hendee, 1830.

Gilbert, Armida. "Emerson in the Context of the Woman's Rights Movement." In *A Historical Guide to Ralph Waldo Emerson*. Edited by Joel Myerson. Oxford: Oxford University Press, 2000, 211–49.

Goethe, Johann Wolfgang von. *Scientific Studies*. Edited and translated by Douglas Miller. New York: Suhrkamp, 1988.

Goodwin, Joan. "Sarah Alden Ripley, Another Concord Botanist." *Concord Saunterer*, n.s., 1 (fall 1993): 76–86.

Gougeon, Len. "Emerson and the Woman Question: The Evolution of His Thought." *New England Quarterly* 71.4 (1998): 570–92.

———. " 'Fortune of the Republic': Emerson, Lincoln, and Transcendental Warfare." *ESQ* 45 (1999): 259–324.

———. *Virtue's Hero: Emerson, Antislavery, and Reform.* Athens: University of Georgia Press, 1990.

Gould, Stephen Jay. *The Mismeasure of Man.* New York: Norton, 1981.

Grant, Edward. "Science and Theology in the Middle Ages." In *God and Nature.* Edited by David C. Lindberg and Ronald L. Numbers. Berkeley: University of California Press, 1986, 49–75.

Grusin, Richard A. *Transcendental Hermeneutics: Institutional Authority and the Higher Criticism of the Bible.* Durham: Duke University Press, 1991.

Gura, Philip F. *The Wisdom of Words: Language, Theology, and Literature in the New England Renaissance.* Middletown, Conn.: Wesleyan University Press, 1981.

Guyot, Arnold. *The Earth and Man: Lectures on Comparative Physical Geography, in Its Relation to the History of Mankind.* Boston: Gould, Kendall, and Lincoln, 1849.

Hacking, Ian. *Representing and Intervening: Introductory Topics in the Philosophy of Natural Science.* Cambridge: Cambridge University Press, 1983.

Halttunen, Karen. *Confidence Men and Painted Women: A Study of Middle-Class Culture in America, 1830–1870.* New Haven: Yale University Press, 1982.

Harding, Walter. *Emerson's Library.* Charlottesville: University Press of Virginia, 1967.

Haüy, René Just. *An Elementary Treatise on Natural Philosophy.* Translated by Olinthus Gregory. 2 vols. London, 1807.

Hedge, Frederic Henry. "Coleridge's Literary Character—German Metaphysics." *Christian Examiner* 14 (n.s. 9, March 1833): 108–29. Reprinted in Cameron, *Emerson the Essayist,* 2:59–69.

Henry, Joseph. *A Scientist in American Life: Essays and Lectures of Joseph Henry.* Washington, D.C.: Smithsonian, 1980.

Herder, Johann Gottfried. *Outlines of a Philosophy of the History of Man* (1784–1791). Translated by T. Churchill. 4 vols. 1800; reprint, New York: Bergman, n.d.

Herschel, John F. W. *Familiar Lectures on Scientific Subjects.* London: Alexander Strahan, 1867.

———. *A Preliminary Discourse on the Study of Natural Philosophy.* 1830; facsimile ed., Chicago: University of Chicago Press, 1987.

———. "Quetelet on Probabilities." *Edinburgh Review* 185 (July 1850): 1–57.

Hodder, Alan H. " 'After a High Negative Way': Emerson's 'Self-Reliance' and the Rhetoric of Conversion." *Harvard Theological Review* 84 (1991): 423–46.

Holmes, Oliver Wendell. *The Poetical Works of Oliver Wendell Holmes.* 3 vols. Boston: Houghton Mifflin, 1891.

Hopkins, Vivian C. "Emerson and Bacon." *American Literature* 29 (1958): 408–30.

Horwitz, Howard. *By the Law of Nature: Form and Value in Nineteenth-Century America.* New York: Oxford University Press, 1991.

Humboldt, Alexander von. *Cosmos: A Sketch of the Physical Description of the Universe.* Translated by Elizabeth C. Otté. 5 vols. New York: Harper and Brothers, 1850–1870. Fascimile ed., vols. 1–2, Baltimore: Johns Hopkins University Press, 1997.

Jacob, Margaret C. "Christianity and the Newtonian Worldview." In *God and Nature.* Edited by David C. Lindberg and Ronald L. Numbers. Berkeley: University of California Press, 1986, 238–55.

Jacobson, David. *Emerson's Pragmatic Vision: The Dance of the Eye.* University Park: Pennsylvania State University Press, 1993.

Jacyna, L. S. "Romantic Thought and the Origins of Cell Theory." In *Romanticism and the Sciences.* Edited by Andrew Cunningham and Nicholas Jardine. Cambridge: Cambridge University Press, 1990, 161–68.

Johnson, Mark. *The Body in the Mind: The Bodily Basis of Meaning, Imagination, and Reason.* Chicago: University of Chicago Press, 1987.

Kant, Immanuel. *Metaphysical Foundations of Natural Science.* 1786. Translated by James Ellington. Indianapolis: Bobbs-Merrill, 1970.

——. *The Philosophy of Kant: Immanuel Kant's Moral and Political Writings.* Edited by Carl J. Friedrich. New York: Random House, 1949.

Kidd, John. *On the Adaptation of External Nature to the Physical Condition of Man.* 1833. 2d ed. Philadelphia: Carey, Lea and Blanchard, 1836.

Kirby, William, and William Spence. *An Introduction to Entomology; or, Elements of the Natural History of Insects.* 4 vols. 5th ed. London: 1828.

Knight, David. *The Age of Science.* New York: Basil Blackwell, 1986.

——. "Romanticism and the Sciences." In *Romanticism and the Sciences.* Edited by Andrew Cunningham and Nicholas Jardine. Cambridge: Cambridge University Press, 1990, 13–24.

Knox, Robert. *The Races of Man: A Fragment.* Philadelphia: Lea and Blanchard, 1850.

Koch, Robert. "The Case of Latour." *Configurations* 3.3 (1995): 319–47.

Kuhn, Thomas S. "Metaphor in Science." In *The Road since Structure.* Chicago: University of Chicago Press, 2000, 196–207.

Lakoff, George, and Mark Johnson. *Metaphors We Live By.* Chicago: University of Chicago Press, 1980.

Lamarck, Jean-Baptiste. *Zoological Philosophy: An Exposition with Regard to the Natural History of Animals.* 1809; reprint, Chicago: University of Chicago Press, 1984.

Laplace, Pierre Simon, Marquis de. *A Philosophical Essay on Probabilities.* 1814. Translated from the 6th French ed. by Frederick Wilson Truscott and Frederick Lincoln Emory. 1902; reprint, New York: Dover, 1952.

Lardner, Dionysius. *A Treatise on Hydrostatics and Pneumatics.* London: Longman, Rees, Orme, Brown, and Green, 1831.

Latour, Bruno. *We Have Never Been Modern.* Translated by Catherine Porter. Cambridge: Harvard University Press, 1993.

Lawlor, Robert. *Sacred Geometry.* New York: Crossroad, 1982.

Lee, Sarah (Wallis) Bowdich. *Memoirs of Baron Cuvier*. New York: J. and J. Harper, 1833.

Levere, Trevor H. "Coleridge and the Sciences." In *Romanticism and the Sciences*. Edited by Andrew Cunningham and Nicholas Jardine. Cambridge: Cambridge University Press, 1990, 295–306.

Levin, Jonathan. *The Poetics of Transition: Emerson, Pragmatism, and American Literary Modernism*. Durham: Duke University Press, 1999.

Levine, George. "One Culture: Science and Literature." In *One Culture: Essays in Science and Literature*. Edited by George Levine. Madison: University of Wisconsin Press, 1987, 3–32.

Liebig, Justus von. *Animal Chemistry, or Organic Chemistry in Its Application to Physiology and Pathology*. 1842; facsimile ed., New York: Johnson Reprint, 1964.

Lindner, Carl M. "Newtonianism in Emerson's Nature." *ESQ* 20 (1974): 260–69.

Linnaeus, Carl. *Lachesis Lapponica, or a Tour in Lapland*. 2 vols. London, 1811.

Lopez, Michael. *Emerson and Power: Creative Antagonism in the Nineteenth Century*. De Kalb: Northern Illinois University Press, 1996.

Lurie, Edward. *Louis Agassiz: A Life in Science*. 1960; reprint, Baltimore: Johns Hopkins University Press, 1988.

Lyell, Charles. *Principles of Geology*. 3 vols. 1830–1833; facsimile edition, Chicago: University of Chicago Press, 1990–1992.

———. Review of *Transactions of the Geological Society of London, 1824*. *Quarterly Review* 34.58 (1826): 507–40.

Marsh, George Perkins. *Man and Nature; or, Physical Geography as Modified by Human Action*. 1864; reprint, Cambridge: Harvard University Press, 1965.

Marsh, James. "Preliminary Essay." In Samuel Taylor Coleridge, *Aids to Reflection*. 1829; reprint, Port Washington, N.Y.: Kennikat, 1971, 9–58.

Matthiessen, F.O. *American Renaissance: Art and Expression in the Age of Emerson and Whitman*. Oxford: Oxford University Press, 1941.

Menand, Louis. *The Metaphysical Club*. New York: Farrar, Straus and Giroux, 2001.

Merchant, Carolyn. "Reinventing Eden: Western Culture as a Recovery Narrative." In *Uncommon Ground: Toward Reinventing Nature*. Edited by William Cronon. New York: Norton, 1995, 132–59.

Miller, Craig W. "Coleridge's Concept of Nature." *Journal of the History of Ideas* 25 (1964): 77–96.

Miller, Perry. *The New England Mind: The Seventeenth Century*. Cambridge: Harvard University Press, 1939.

Modiano, Raimonda. *Coleridge and the Concept of Nature*. Tallahassee: Florida State University Press, 1985.

Moore, James R. "Geologists and Interpreters of Genesis in the Nineteenth Century." In *God and Nature*. Edited by David C. Lindberg and Ronald L. Numbers. Berkeley: University of California Press, 1986, 322–50.

Morris, Saundra. "The Threshold Poem, Emerson, and 'The Sphinx.' " *American Literature* 69.3 (1997): 547–70.

Mott, Wesley T. *"The Strains of Eloquence": Emerson and His Sermons*. University Park: Pennsylvania State University Press, 1989.

Neufeldt, Leonard N. *The House of Emerson*. Lincoln: University of Nebraska Press, 1982.

——. "The Science of Power: Emerson's Views on Science and Technology in America." *Journal of the History of Ideas* 38 (April–June 1977): 329–44.

Newfield, Christopher. *The Emerson Effect: Individualism and Submission in America*. Chicago: Chicago University Press, 1996.

Nicoloff, Philip L. *Emerson on Race and History: An Examination of "English Traits."* New York: Columbia University Press, 1961.

Oegger, Guillaume. "The True Messiah." 1829. Reprinted in Cameron, *Emerson the Essayist*, 2:83–99.

Owen, Richard. *Palaeontology; or, a Systematic Summary of Extinct Animals and Their Geological Relations*. Edinburgh: Adam and Charles Black, 1860.

Packer, Barbara. "Emerson and the Terrible Tabulations of the French." In *Transient and Permanent: The Transcendentalist Movement and Its Contexts*. Edited by Charles Capper and Conrad E. Wright. Boston: Northeastern University Press, 1999, 148–67.

——. *Emerson's Fall: A New Interpretation of the Major Essays*. New York: Continuum, 1982.

Paley, William. *Natural Theology; or, Evidences of the Existence and Attributes of the Deity, Collected from the Appearances of Nature*. 1802. In *Works*. Vol. 5. London: C. and J. Rivington et al., 1825.

Parker, Theodore. Review of William Whewell, *The Philosophy of the Inductive Sciences, founded upon their History. The Dial* 2.4 (1842): 529–30.

Patell, Cyrus R. "Emersonian Strategies: Negative Liberty, Self-Reliance, and Democratic Individuality." *Nineteenth-Century Literature* 48 (March 1994): 440–79.

Peabody, William B. O. "Study of Natural History." Review of William Swainson, *A Preliminary Discourse on the Study of Natural History* (London, 1834). *North American Review* 41 (October 1835): 406–30.

Pease, Donald. *Visionary Compacts: American Renaissance Writings in Cultural Context*. Madison: University of Wisconsin Press, 1987.

Pelikan, Jaroslav. "Natural History Married to Human History: Ralph Waldo Emerson and the 'Two Cultures.'" In *The Rights of Memory: Essays on History, Science, and American Culture*. Edited by Taylor Littleton. University: University of Alabama Press, 1986, 35–75.

Pesmen, Dale. "Reasonable and Unreasonable Worlds: Some Expectations of Coherence in Culture Implied by the Prohibition of Mixed Metaphor." In *Beyond Metaphor*. Edited by James W. Fernandez. Stanford: Stanford University Press, 1991, 213–43.

Pickering, Charles. *The Races of Man: and Their Geographical Distribution*. Boston: Little and Brown, 1848.

Plato. *The Collected Dialogues*. Edited by Edith Hamilton and Huntington Cairns. Princeton: Princeton University Press, 1961.

Playfair, John. *Dissertation Second: Exhibiting a General View of the Progress of Mathematical and Physical Science, since the Revival of Letters in Europe*.

1817. Reprinted in *Dissertations on the Progress of Knowledge*. New York: Arno Press, 1975.

——. *Illustrations of the Huttonian Theory of the Earth*. 1802; facsimile ed., Urbana: University of Illinois Press, 1956.

Poovey, Mary. *Making a Social Body: British Cultural Formation, 1830–1864*. Chicago: University of Chicago Press, 1995.

Porte, Joel. *Emerson and Thoreau: Transcendentalists in Conflict*. Middletown, Conn.: Wesleyan University Press, 1966.

Porter, David. *Emerson and Literary Change*. Cambridge: Harvard University Press, 1978.

Porter, Theodore M. *The Rise of Statistical Thinking, 1820–1900*. Princeton: Princeton University Press, 1986.

Prichard, James Cowles. *The Natural History of Man*. 2 vols. 4th ed. London: H. Baillière, 1855.

Priestley, Joseph. *Lectures on History and General Policy*. 2 vols. Philadelphia, 1803.

Quetelet, Adolphe M. A. *A Treatise on Man and the Development of His Faculties*. 1835. Edinburgh, 1842; reprint, New York: Burt Franklin, 1968.

Quinn, Naomi. "The Cultural Basis of Metaphor." In *Beyond Metaphor*. Edited by James W. Fernandez. Stanford: Stanford University Press, 1991, 56–93.

Reed, Sampson. *Observations on the Growth of the Mind*. 1821. Reprinted in Cameron, *Emerson the Essayist*, 2:12–31.

Rehbock, Philip F. "Transcendental Anatomy." In *Romanticism and the Sciences*. Edited by Andrew Cunningham and Nicholas Jardine. Cambridge: Cambridge University Press, 1990, 144–60.

Richardson, Robert D., Jr. *Emerson: The Mind on Fire*. Berkeley: University of California Press, 1995.

Robinson, David. *Apostle of Culture: Emerson as Preacher and Lecturer*. Philadelphia: University of Pennsylvania Press, 1982.

——. *Emerson and the Conduct of Life: Pragmatism and Ethical Purpose in the Later Work*. Cambridge: Cambridge University Press, 1993.

——. "Emerson's Natural Theology and the Paris Naturalists: Toward a Theory of Animated Nature." *Journal of the History of Ideas* 41 (1980): 69–88.

——. "Fields of Investigation: Emerson and Natural History." In *American Literature and Science*. Edited by Robert J. Scholnick. Lexington: University Press of Kentucky, 1992, 94–109.

——. "*The Method of Nature* and Emerson's Period of Crisis." In *Emerson Centenary Essays*. Edited by Joel Myerson. Carbondale: Southern Illinois University Press, 1982, 74–92.

Roget, Peter Mark. *Animal and Vegetable Physiology, Considered with Reference to Natural Theology*. 2 vols. Philadelphia: Carey, Lea, and Blanchard, 1836.

Ross, Sydney. "*Scientist*: Story of a Word." *Annals of Science* 18.2 (1962): 65–85.

Rossi, William. "Emerson, Nature, and the Natural Sciences." In *A Historical Guide to Ralph Waldo Emerson*. Edited by Joel Myerson. New York: Oxford University Press, 2000, 101–50.

Rothman, Tony. "Irreversible Differences." *The Sciences* 37 (July/August 1997): 26–31.

Schaffer, Simon. "Genius in Romantic Natural Philosophy." In *Romanticism and the Sciences*. Edited by Andrew Cunningham and Nicholas Jardine. Cambridge: Cambridge University Press, 1990, 82–98.

Schelling, Friedrich Wilhelm Joseph von. *Ideas for a Philosophy of Nature*. 2d ed. Edited by Robert Stern. Translated by Errol E. Harris and Peter Heath. Cambridge: Cambridge University Press, 1988.

Schneider, Richard. " 'Climate Does Thus React on Man': Wildness and Geographic Determinism in Thoreau's 'Walking.' " In *Thoreau's Sense of Place: Essays in Environmental Writing*. Edited by Richard Schneider. Iowa City: University of Iowa Press, 2000, 44–60.

Secord, James A. *Victorian Sensation: The Extraordinary Publication, Reception, and Secret Authorship of "Vestiges of the Natural History of Creation."* Chicago: University of Chicago Press, 2000.

Shaffer, Elinor S. "Romantic Philosophy and the Organization of the Disciplines: The Founding of the Humboldt University of Berlin." In *Romanticism and the Sciences*. Edited by Andrew Cunningham and Nicholas Jardine. Cambridge: Cambridge University Press, 1990, 38–54.

Shapin, Steven. *A Social History of Truth: Civility and Science in Seventeenth-Century England*. Chicago: University of Chicago Press, 1994.

Simmons, Nancy Craig. "Man without a Shadow: The Life and Work of James Elliot Cabot, Emerson's Biographer and Literary Executor." Ph.D. diss., Princeton University, 1980.

Smith, Jonathan. *Fact and Feeling: Baconian Science and the Nineteenth-Century Literary Imagination*. Madison: University of Wisconsin Press, 1994.

Snow, C. P. *The Two Cultures and a Second Look*. Cambridge: Cambridge University Press, 1964.

Stallo, Johann Bernhard. *General Principles of the Philosophy of Nature, with an outline of some of its recent developments among the Germans, embracing the philosophical systems of Schelling and Hegel, and Oken's System of Nature*. Boston: Crosby and Nichols, 1848.

Stanton, William. *The Leopard's Spots: Scientific Attitudes toward Race in America, 1815–1859*. Chicago: University of Chicago Press, 1960.

Stephens, Lester D. *Science, Race, and Religion in the American South: John Bachman and the Charleston Circle of Naturalists, 1815–1895*. Chapel Hill: University of North Carolina Press, 2000.

Sterrenburg, Lee. "A Narrative Overview: The Making of the Concept of the Global 'Environment' in Literature and Science." Manuscript.

Stewart, Dugald. *Dissertation: Exhibiting the Progress of Metaphysical, Ethical, and Political Philosophy since the Revival of Letters in Europe*. 1815. In *The Collected Works of Dugald Stewart*. Edited by William Hamilton. Vol. 1. Edinburgh: Thomas Constable, 1854.

———. *Elements of the Philosophy of the Human Mind*. 1792–1827. In *The Collected Works of Dugald Stewart*. Edited by William Hamilton. Vols. 2–4. Edinburgh: Thomas Constable, 1854.

Stewart, Susan. *On Longing: Narratives of the Miniature, the Gigantic, the Souvenir, the Collection.* Durham: Duke University Press, 1993.

Thompson, D'Arcy. *On Growth and Form.* Edited by John Tyler Bonner. Cambridge: Cambridge University Press, 1961.

Thoreau, Henry David. *The Writings of Henry David Thoreau: Journal.* Edited by John C. Broderick et al. 7 vols. to date. Princeton: Princeton University Press, 1981–.

Turner, Frank Miller. *Contesting Cultural Authority: Essays in Victorian Intellectual Life.* Cambridge: Cambridge University Press, 1993.

Tyndall, John. *Fragments of Science.* 6th ed. 2 vols. New York: Collier, 1902.

Van Cromphout, Gustaaf. *Emerson's Modernity and the Example of Goethe.* Columbia: University of Missouri Press, 1990.

Van Leer, David. *Emerson's Epistemology: The Argument of the Essays.* Cambridge: Cambridge University Press, 1986.

von Frank, Albert J. *The Trials of Anthony Burns: Freedom and Slavery in Emerson's Boston.* Cambridge: Harvard University Press, 1998.

Walls, Laura Dassow. "The Anatomy of Truth: Emerson's Poetic Science." *Configurations* 5 (1997): 425–61.

———. "Believing in Nature: Wilderness and Wildness in Thoreauvian Science." In *Thoreau's Sense of Place: Essays in Environmental Criticism.* Edited by Richard Schneider. Iowa City: University of Iowa Press, 2000, 15–27.

———. *Seeing New Worlds: Henry David Thoreau and Nineteenth-Century Natural Science.* Madison: University of Wisconsin Press, 1995.

Watson, James D. *The Double Helix: A Personal Account of the Discovery of the Structure of DNA.* New York: Norton, 1980.

Waugh, Albert E. *Sundials: Their Theory and Construction.* New York: Dover, 1973.

West, Cornel. *The American Evasion of Philosophy: A Genealogy of Pragmatism.* Madison: University of Wisconsin Press, 1989.

Wetzels, Walter D. "Johann Wilhelm Ritter: Romantic Physics in Germany." In *Romanticism and the Sciences.* Edited by Andrew Cunningham and Nicholas Jardine. Cambridge: Cambridge University Press, 1990, 199–212.

Whewell, William. *Astronomy and General Physics, Considered with Reference to Natural Theology.* 1833; reprint, Philadelphia: Carey, Lea and Blanchard, 1836.

———. Review of John Herschel, *A Preliminary Discourse on the Study of Natural Philosophy. Quarterly Review* 45 (July 1831): 374–407.

———. Review of Mary Somerville, *On the Connexion of the Physical Sciences. Quarterly Review* 51 (March 1834): 54–68.

———. *Selected Writings on the History of Science.* Edited by Yehuda Elkana. Chicago: University of Chicago Press, 1984.

Whicher, Stephen E. *Freedom and Fate: An Inner Life of Ralph Waldo Emerson.* Philadelphia: University of Pennsylvania Press, 1953.

Whitehead, Alfred North. *Science and the Modern World.* 1925; reprint, New York: Macmillan, 1953.

Wider, Sarah Ann. *The Critical Reception of Emerson: Unsettling All Things.* Rochester, N.Y.: Camden House, 2000.

Williams, Raymond. "Culture and Civilization." In *Encyclopedia of Philosophy*. Edited by Paul Edwards. Vol. 2. New York: Macmillan, 1967, 273–76.

———. *Keywords: A Vocabulary of Culture and Society*. 1976; reprint, New York: Oxford University Press, 1983.

Wilson, Eric. *Emerson's Sublime Science*. New York: St. Martin's, 1999.

———. *Romantic Turbulence: Chaos, Ecology, and American Space*. New York: St. Martin's, 2000.

Winsor, Mary P. "Louis Agassiz and the Species Question." *Studies in History of Biology* 3 (1979): 89–117.

———. *Reading the Shape of Nature: Comparative Zoology at the Agassiz Museum*. Chicago: University of Chicago Press, 1991.

Winthrop, John. *Winthrop's Journal, "History of New England," 1630–1649*. Edited by James Kendall Hosmer. 2 vols. New York: Charles Scribner's Sons, 1908.

Wollaston, William. *The Religion of Nature Delineated*. 1724; reprint, Delmar, N.Y.: Scholar's Facsimiles and Reprints, 1975.

Yeo, Richard. *Defining Science: William Whewell, Natural Knowledge, and Public Debate in Early Victorian Britain*. Cambridge: Cambridge University Press, 1993.

———. "An Idol of the Market-Place: Baconianism in Nineteenth Century Britain." *History of Science* 23 (1985): 251–98.

Young, Robert M. *Darwin's Metaphor: Nature's Place in Victorian Culture*. Cambridge: Cambridge University Press, 1985.

Index